INTERNATIONAL

REVIEW OF CYTOLOGY

VOLUME 99

Molecular Approaches to the Study of Protozoan Cells

INTERNATIONAL

Review of Cytology

EDITED BY

G. H. BOURNE
St. George's University School of Medicine
St. George's, Grenada
West Indies

J. F. DANIELLI
(Deceased April 22, 1984)

ASSISTANT EDITOR

K. W. JEON
Department of Zoology
University of Tennessee
Knoxville, Tennessee

VOLUME 99

Molecular Approaches to the Study of Protozoan Cells

EDITED BY

K. W. JEON
Department of Zoology
University of Tennessee
Knoxville, Tennessee

ACADEMIC PRESS, INC. 1986

Harcourt Brace Jovanovich, Publishers
Orlando   San Diego   New York   Austin
London   Montreal   Sydney   Tokyo   Toronto

QH573
I5
vol. 99
1986

ACADEMIC PRESS, INC.
Orlando, Florida 32887

United Kingdom Edition published by
ACADEMIC PRESS INC. (LONDON) LTD.
24–28 Oval Road, London NW1 7DX

LIBRARY OF CONGRESS CATALOG CARD NUMBER: 52-5203

ISBN 0–12–364499–2

PRINTED IN THE UNITED STATES OF AMERICA

86 87 88 89   9 8 7 6 5 4 3 2 1

Contents

The Genome of Hypotrichous Ciliates

H. KRAUT, H. J. LIPPS, AND D. M. PRESCOTT

Structure and Formation of Telomeres in Holotrichous Ciliates

ELIZABETH H. BLACKBURN

Genome Reorganization in *Tetrahymena*

CLIFFORD F. BRUNK

The Molecular Biology of Antigenic Variation in Trypanosomes: Gene Rearrangements and Discontinuous Transcription

TITIA DE LANGE

Kinetoplast DNA in Trypanosomid Flagellates

LARRY SIMPSON

Chlamydomonas reinhardtii: A Model System for the Genetic Analysis of Flagellar Structure and Motility

BESSIE PEI-HSI HUANG

Genetic, Biochemical, and Molecular Approaches to *Volvox* Development and Evolution

David L. Kirk and Jeffrey F. Harper

The Ribosomal Genes of *Plasmodium*

Thomas F. McCutchan

Molecular Biology of DNA in *Acanthamoeba*, *Amoeba*, *Entamoeba*, and *Naegleria*

Thomas J. Byers

Contributors

Numbers in parentheses indicate the pages on which the authors' contributions begin.

ELIZABETH H. BLACKBURN (29), *Department of Molecular Biology, University of California, Berkeley, California 94720*

CLIFFORD F. BRUNK (49), *Biology Department and Molecular Biology Institute, University of California, Los Angeles, California 90024*

THOMAS J. BYERS (311), *Department of Microbiology, The Ohio State University, Columbus, Ohio 43210-1292*

TITIA DE LANGE[1] (85), *Division of Molecular Biology, Antoni van Leeuwenhoekhuis, The Netherlands Cancer Institute, 1066 CX Amsterdam, The Netherlands*

JEFFREY F. HARPER (217), *Department of Biology, Washington University, St. Louis, Missouri 63130*

BESSIE PEI-HSI HUANG (181), *Department of Cell Biology, Baylor College of Medicine, Houston, Texas 77030*

DAVID L. KIRK (217), *Department of Biology, Washington University, St. Louis, Missouri 63130*

H. KRAUT (1), *Institut für Biologie III, Abt. Zellbiologie, University of Tübingen, Federal Republic of Germany*

H. J. LIPPS (1), *Institut für Biologie III, Abt. Zellbiologie, University of Tübingen, Federal Republic of Germany*

THOMAS F. MCCUTCHAN (295), *Laboratory of Parasitic Diseases, National Institute of Allergy and Infectious Diseases, National Institutes of Health, Bethesda, Maryland 20205*

---

[1] Present address: Department of Microbiology and Immunology, University of California, San Francisco, California 94143.

D. M. Prescott (1), *Department of Molecular, Cellular and Developmental Biology, University of Colorado, Boulder, Colorado 80309-0347*

Larry Simpson (119), *Department of Biology and Molecular Biology Institute, University of California, Los Angeles, California 90024*

Preface

Protozoans have long been model cells for biological research as well as being choice material for teaching at all educational levels [see for example, "Protozoa in Biological Research" by G. N. Calkins and F. M. Summers (1941) and "Progress in Protozoology," four volumes edited by T. T. Chen (1967–1969)]. In recent years, however, the use of protozoan cells for basic biomedical research appears to have waned, possibly due to lack of research funds resulting from emphasis on more mission orientation and application with the perception that studies on protozoans are not relevant to human physiology and welfare. However, since all living cells display the same basic life processes, it is still true that protozoans are ideal tools for solving fundamental biological problems of general interest. They offer many advantages over cultured tissue cells or microbial cells, such as larger size, greater ease with which they can be cultured and manipulated, and the wide range of biological phenomena that can be studied with each cell type. In addition, many parasitic and pathogenic protozoans continue to plague much of the world's human population and livestock. New molecular approaches in studying the physiology of these cells and host–parasite interactions would enhance the solution of world health problems.

The purpose of this present volume is twofold: first to stimulate the protozoologists to apply modern molecular–biological approaches and techniques in their research, and second to inform molecular biologists of the various advantages protozoa offer as experimental systems. This volume represents an attempt to bring together recently obtained results from such molecular–biological studies on representative protozoans in one place.

Articles have been arranged to describe representative samples of ciliates, flagellates, plasmodium, and amoebas in that order. In the first three articles, the genomic structure and expression in both holotrichous and hypotrichous ciliates are reviewed as follows: Kraut, Lipps, and Prescott describe the structure of the macronucleus and molecular events occurring during macronuclear differentiation; Blackburn, the structure and formation of telomeres; and Brunk, reorganization of the genome. In the next four articles, three kinds of flagellates are covered from the following different points of view: the basis for antigenic variation in trypanosomes by De Lange, kinetoplast DNA by Simpson, genetic analysis of flagellar structure and function by Huang, and molecular and genetic analysis of *Volvox* development by Kirk and Harper. McCutchan describes the ribosomal genes of *Plasmodium* in the next article and, finally, an article by

Byers covers the DNA studies in four groups of amoebas. Amoebas have previously not been favored material in nucleic acid studies, but their potential usefulness as an experimental system in this area of molecular biology seems to be gaining recognition.

The organisms included in this volume are by no means all inclusive, and it must be stated that the material covered represents only a fraction of what has been learned about the molecular biology of protozoa. Even during the short period of time this volume has been in production, many new results have been reported, some being highlighted in science news magazines. I regret that a few other articles that had been planned for this volume were not complete at press time and could not be included here. They may appear as regular articles in future volumes of the *International Review of Cytology*. I hope that this volume will stimulate the interests of both protozoologists and molecular biologists in each others' fertile fields. I thank the authors for their superb cooperation and the production editors at Academic Press for their skillful work.

KWANG W. JEON

The Genome of Hypotrichous Ciliates

H. Kraut,* H. J. Lipps,* and D. M. Prescott†

*Institut für Biologie III, Abt. Zellbiologie, University of Tübingen, Federal
Republic of Germany, and †Department of Molecular, Cellular and
Developmental Biology, University of Colorado, Boulder, Colorado

I. Introduction

Like the majority of ciliated protozoa, hypotrichous ciliates are characterized by their nuclear dimorphism (Raikov, 1982). During their vegetative life cycles the cells contain diploid micronuclei and DNA-rich macronuclei of varying number and shape (Grell, 1973). The micronuclei show little or no transcriptional activity, and in some hypotrichous ciliates, like *Stylonychia,* can be removed from the cell without significant loss of viability of the clones (Ammermann, 1971). Their high-molecular-weight DNA is arranged in chromosomes. During mitosis these chromosomes are separated in an organized way with the help of a spindle apparatus.

The macronucleus is the major site of RNA transcription and synthesizes all RNA necessary for vegetative growth of the cell. Its DNA is organized in short, gene-sized molecules, which are distributed during cell division by a process called amitosis (Grell, 1973). Vegetative growth usually continues for a limited number of generations that is characteristic for each species. Without sexual reproduction clones of most species die out. Sexual reproduction occurs when two competent cells, which are morphologically identical but of different mating type, meet. In the course of conjugation all micronuclei undergo a series of divisions that finally lead to differentiation into a haploid stationary and migratory micronucleus. This process is described in detail for *Euplotes* (Heckman, 1963,

1

1964) but is, with slight modifications, characteristic for other hypotrichous ciliates as well. At the beginning of conjugation, the micronucleus
divides mitotically (premeiotic division) followed by a meiotic division of
the two resulting diploid nuclei. Six out of eight generated haploid nuclei
become pycnotic and are resorbed in the cytoplasm. The remaining two
nuclei undergo a postmeiotic division. Two of the four nuclei are again
resorbed; the other two differentiate into a migratory and a stationary
nucleus. The migratory nuclei are exchanged between the conjugating
cells and fuse with the stationary nucleus in the recipient cell to form a
diploid synkaryon. The synkaryon divides mitotically, and one of the
daughter nuclei becomes the new micronucleus; the other develops in a
series of morphological and molecular changes into a new vegetative
macronucleus.

Figure 1 illustrates schematically macronuclear development of hypo-

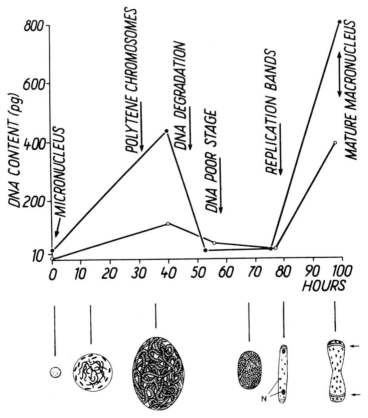

FIG. 1. Schematic illustration of macronuclear development of hypotrichous ciliates.
(●), *Stylonychia lemnae*; (○), *Euplotes aediculatus* (Kraut and Lipps, 1983).

trichous ciliates and gives a rough impression of the cytological events taking place during this nuclear differentiation process (Ammermann, 1964; Alonso and Perez-Silva, 1965; Rao and Ammermann, 1970; Kloetzel, 1970; Prescott *et al.*, 1971, 1973, 1979; Ammermann *et al.*, 1974; Prescott and Murti, 1974; Spear and Lauth, 1976; Kraut and Lipps, 1983). A few hours after the separation of the two exconjugants, chromosomes become visibly arranged at the inner surface of the already enlarged nuclei. A rapid DNA synthesis converts these chromosomes into polytene chromosomes during the next 20 hours. About 40 hours after separation, at the peak of polyteny, the giant chromosomes start to disintegrate, and most of their DNA is degraded, leading to a DNA-poor stage. At the beginning of a second phase of DNA synthesis, replication bands move toward the center of the macronuclear anlagen. After several rounds of replication, the new macronucleus reaches its mature stage and highest DNA content.

In this article we describe the structure of the macronucleus as well as the cytological and molecular events during macronuclear differentiation.

II. Structure of the Macronucleus

A. DNA STRUCTURE IN MACRONUCLEI

1. General Properties

As early as 1971, Prescott *et al.* (1971) reported that the DNA of macronuclei isolated from *Stylonychia* is of very low molecular weight and of relative uniform length when investigated by electron microscopy and sedimentation analysis. While the DNA of micronuclei and from polytene-stage nuclei (Spear and Lauth, 1976) sediments rapidly, the DNA from macronuclei sediments slowly in a broad peak when centrifuged through a sucrose gradient. The chromosomal micronuclear DNA, which has an estimated molecular weight of several hundred $\times 10^6$ in *Oxytricha* (Spear and Lauth, 1976; Prescott, 1983), is obviously processed during macronuclear development to DNA molecules with reduced molecular weight [averaging 0.8–1 μm in length or 2×10^6 (Prescott *et al.*, 1971; Wesley, 1975)] of a physical size that is very likely equivalent to one or a few genes only (Prescott *et al.*, 1973). This processing changes the physical and chemical properties of the DNA to a dramatic extent, as expected from earlier cytological studies. When micronuclear DNA of *Stylonychia mytilus*[1] (Ammermann *et al.*, 1974) and *Oxytricha* sp. (Bostock and Pres-

---

[1] 1983 *Stylonychia mytilus* syngen I was renamed *Stylonychia lemnae* (Ammermann and Schlegel, 1983).

cott, 1972; Spear and Lauth, 1976) is centrifuged to equilibrium in CsCl, several satellites are detectable, whereas DNA prepared from macronuclei bands in a single peak fraction.

A reduction in density satellite DNA content already takes place during polytene chromosome formation. By reassociation kinetics after thermal dissociation, DNA from nuclei in different developmental stages shows striking differences, which are summarized in Table I (Ammermann et al., 1974; Lauth et al., 1976). Micronuclear DNA of Oxytricha and Stylonychia reassociates as multiple DNA components, indicating the presence of repeated and nonrepeated sequences. The kinetic complexity is in the order of $0.2–1.5 \times 10^{12}$ Da. The reassociation of macronuclear DNA, however, follows simple second-order reaction kinetics and in this respect resembles the DNA of Escherichia coli. No repetitive DNA could be found in macronuclei, which thus is different from almost all other eukaryotic nuclei. The kinetic complexity of macronuclear DNA is in the range of 3×10^{10} Da, about 10 times the kinetic complexity of E. coli DNA (Lauth et al., 1976; Steinbrück et al., 1981).

Taken together, the results suggest that only part of the micronuclear DNA is present in the macronuclear genome and that most of the chromosomal DNA is not essential during the vegetative life cycles of the cells. The reduction of sequence complexity together with the degradation of most of the spacer DNA in the vesicle stage at the end of polytenization (Prescott and Murti, 1974) finally results in low-molecular-weight DNA molecules (Wesley, 1975; Lipps and Steinbrück, 1978; Elsevier et al., 1978; Swanton et al., 1980; Steinbrück et al., 1981; Steinbrück, 1983).

This macronuclear DNA of Oxytricha and other closely related hypotrichs exists in the form of distinct molecules as shown by agarose gel electrophoresis (Fig. 2). The sizes of the fragments range from 0.3 to about 20 kb, with an average size of 2–3 kb. Some fragments are seen as

TABLE I
AMOUNTS AND COMPLEXITIES OF DNAs OF THREE HYPOTRICHS

	DNA content (pg)		Kinetic complexity (GC corrected)	
	Macro-nucleus	Micro-nucleus	Macro-nucleus	Micro-nucleus
Stylonychia mytilus	788	13	3.1×10^{10}	1.45×10^{12}
Euplotes aediculatus	380	—	2.8×10^{10}	—
Oxytricha sp.	116	1.32	3.6×10^{10}	1.5×10^{12}
Escherichia coli				2.8×10^{9}

kb

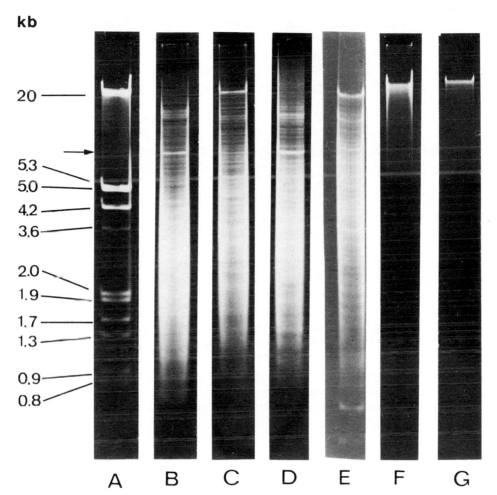

FIG. 2. Electrophoretic patterns of macronuclear DNA separated on agarose gels. (A) λ-DNA digested with *Eco*RI and *Hin*dIII. (B) *Euplotes aediculatus.* (C) *Paraurostyla weissei.* (D) *Oxytricha similis.* (E) *Stylonychia lemnae.* (F) *Paramecium bursaria.* (G) Micronuclear DNA from *Stylonychia lemnae* (Steinbrück *et al.,* 1981).

prominent bands, indicating that certain DNA molecules occur in higher copy number. This banding pattern is consistent and characteristic for different species and can be used as a systematic criterion to identify hypotrichous ciliates (Swanton *et al.,* 1980; Steinbrück *et al.,* 1981; Steinbrück, 1983; Steinbrück and Schlegel, 1983).

Knowing the total complexity of macronuclear DNA and the DNA content of a single macronucleus, the number of different gene-sized

DNA pieces can be calculated to be between 10,000 in *Stylonychia* and 24,000 in *Oxytricha* (Lipps *et al.*, 1978; Prescott *et al.*, 1979; Nock, 1981; Prescott, 1983), respectively. Because not all genes are present in the same copy number, this is a rough estimation. It is known that some of the genes are highly amplified, as in the case of rDNA (Lipps and Steinbrück, 1978; Spear, 1980), which is present in about 600,000 copies per macronucleus in *Stylonychia* and in about 100,000 copies in *Oxytricha* (M. T. Swanton, unpublished). Other genes may be underrepresented or over-amplified with age of a clone (Steinbrück, 1983). A macronuclear molecule of 1.23 kb becomes overamplified about 1 year after conjugation from 30,000 to 2.8×10^6 copies per macronucleus (E. Helftenbein, personal communication). This gene is transcribed into an mRNA of 0.7 kb and is translated. The protein has not yet been identified.

⠀Previous investigations with macronuclear DNA from *Oxytricha, Euplotes,* and *Paraurostyla* (Wesley, 1975; Lawn *et al.*, 1978; Herrick and Wesley, 1978) showed that these DNA molecules contain inverted repeats at or close to the ends of the gene-sized fragments. Single-stranded DNA (after denaturation with heat or high pH) is able to form circles upon reassociation, as shown by electron microscopy (Fig. 3). It became clear from sequencing data that all the DNA molecules share a short inverted repeat at their ends, which has the structure (Fig. 4) as shown for *Oxytricha* sp., *Oxytricha nova, Stylonychia pustulata* (Oka *et al.*, 1980; Klobutcher *et al.*, 1981), and *S. mytilus*. The sequence is $5'$-$C_4A_4C_4A_4C_4A_4C_4$ for *Euplotes* (Klobutcher *et al.*, 1981). In addition, each molecule has an extended $3'$ single-stranded end that is $3'$-$G_4T_4G_4T_4$ in *Oxytricha* and *Stylonychia* and $G_2T_4G_4T_4$ in *Euplotes*. These sequences are the evolutionarily most conserved sequences known among the different macronuclear DNA molecules of one species and among different species of hypotrichous ciliates.

Similar terminal repeats have been found at the ends of extrachromosomal rDNA and other macronuclear DNA molecules of *Tetrahymena* (Blackburn and Gall, 1978; Katzen *et al.*, 1981; Yao and Yao, 1981). *Tetrahymena* and *Glaucoma chattoni* are both holotrichous ciliates and have terminal $5'$-C_4A_2 sequence repeats, as do *Paramecium* and *Colpidium* (Yao and Yao, 1981; M. C. Yao, personal communication). A $5'$-CCCTAA sequence repeat occurs in flagellates (Blackburn and Challoner, 1984; Van der Ploeg *et al.*, 1984), a $5'$-$(CCCTA)_n$ sequence in the slime mold *Physarum* (Johnson, 1980), and a $5'$-$(C_{1-8}T)_n$ sequence in *Dictyostelium* DNA (Emery and Weiner, 1981). A tandemly repeated telomeric DNA sequence of $5'$-$C_{2-3}A(CA)_{1-3}$ also occurs in the yeast *Saccharomyces cerevisiae* (Shampay *et al.*, 1984). The accuracy with which these sequences are maintained in ciliates and other lower eukaryotes

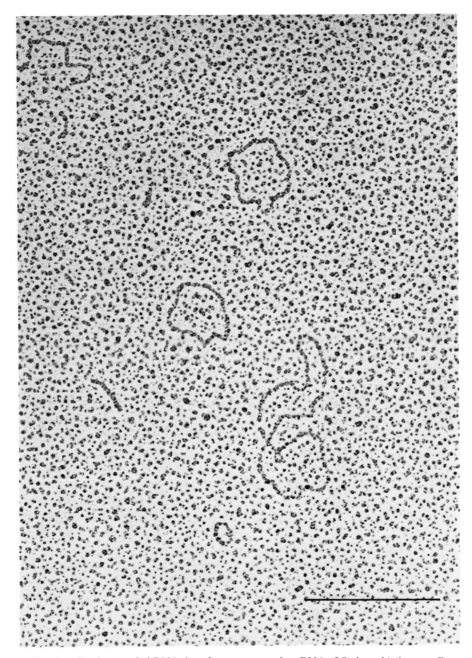

FIG. 3. Single-stranded DNA rings from macronuclear DNA of *Stylonychia lemnae*. Bar represents 0.5 μm.

$$5'\text{-}C_4A_4C_4A_4C_4 \text{ ---} \wedge\!\wedge \text{--- } G_4T_4G_4T_4G_4T_4G_4T_4G_4\text{-}3'$$
$$3'\text{-}G_4T_4G_4T_4G_4T_4G_4T_4G_4 \text{ ---} \wedge\!\wedge \text{--- } C_4A_4C_4A_4C_4\text{- }5'$$

FIG. 4.   Schematic diagram of one macronuclear gene-sized fragment (Lipps *et al.*, 1982).

suggests that they have an essential function presumably as telomeric structures (Blackburn *et al.*, 1983; Blackburn, 1984). This is discussed in more detail below.

2. *Specific Genes*

Prescott *et al.* (1979) proposed that each small DNA molecule of the macronucleus might represent one gene plus all control elements for its own replication and transcription (Lawn *et al.*, 1978; Lipps and Steinbrück, 1978). Most of the gene-sized macronuclear DNA molecules are not yet identified as specific genes, with a few exceptions. In *Oxytricha* sp., *Oxytricha fallax, O. nova,* and *S. mytilus,* a molecule of 7–8 kb (6.8 kb in *Stylonychia,* 7.4 kb in *O. nova*) codes for 25 S, 19 S, and 5.8 S rRNAs (Lipps and Steinbrück, 1978; Rae and Spear, 1978; Spear, 1980; Swanton *et al.*, 1980), whereas a much smaller molecule (690 bp in *Oxytricha,* 200–300 bp in *Stylonychia*) codes for the 5 S rRNA (Lipps and Steinbrück, 1978; Rae and Spear, 1978). Several DNA molecules of different size (400–3200 bp) code for tRNAs (C. L. Jahn, personal communication).

By probing with labeled cloned histone genes from sea urchin (*Psammechinus miliaris*), the five genes coding for histone proteins could be identified in *S. mytilus* macronuclear DNA that was fractionated by agarose gel electrophoresis. Two to six different macronuclear molecules hybridized with individual histone gene probes (Fig. 5a). The five histone genes are not clustered, but instead sequences homologous to the histone genes are scattered among 18 gene-sized molecules and are not transcribed as one polycistronic unit (Elsevier *et al.*, 1978). The actin and tubulin genes exist in multigene families as described for most other eukaryotes. Two molecules of 1.6 and 1.4 kb contain actin genes (Kaine and Spear, 1980, 1982) in *O. fallax* and *O. nova,* and two molecules of 2.03 and 1.88 kb carry single genes for β-tubulin (B. B. Spear, personal communication). Two gene-sized molecules of about 2 kb code for α-tubulin genes (E. Helftenbein, personal communication) in *O. fallax,* and two size classes of DNA (1.85 and 1.73 kb code for α-tubulin in *S. lemnae* (Helftenbein, 1985) (Fig. 5b). The 1.85-kb molecules exist in 55,000 copies and the 1.73-kb molecules in 17,000 copies per macronucleus. The copy numbers are found to be very specific for the particular genes in general and are regulated in a way which is not understood. The sequences of the 1.6-

kb actin gene from *O. fallax* and the 1.85-kb α-tubulin gene from *S. lemnae* are not available. There are no intervening sequences in either gene. By comparing the codon usage in *Stylonychia* as well as in *Oxytricha, Paramecium,* and *Tetrahymena,* unique features were deduced. Open reading frames of the α-tubulin gene in *Stylonychia* (Helftenbein, 1985), the i antigen *A* gene in *Paramecium* (Preer *et al.,* 1985), and the *H3* gene in *Tetrahymena* (Horowitz and Gorovsky, 1985) are interrupted by TAA and TAG stop codons that very likely code for glutamine. TGA seems to be the only translational stop codon used in ciliates. This usage of TAA and TAG to code for amino acids differs from the universal code, as does the usage of TGA in mitochondria of *Paramecium,* fungi, and mammals. Besides this unconventional usage of stop codons, ciliates prefer codons that are very seldom used in other eukaryotes to specify amino acids.

The noncoding regions have a very low G + C content, 25 compared to 49% in the coding region. They contain, besides the already described terminal repeats, sequences that are presumably involved in DNA replication and sequences that are supposed to be signals for the initiation and

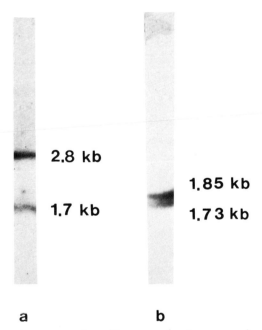

2.8 kb

1.85 kb

1.7 kb

1.73 kb

a b

FIG. 5. Hybridization pattern of specific gene probes to macronuclear DNA of *Stylonychia lemnae.* (a) Histone H4 (courtesy of I. Wefes). (b) α-Tubulin. (Courtesy of E. Helftenbein.)

control of transcription and translation (Helftenbein, 1985). The Pribnow box that occurs at position -20 in the actin gene in *O. fallax* (Kaine and Spear, 1982) is not found in the *Stylonychia* α-tubulin gene. These sequencing data, together with the data from hybridizations with specific gene probes to macronuclear DNA and from transformation experiments in *Stylonychia,* allow the assumption that each macronuclear DNA molecule functions as a replication and transcription unit. It was possible to transform *Stylonychia* stably with a recombinant DNA molecule where the neomycin resistance gene from *E. coli, Tn 5,* was integrated into gene-sized DNA molecules. Further experiments with specifically modified Neo/Ma recombinant DNA will provide more information about the function of the terminal C_4A_4 sequences as well as other sequences that are essential for the replication and transcription of these eukaryotic cells (Wünning and Lipps, 1983).

3. *Function of the Terminal Sequence*

The occurrence of a simple sequence on all the ends of macronuclear gene-sized DNA molecules suggests that it must be an important biological sequence. It is nearly identical in all hypotrichous ciliates examined so far (Oka *et al.,* 1980; Klobutcher *et al.,* 1981; Lipps and Erhardt, 1981) and very similar to terminal sequences found in other lower eukaryotes (Blackburn and Gall, 1978; Johnson, 1980; Emery and Weiner, 1981; Katzen *et al.,* 1981; Yao and Yao, 1981; Blackburn and Challoner, 1984; Shampay *et al.,* 1984; Van der Ploeg *et al.,* 1984).

When the terminal restriction fragment of *Tetrahymena* rDNA was ligated to the ends of a linearized yeast plasmid, this plasmid was maintained indefinitely and allowed its autonomous replication in yeast. Thus, it is able to fulfill at least some requirements of a telomere in yeast (Szostak and Blackburn, 1982; Dani and Zakian, 1983; Murray and Szostak, 1983). However, after replication in yeast, yeast-specific telomeric repeats are added onto the ends of *Tetrahymena* terminal repeats. When *Tetrahymena* rDNA was microinjected into *Xenopus* eggs, the molecules replicated as linear molecules, but grew longer by the addition of DNA onto their ends (Berg *et al.,* 1982).

Similarly, when the ends of rDNA from *Oxytricha* were ligated to the ends of a linearized yeast plasmid, addition of yeast telomeric sequences occurred (Pluta *et al.,* 1984). Ciliate terminal sequences may function in yeast because they are recognized as correct ends for the addition of yeast-specific telomeric repeats (Blackburn, 1984). Whether the terminal C_4A_4 sequences in hypotrichous macronuclear DNA function as telomeric sequences as such or whether additional telomeric sequences are added during replication of these molecules still has to be examined.

In micronuclear DNA of *Oxytricha,* most of the C_4A_4 sequences form large clusters at the ends of the chromosomes and may be involved in the replication of micronuclear chromosomes (Boswell *et al.,* 1982; Dawson and Herrick, 1982, 1984a; Pluta *et al.,* 1982; Boswell *et al.,* 1983; Oka and Honjo, 1983).

This sequence has another remarkable property. When macronuclear DNA is incubated under increased ionic strength (above 100 mM NaCl), it forms long aggregates. The formation of these aggregates is very slow, and the aggregates can be destroyed either by high temperature, high pH (>9.0), or low salt, suggesting that they are stabilized by additional hydrogen bonds. When the terminal sequences are removed by exonuclease digestion, no aggregation occurs, showing that the terminal sequences are responsible for this process. When DNA is incubated in high salt under conditions where ring formation is favored and then subjected to centrifugation on CsCl–ethidium bromide gradients, supercoiled rings can be recovered. In this form no free rotation of DNA strands is possible, showing that aggregation is not just due to cohesion processes via single-stranded DNA. A number of models have been proposed for the structure formed, including three- or four-stranded DNA complexes; however, the exact structure still has to be determined. An indication that such interactions may also occur *in vivo* comes from experiments on chromatin structure of *Oxytricha* (see below), and interactions of this type may also have an impact on eukaryotic chromosome structure in general (Lipps, 1980).

4. *Replication of Macronuclei*

Early autoradiographic studies showed that DNA replication in the macronucleus of hypotrichous ciliates occurs in a cytological structure called the replication band (Gall, 1959; Prescott and Kimball, 1961; Kimball and Prescott, 1962; Prescott, 1966). Replication bands are formed at specific initiation regions and migrate along the macronucleus. In *Euplotes,* replication bands originate at each end of the macronucleus and move toward one another. In *Oxytricha* and *Stylonychia,* a single replication band is formed at one end of each macronucleus.

A great number of ultrastructural investigations have been made on the replication band (Fauré-Fremiet *et al.,* 1957; Kluss, 1962; Inaba and Suganuma, 1966; Chakraborty, 1967; Pheghan and Moses, 1967; Ringertz *et al.,* 1967; Evenson and Prescott, 1970; Ruffolo, 1978; Olins *et al.,* 1981). The four major ultrastructural zones of the replication band are the pre-replication chromatin, the forward zone, the rear zone, and postreplication chromatin (Fig. 6). While DNA synthesis occurs in the rear zone of the replication band (Prescott and Kimball, 1961; Kimball and Prescott, 1962; Ringertz *et al.,* 1967; Evenson and Prescott, 1970), the forward zone

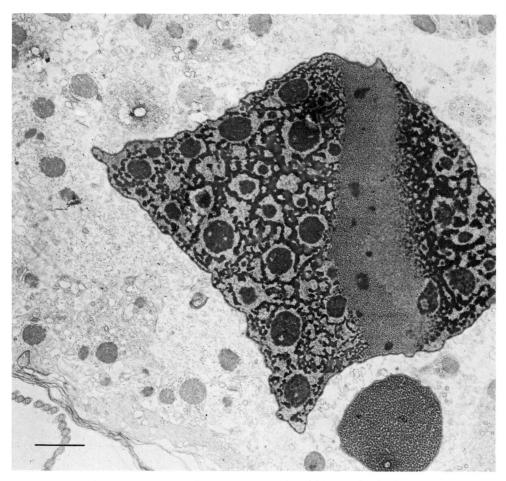

Fɪɢ. 6. Survey view of a *Stylonychia* macronucleus with a replication band. The right portion of the replication band is the forward zone; the portion on the left side is the rear zone. Bar represents 2 μm. (Courtesy of A. L. Olins and D. E. Olins.)

appears to be the region where chromatin reorganization preparatory to DNA synthesis occurs. The chromatin fibers of the forward zone are among the most regular chromatin fibers observed. They have a diameter of 40–50 nm and can be observed to be as long as 400–500 nm (Olins *et al.*, 1981). The longest forward zone chromatin fibers could contain 200–250 nucleosomes containing 40–50 kb of DNA. Since macronuclear DNA molecules have an average length of about 3 kb, this observation suggests that DNA continuity is not required for the length and regularity of the 40- to 50-nm chromatin fibers. These fibers are not observed in the rear zone where a rather diffuse meshwork of 10-nm fibers is observed. Thus, any higher order chromatin structure seems to be absent in the actual zone of DNA replication. This may also be reflected in the absence of Z-DNA in the replication band. While macronuclei react very strongly with antibodies against Z-DNA, no reaction is seen with the replication band (Lipps *et al.*, 1983).

Very little is known about the replication of single gene-sized DNA molecules. Electron microscopic analysis of replicating DNA molecules suggests that replication origins are present near each end of the gene-sized DNA molecules internal to the terminal inverted repeat sequence (Murti and Prescott, 1983). Indeed, sequences resembling ARS sequences of other eukaryotes are found 50–150 bp from one end in the actin gene of *Oxytricha* (Kaine and Spear, 1982) and the α-tubulin gene from *Stylonychia* (Helftenbein, 1985). As shown by transformation experiments, plasmids carrying an ARS sequence from *Dictyostelium* (Hirth *et al.*, 1982) are efficiently replicated in the macronucleus from *Stylonychia* (I. Wünning and H. J. Lipps, unpublished). Sequences from macronuclei and micronuclei from *Oxytricha* that allow autonomously replicating activity in *Saccharomyces* have been cloned. However, the nucleotide sequence of the different ARS regions as well as their location within the gene-sized DNA molecules still has to be determined (Colombo *et al.*, 1984).

B. CHROMATIN STRUCTURE IN MACRONUCLEI

Several important questions can be asked about the chromatin organization of the ciliate nuclei. The occurrence of two functionally different nuclei within the same cell provides the opportunity to compare the chromatin structure of the transcriptionally very active macronucleus with that of the inactive micronucleus. The organization of DNA in short gene-sized DNA molecules in the macronucleus of hypotrichous ciliates may allow the study of the chromatin structures of complete defined eukaryotic genes. Finally, the question arises how these gene-sized DNA molecules are organized *in vivo* in macronuclear chromatin.

All five histone proteins are found in macronuclei and micronuclei. While the four core histones, H2a, H2b, H3, and H4, are very similar to the core histones of other eukaryotes (Lipps *et al.*, 1974; Caplan, 1975; Lipps and Hantke, 1975; Butler *et al.*, 1984), histone H1 is present only in a small quantity in the macronucleus and may have some special properties. Histone H1 from *Oxytricha* macronuclei migrates on SDS–polyacrylamide gels as a protein with a molecular weight around 20,000, very much like histone H1 of other eukaryotes. However, on polyacrylamide urea gels, it migrates in front of the four core histones, although it has an amino acid composition very similar to calf thymus histone H1 (Caplan, 1975). No explanation for this peculiar behavior has been provided, and it is not clear whether H1 histones from other hypotrichous ciliates have similar properties. Like the active chromatin fractions from other eukaryotes, histone H3 and H4 are highly acetylated in macronuclear chromatin, but not in micronuclear chromatin (Lipps, 1975).

Limited digestion of macronuclear and micronuclear chromatin with micrococcal nuclease revealed a nucleosome repeat of 220 bp in macronuclei of *Stylonychia* and *Oxytricha* and a repeat length of 200 bp in micronuclei of *Stylonychia* and 180 bp in micronuclei of *Oxytricha* (Lipps and Morris, 1977; Lawn *et al.*, 1978; Lipps *et al.*, 1978; Butler *et al.*, 1984). In all nuclei, the core particle comprises about 140 bp.

Lysis of macronuclei under low ionic strength releases short chromatin fragments (Fig. 7). The length distribution of these chromatin fragments corresponds well to the size distribution of macronuclear DNA molecules. Analysis of the soluble chromatin by sucrose gradients followed by agarose electrophoresis of the DNA from the different fractions and probing with specific gene probes showed that each chromatin fragment corresponds to one gene-sized DNA molecule (Lipps *et al.*, 1978; Butler *et al.*, 1984). Thus, under these conditions, it should be possible to isolate and characterize specific macronuclear genes in their chromatin configuration.

Observations on the chromatin structure obtained by electron microscopy suggest that the DNA of the macronucleus is organized in a continuous chromatin structure. When macronuclei are gently lysed on a deionized water hypophase, only long chromatin fibers consisting of closely spaced 300 Å superbeads are observed (Meyer and Lipps, 1981). This discrepancy raised the question how the short DNA molecules could be integrated into such a higher structure.

Lysis of macronuclei at salt concentration of 0.5 *M* and above releases high-molecular-weight chromatin. Analysis of this chromatin by sucrose gradient centrifugation and agarose gel electrophoresis showed that the chromatin occurs in long aggregates under these ionic conditions. The aggregate structure can be destroyed either by treatment with low salt, high temperature (65°C), high pH (>9.0), or by treatment with 1% SDS or pronase. These results suggest that the aggregate structure depends on both nucleic acid–nucleic acid and nucleic acid–protein interactions. When macronuclei are lysed in 2 *M* NaCl and the DNA–protein complex is isolated, one major nuclear protein with a molecular weight of about 50,000 copurifies with the DNA. Although this protein has a high affinity for DNA, no sequence-specific binding to purified macronuclear DNA was found. The DNA–protein complex is stable under physiological ionic

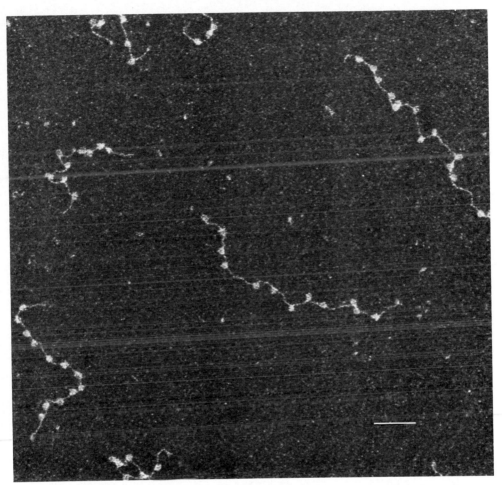

FIG. 7. Darkfield electron micrograph of gene-sized *Euplotes* macronuclear chromatin. Bar represents 50 nm. (Courtesy of C. L. Cadilla, J. M. Harp, A. L. Olins, and D. E. Olins.)

strength. Analysis of the chromatin aggregate by electron microscopy showed structures similar to those observed after *in vitro* aggregation of macronuclear DNA. A model was suggested in which the terminal sequences interact with each other, and this interaction is stabilized under physiological conditions by a nuclear protein (Lipps *et al.*, 1982).

The organization of the terminal C_4A_4 sequence in macronuclear chromatin was examined in more detail. This sequence occurs in the macronucleus in a DNase I hypersensitive configuration (Lipps and Erhardt, 1981). Probing the ends of the molecules with micrococcal nuclease and methidium propyl–ethylene diaminetetraacetic acid (EDTA)–Fe(II) showed a terminal 100-bp DNA–protein complex with nucleosomes phased inward from this complex (Gottschling and Cech, 1984). However, the exact nature of this DNA–protein complex is not known.

III. Macronuclear Development

A. Morphological Events during Macronuclear Development

The differentiation into macronuclei and micronuclei starts with the mitotic division of the synkaryon. However, in hypotrichous ciliates it is not known whether the position of the two nuclei within the cell decides their further differentiation, as in suctoria (Lanners, 1978).

In *Stylonychia,* the first DNA reduction event occurs during the first DNA synthesis phase. Only 30% of the micronuclear chromosomes enter the polytenization stage; the rest becomes pycnotic and eventually disintegrates (Ammermann, 1971; Ammermann *et al.,* 1974). The remaining chromosomes undergo several rounds of replication, resulting in polytene chromosomes with a 100–200 polyteny (Ammermann, 1971; Ammermann *et al.,* 1974). This stage is reached about 40 hours after conjugation. The number of polytene chromosomes has been estimated to be 50–100, and the total number of bands is around 10,000–15,000 (Ammermann *et al.,* 1974). In *Oxytricha,* all micronuclear chromosomes enter the polytenization phase. The macronuclear anlage contains about 120 giant chromosomes. Each chromosome has on average 81 bands, in total about 10,000 bands (Spear and Lauth, 1976). However, when the DNA of polytene chromosomes was analyzed on CsCl gradients, a shift in buoyant density compared to micronuclear DNA was observed. This change in buoyant density has been interpreted as a result of differential underreplication of some DNA sequences (presumably repetitious) during polytene chromosome formation (Spear and Lauth, 1976).

The structure of the polytene chromosomes has been studied in great detail by light microscopy, electron microscopical thin sections, and spreading procedures (Ammermann, 1964, 1965, 1971; Alonso and Perez-Silva, 1965; Kloetzel, 1970; Rao and Ammermann, 1970; Ammermann *et al.,* 1974; Prescott and Murti, 1974; Murti, 1976; Spear and Lauth, 1976; Meyer and Lipps, 1980, 1981, 1984). The polytene chromosomes of hypotrichous ciliates have a diversity of banding patterns very much like those of Diptera. However, in addition they contain heterochromatic blocks. The amount of heterochromatin is different in the different species (Rao and Ammermann, 1970; Spear and Lauth, 1976) (Fig. 8). Typical puffs have never really been demonstrated, and it is unclear whether polytene

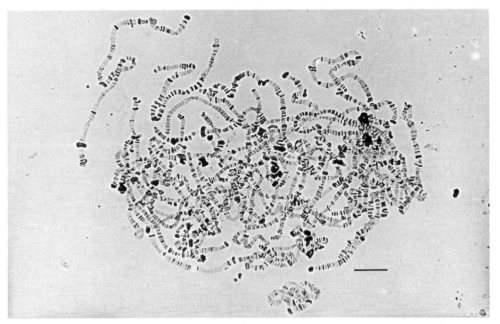

FIG. 8. Polytene chromosomes from *Stylonychia lemnae*. Bar represents 20 μm. (Courtesy of D. Ammermann.)

chromosomes of hypotrichous ciliates have any transcriptional activity (Alonso and Perez-Silva, 1965, Ammermann, 1971; Sapra and Ammermann, 1973; Ammermann *et al.,* 1974; Spear and Lauth, 1976). Bands of different size connected by interband fibers are also clearly visible in the electron microscope in thin sections (Fig. 9). In spread preparations of the developing macronucleus, it became apparent that during polytene chromosome formation a reorganization of chromatin occurs. The micronuclear chromatin and the chromatin of the early macronuclear anlage appear as coiled 30-nm fibers interconnected by 10-nm fibers. In later stages of polytene chromosome formation 30-nm fibers become organized into looplike configurations. Finally, in the polytene chromosomes the bands appear as aggregates of 30-nm loops. These loops contain between 10 and 200 kb of DNA, with an average of about 40 kb, very similar to the DNA content found in a domain (probably corresponding to a chromomere) of *Drosophila* (Worcel, 1978). The interband fibers contain between 1 and 10 kb of DNA (Meyer and Lipps, 1980, 1981, 1984). In these preparations, the organization of the chromatid of *Stylonychia* polytene chromosomes resembles the lampbrush pattern of meiotic prophase chromosomes as described, e.g., in the spermatocytes of *Chironomus* (Keyl, 1975) (Fig. 10).

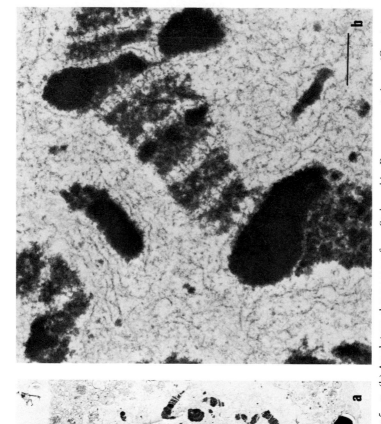

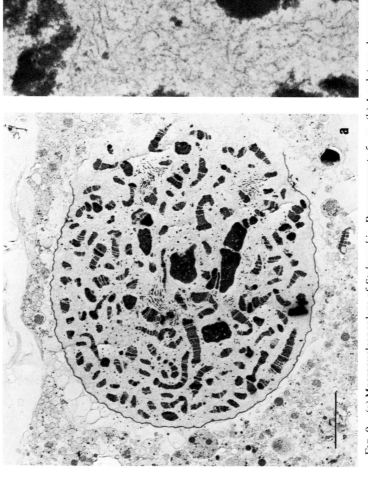

Fig. 9. (a) Macronuclear anlage of *Stylonychia*. Bar represents 5 μm. (b) A polytene chromosome from *Stylonychia*. Bar represents 1 μm. (Courtesy of A. L. Olins and D. E. Olins.)

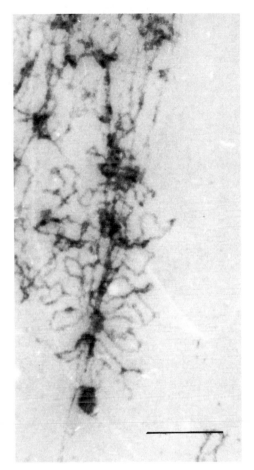

FIG. 10.   Spread preparation of *Stylonychia* polytene chromosomes. Bar represents 0.5 μm.

The degradation of the polytene chromosomes becomes apparent in electron microscopic thin sections by the appearance of membranous material surrounding the bands of the chromosomes. In later stages, bands are completely enclosed in vesicles in which DNA degradation occurs (Fig. 11). However, from these investigations it is not clear whether only the DNA to be degraded is enclosed in vesicles or whether all sequences are contained in the vesicles. The latter would imply a selective digestion of sequences occurring in the vesicles (Kloetzel, 1970; Murti, 1976). A clear answer could only be obtained by isolation of the vesicles and analysis of their DNA. In *Stylonychia,* the number of vesi-

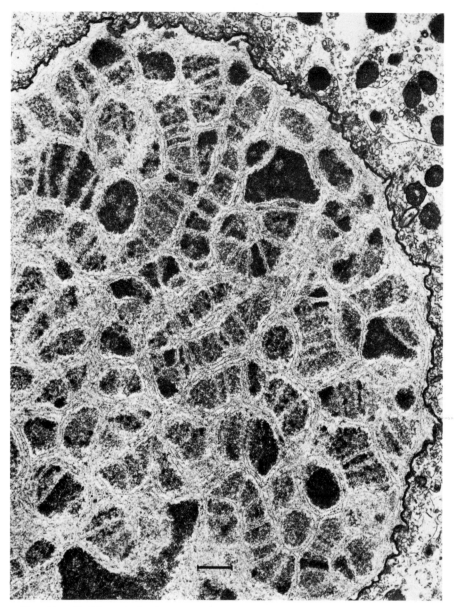

Fig. 11.   Vesicle stage in the macronuclear anlage of *Stylonychia* (Kloetzel, 1970). Bar represents 1 μm.

cles is identical to the number of bands (Ammermann, 1971; Ammermann *et al.*, 1974); in *Oxytricha,* the number of vesicles is only 2700, therefore, several bands would have been enclosed in the same vesicle (Spear and Lauth, 1976).

A different picture is obtained when chromatin elimination is examined in spread preparations. When elimination starts, the 30-nm loops, representing the bands of the chromosome, are released from the axis in the form of chromatin rings. In later stages, only the 10-nm fiber is retained. The 30-nm loops are subsequently degraded into superbeads and eventually completely degraded (Fig. 12). From this observation, it was concluded that macronuclear sequences are contained in the interbands of the chromosomes, whereas the sequences to be degraded are mainly localized in the bands and heterochromatin blocks of the polytene chromosomes. Furthermore, the reorganization of chromatin during polytene chromosome formation may reflect a reorganization of DNA sequences. Thus, the main function of polytene chromosomes in hypotrichous ciliates may be related to the subsequent specific elimination events. It should be noted that elimination in the form of chromatin rings has also been described in the copepoda *Cyclops* (Beermann and Meyer, 1980).

The percentage of eliminated DNA is correlated with the DNA content of the micronucleus. In *Stylonychia* and *Oxytricha,* 95% of the DNA is degraded, leading to a DNA-poor stage with a DNA content of 2C (the DNA content of the diploid micronucleus). In *Euplotes,* whose micronuclear DNA content is much lower, about 80% of the DNA is decomposed and the DNA content of the DNA-poor stage is 20C (Ammermann, 1971; Ammermann *et al.*, 1974; Steinbrück *et al.*, 1981). The DNA-poor stage lasts for about 40 hours. With the beginning of the second DNA synthesis phase, replication bands move toward the center of the macronuclear anlage. At this stage the first nucleoli also appear. After several rounds of replication, the new macronucleus then reaches its final DNA content.

B. Gene Processing during Macronuclear Development

By investigating the morphological events during macronuclear development from micronuclei, it became obvious that an extensive reorganization of the micronuclear genome takes place.

As described in Section II,A, the macronuclear DNA molecules of all ciliates carry a tandemly repeated DNA sequence on each end that is arranged as an inverted repeat. These terminal repeat sequences do not flank the prospective macronuclear genes while they are still organized within the chromosomal DNA of the micronuclei. This is shown for the hypotrichous ciliates *O. nova* (Boswell *et al.*, 1982; Klobutcher *et al.*,

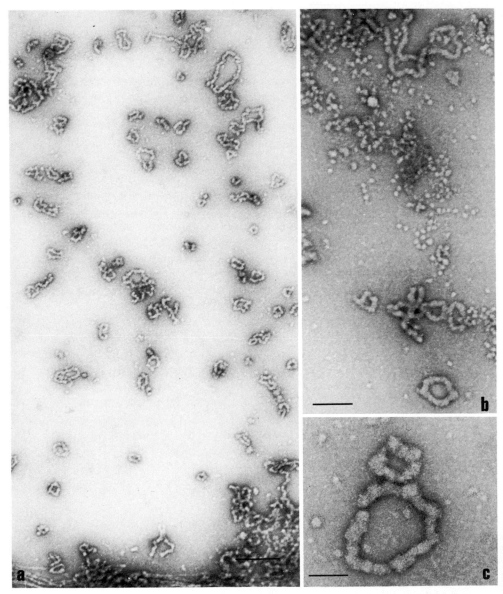

FIG. 12. Eliminated chromatin loops in the macronuclear anlage from *Stylonychia*. Bar represents (a) 0.5 μm; (b) 0.2 μm; (c) 0.1 μm (Meyer and Lipps, 1980).

1984) and *O. fallax* (Dawson and Herrick, 1982, 1984a,b). Hybridization experiments with labeled C_4A_4 containing the macronuclear inverted terminal repeat (Ma-ITR) or the C_4A_4 containing fragment in a recombinant plasmid (pOH7) to restriction digests of chromosomal DNA revealed the existence of C_4A_4 sequences in the micronuclear genome. The overwhelming amount of the homologous sequences is arranged in large blocks without restriction sites. The blocks are heterogeneous in size after digestion with different restriction enzymes. The main DNA fraction that hydridizes to the Ma-ITR in *O. fallax* is found to be about 9 kb following a complete *Hin*dIII digestion, while a minor fraction can be detected at around 5 kb. Restriction digests with other enzymes like *Pst*I and *Bam*H1 generated very large hybridizing fragments of 20 kb or larger. Blocks of C_4A_4 repeats do not contain any restriction sites for known restriction enzymes. In *O. nova,* after digestion with *Eco*RI, *Hin*dIII, and *Alu*I, a large-sized fraction of micronuclear DNA (9–23 kb) hybridized to a C_4A_4-specific probe isolated from the recombinant plasmid pOH7 (Boswell *et al.,* 1982). This probe did not hybridize to cloned micronuclear DNA fragments that had been shown to contain macronuclear DNA fragments.

The large C_4A_4 clusters are located at the micronuclear telomeres, since they are rapidly destroyed after *Bal*31 digestion of total DNA (Dawson and Herrick, 1982, 1984a). The function of the large terminal repeats and the recently detected, rare (less than 5% of the micronuclear C_4A_4 sequences) internally localized, shorter C_4A_4 repeats (20–40 bp) is not known (Dawson and Herrick, 1984b). Since the telomeric DNA sequences are not located adjacent to the macronuclear genes in chromosomes, they must be acquired after gene excision during macronuclear development. There are two ways these genes could receive their unique ends. Ends could be synthesized *de novo* on all macronuclear gene-sized molecules, or they could be transposed from micronuclear DNA in the course of the macronuclear development. There exists enough C_4A_4 DNA repeats in the micronuclear genome to allow a rearrangement onto the macronuclear DNA ends.

Investigating micronuclear recombinant plasmids, Dawson *et al.* (1983) demonstrated the existence of short tandemly arranged high-repetitious sequences in *S. pustulata* (Dawson *et al.,* 1983). The approximately 160-bp-long repeat sequences can be classified into two families. They obviously are part of the highly repetitious sequences that are part of the more rapidly renaturing chromosomal DNA in reassociation kinetics. Repeats of a similar length (180 and 370 bp) have been detected in *O. fallax* (Dawson *et al.,* 1984). Large 11-kb DNA fragments after *Eco*RI digestion which represent a repetitive sequence family of the chromosomal DNA of

O. nova are also found in recombinant phages along with phages that contain repetitive plus unique or unique DNA only (Boswell *et al.*, 1983). These latter repetitive sequences, which in total comprise up to 10% of the micronuclear sequences, obviously are interrupted by macronuclear genes or gene families, which are clusters of two to five genes.

More information on the splicing events is provided by comparing macronuclear and micronuclear gene pairs by sequencing (Oka and Honjo, 1983; Klobutcher *et al.*, 1984). There are no C_4A_4 repeats adjacent to the micronuclear copies of genes, in keeping with earlier hybridization experiments. A full-length, cloned, macronuclear gene called *C2* and its micronuclear counterpart were studied in detail (Klobutcher *et al.*, 1984). Three versions differing by ~3% base-pair nonmatching were identified in micronuclear DNA; only two of these versions could be found in macronuclear DNA, suggesting that one of the *C2* genes did not survive macronuclear development. The mismatching was confined almost entirely to regions outside the open reading frame of the gene. The mismatching within the coding region did not change the amino acid sequence specified by the gene.

Comparison of the *C2* gene in a matched pair of macronuclear–micronuclear genes showed that the micronuclear copy contains three short (49-, 49-, and 32-bp) segments that are removed from the gene during macronuclear development. These sequences, called internal eliminated sequences (IESs), resemble introns, but their relationship to introns is not known. One IES is located in a 3′-nontranscribed region of the gene, another is in the 3′-noncoding transcribed region, and the third occurs in the polypeptide coding region. It is clear that they are removed during macronuclear development, showing that DNA processing includes cutting and splicing of micronuclear DNA to produce a macronuclear gene. The IESs have some structural features in common. Each IES is flanked by a short direct repeat (3–5 bp) and has adjacent, nearly perfect, inverted repeats, a structure similar to transposons. This is very likely, the structure that allows—in the case of hypotrichous ciliates—a precise splicing out of the internal sequences in the course of the development of the macronucleus.

Splicing events that specifically eliminate micronuclear DNA and rearrange DNA have also been described for *Tetrahymena*, a holotrichous ciliate (Blackburn and Gall, 1978; Blackburn *et al.*, 1983; Callahan *et al.*, 1984; Yao *et al.*, 1981, 1984; Yao and Yao, 1981). A precise breakage and rejoining of DNA is found to occur very frequently and finally leads to the formation of a somatic macronucleus from the germinal micronucleus.

Intermediate stages in the processing of gene-sized molecules of DNA during macronuclear development have been observed in the marine hypotrich *Euplotes crassus* (Roth and Prescott, 1985). Immediately after transection of the polytene chromosomes, one of three genes studied appeared in an intermediate DNA fragment about twice as long as the macronuclear version of the gene. After some hours, this intermediate disappeared and was replaced by a second smaller intermediate consisting of the gene with oversized telomeric sequences (C_4A_4). For two other genes, the only intermediates detected were of the latter type, i.e., gene-sized molecules with oversized telomeres. These penultimate intermediates were succeeded late in macronuclear development with the mature form macronuclear gene-sized molecules; i.e., the telomeres had been trimmed.

All of these observations on various hypotrichs have provided a general outline of macronuclear development in which genes are excised from chromosomes, large amounts of unique and all repetitious sequences are eliminated, genes are cut and spliced to remove intron-like sequences, oversized telomeric sequences are added and then trimmed, and a few genes (such as rDNA) are finally differentially amplified. These complicated steps provide an unusual opportunity to study the ways in which DNA sequences can be manipulated by an organism.

ACKNOWLEDGMENTS

This work was supported in part by NIGMS Research Grant GM19199 to D. M. Prescott.

REFERENCES

Alonso, P., and Perez-Silva, J. (1965). *Nature (London)* **205**, 313–314.
Ammermann, D. (1964). *Naturwissenschaften* **51**, 249.
Ammermann, D. (1965). *Arch. Protistenkd.* **108**, 109–152.
Ammermann, D. (1971). *Chromosoma* **33**, 209–238.
Ammermann, D., and Schlegel, M. (1983). *J. Protozool.* **30**, 290–294.
Ammermann, D., Steinbrück, G., von Berger, L., and Hennig, W. (1974). *Chromosoma* **45**, 401–419.
Beermann, S., and Meyer, G. F. (1980). *Chromosoma* **77**, 277–283.
Berg, C. A., Gans, E. R., and Gall, J. G. (1982). *J. Cell Biol.* **95**, 220a.
Blackburn, E. H. (1984). *Cell* **37**, 7–8.
Blackburn, E. H., and Challoner, P. B. (1984). *Cell* **36**, 447–457.
Blackburn, E. H., and Gall, J. G. (1978). *J. Mol. Biol.* **120**, 33–53.
Blackburn, E. H., Budarf, M. L., Challoner, P. B., Cherry, J. M., Howard, E. A., Katzen, A. L., Pan, W. C., and Ryan, T. (1983). *Cold Spring Harbor Symp. Quant. Biol.* **47**, 1195–1207.
Bostock, C. J., and Prescott, D. M. (1972). *Proc. Natl. Acad. Sci. U.S.A.* **69**, 139–142.
Boswell, R. E., Klobutcher, L. A., and Prescott, D. M. (1982). *Proc. Natl. Acad. Sci. U.S.A.* **79**, 3255–3259.

Boswell, R. E., Jahn, C. L., Greslin, A. F., and Prescott, D. M. (1983). *Nucleic Acids Res.* **11**, 3651–3663.

Butler, A. P., Laughlin, T. J., Cadilla, C. L., Henry, J. M., and Olins, D. E. (1984). *Nucleic Acids Res.* **12**, 3201–3217.

Callahan, R. M., Shalke, G., and Gorovsky, M. A. (1984). *Cell* **36**, 441–445.

Caplan, E. B. (1975). *Biochim. Biophys. Acta* **407**, 109–113.

Chakraborty, J. (1967). *J. Protozool.* **14**, 59–64.

Colombo, M. M., Swanton, M. T., Donini, P., and Prescott, D. M. (1984). *Mol. Cell. Biol.* **4**, 1725–1729.

Dani, G. M., and Zakian, V. A. (1983). *Proc. Natl. Acad. Sci. U.S.A.* **80**, 3406–3410.

Dawson, D., and Herrick, G. (1982). *Nucleic Acids Res.* **10**, 2911–2923.

Dawson, D., and Herrick, G. (1984a). *Cell* **36**, 171–177.

Dawson, D., and Herrick, G. (1984b). *Mol. Cell. Biol.* **4**, 2661–2667.

Dawson, D., Stetler, D. J., Swanton, M. T., and Herrick, G. (1983). *J. Protozool.* **30**, 629–635.

Dawson, D., Buckley, B., Cartinhour, S., Myers, R., and Herrick, G. (1984). *Chromosoma* **90**, 289–294.

Elsevier, S. M., Lipps, H. J., and Steinbrück, G. (1978). *Chromosoma* **69**, 291–306.

Emery, H. S., and Weiner, A. M. (1981). *Cell* **26**, 411–419.

Evenson, D. P., and Prescott, D. M. (1970). *Exp. Cell Res.* **63**, 245–252.

Fauré-Fremiet, E., Rouiller, C., and Gauchery, M. (1957). *Exp. Cell Res.* **12**, 135–144.

Gall, J. G. (1959). *J. Biophys. Biochem. Cytol.* **9**, 295–308.

Gottschling, D. E., and Cech, T. R. (1984). *Cell* **38**, 501–510.

Grell, K. G. (1973). "Protozoology." Springer-Verlag, Berlin and New York.

Heckmann, K. (1963). *Arch. Protistenkd.* **106**, 393–421.

Heckmann, K. (1964). *Z. Vererbungsl.* **95**, 114–124.

Helftenbein, E. (1985). *Nucleic Acids Res.* **13**, 415–432.

Herrick, G., and Wesley, R. D. (1978). *Proc. Natl. Acad. Sci. U.S.A.* **75**, 2626–2630.

Hirth, K. P., Edwards, C., and Firtel, R. A. (1982). *Proc. Natl. Acad. Sci. U.S.A.* **79**, 7356–7360.

Horowitz, S., and Gorovsky, M. A. (1985). *Proc. Natl. Acad. Sci. U.S.A.* **82**, 2452–2455.

Inaba, F., and Suganuma, Y. (1966). *J. Protozool.* **13**, 137–147.

Johnson, E. M. (1980). *Cell* **22**, 875–886.

Kaine, B. P., and Spear, B. B. (1980). *Proc. Natl. Acad. Sci. U.S.A.* **77**, 5336–5340.

Kaine, B. P., and Spear, B. B. (1982). *Nature (London)* **295**, 430–432.

Katzen, A. L., Cann, G. M., and Blackburn, E. H. (1981). *Cell* **24**, 313–320.

Keyl, H. G. (1975). *Chromosoma* **51**, 75–91.

Kimball, R. F., and Prescott, D. M. (1962). *J. Protozool.* **9**, 88–92.

Klobutcher, L. A., Swanton, M. T., Donini, P., and Prescott, D. M. (1981). *Proc. Natl. Acad. Sci. U.S.A.* **78**, 3015–3019.

Klobutcher, L. A., Jahn, C. L., and Prescott, D. M. (1984). *Cell* **36**, 1045–1055.

Kloetzel, J. A. (1970). *J. Cell Biol.* **47**, 395–407.

Kluss, B. C. (1962). *J. Cell Biol.* **13**, 462–465.

Kraut, H., and Lipps, H. J. (1983). *In* "Advances in Invertebrate Reproductions 3" (W. Engels, eds.), pp. 535–540. Elsevier, Amsterdam.

Lanners, H. N. (1978). *Arch. Protistenkd.* **115**, 370–385.

Lauth, M. R., Spear, B. B., Heumann, J., and Prescott, D. M. (1976). *Cell* **7**, 67–74.

Lawn, R. M., Heumann, J. M., Herrick, G., and Prescott, D. M. (1978). *Cold Spring Harbor Symp. Quant. Biol.* **42**, 483–492.

Lipps, H. J. (1975). *Cell Diff.* **4**, 123–129.

Lipps, H. J. (1980). *Proc. Natl. Acad. Sci. U.S.A.* **77**, 4104–4107.
Lipps, H. J., and Erhardt, P. (1981). *FEBS Lett.* **126**, 219–222.
Lipps, H. J., and Hantke, K. G. (1975). *Chromosoma* **49**, 309–320.
Lipps, H. J., and Morris, N. R. (1977). *Biochem. Biophys. Res. Commun.* **74**, 230–234.
Lipps, H. J., and Steinbrück, G. (1978). *Chromosoma* **69**, 21–26.
Lipps, H. J., Sapra, G. R., and Ammermann, D. (1974). *Chromosoma* **45**, 273–280.
Lipps, H. J., Nock, A., Riewe, M., and Steinbrück, G. (1978). *Nucleic Acids Res.* **5**, 4699–4709.
Lipps, H. J., Gruissem, W., and Prescott, D. M. (1982). *Proc. Natl. Acad. Aci. U.S.A.* **79**, 2495–2499.
Lipps, H. J., Nordheim, A., Lafer, E. M., Ammermann, D., Stollar, B. C., and Rich, A. (1983). *Cell* **32**, 435–441.
Meyer, G. F., and Lipps, H. J. (1980). *Chromosoma* **77**, 285–297.
Meyer, G. F., and Lipps, H. J. (1981). *Chromosoma* **82**, 309–314.
Meyer, G. F., and Lipps, H. J. (1984). *Chromosoma* **84**, 107–110.
Murray, A. W., and Szostak, J. W. (1983). *Nature (London)* **305**, 189–193.
Murti, K. G. (1976). *Handb. Genet.* **5**, 113–137.
Murti, K. G., and Prescott, D. M. (1983). *Mol. Cell. Biol.* **3**, 1562–1566.
Nock, A. (1981). *Chromosoma* **83**, 209–220.
Oka, Y., and Honjo, T. (1983). *Nucleic Acids Res.* **11**, 4325–4333.
Oka, Y., Shiota, S., Nakai, S., Nishida, Y., and Okubo, S. (1980). *Gene* **10**, 301–306.
Olins, A. L., Olins, D. E., Franke, W. W., Lipps, H. J., and Prescott, D. M. (1981). *Eur. J. Cell Biol.* **25**, 120–130.
Pheghan, W. D., and Moses, M. J. (1967). *J. Cell Biol.* **35**, 103a.
Pluta, A. F., Kaine, B. P., and Spear, B. B. (1982). *Nucleic Acids Res.* **10**, 8145–8154.
Pluta, A. F., Dani, G. M., Spear, B. B., and Zakian, V. A. (1984). *Proc. Natl. Acad. Sci. U.S.A.* **81**, 1475–1479.
Preer, J. R., Jr., Preer, L. B., Rudman, B. M., and Barnett, A. J. (1985). *Nature (London)* **314**, 188–190.
Prescott, D. M. (1966). *J. Cell Biol.* **31**, 1–9.
Prescott, D. M. (1983). *In* "Modern Cell Biology" (J. R. McIntosh, ed.), Vol. 2, pp. 329–352. Liss, New York.
Prescott, D. M., and Kimball, R. F. (1961). *Proc. Natl. Acad. Sci. U.S.A.* **47**, 686–693.
Prescott, D. M., and Murti, K. G. (1974). *Cold Spring Harbor Symp. Quant. Biol.* **38**, 609–618.
Prescott, D. M., Bostock, C. J., Murti, K. G., Lauth, M. R., and Gamow, E. (1971). *Chromosoma* **34**, 355–366.
Prescott, D. M., Murti, K. G., and Bostock, C. J. (1973). *Nature (London).* **242**, 596–600.
Prescott, D. M., Heumann, J., Swanton, M., and Boswell, R. E. (1979). *In* "Specific Eukaryotic Genes" (J. Engberg, H. Klenow, and V. Leick, eds.), Vol. 13, pp. 85–99. Munksgaard, Copenhagen.
Rae, P. M. M., and Spear, B. B. (1978). *Proc. Natl. Acad. Sci. U.S.A.* **75**, 4992–4996.
Raikov, J. B. (1982). "The Protozoan Nucleus." Springer-Verlag, Berlin and New York.
Rao, M. V. N., and Ammermann, D. (1970). *Chromosoma* **29**, 246–254.
Ringertz, N. R., Ericsson, J. L. E., and Nilsson, O. (1967). *Exp. Cell Res.* **48**, 97–117.
Roth, M., and Prescott, D. M. (1985). *Cell* **41**, 411–417.
Ruffolo, J. J. (1978). *J. Morphol.* **157**, 211–222.
Sapra, G. R., and Ammermann, D. (1973). *Exp. Cell Res.* **78**, 168–174.
Shampay, J., Szostak, J. W., and Blackburn, E. H. (1984). *Nature (London)* **310**, 154–157.
Spear, B. B. (1980). *Chromosoma* **77**, 193–202.

Spear, B. B., and Lauth, M. R. (1976). *Chromosoma* **54,** 1–13.

Steinbrück, G. (1983). *Chromosoma* **88,** 156–163.

Steinbrück, G., and Schlegel, M. (1983). *J. Protozool.* **30,** 294–300.

Steinbrück, G., Haas, I., Hellmer, K. H., and Ammermann, D. (1981). *Chromosoma* **83,** 199–208.

Swanton, M. T., Greslin, A. F., and Prescott, D. M. (1980). *Chromosoma* **77,** 203–215.

Szostak, J. W., and Blackburn, E. H. (1982). *Cell* **29,** 245–255.

van der Ploeg, L. T. H., Lin, A. Y. C., and Borst, P. (1984). *Cell* **36,** 459–468.

Wesley, R. D. (1975). *Proc. Natl. Acad. Sci. U.S.A.* **72,** 678–682.

Worcel, A. (1978). *Cold Spring Harbor Symp. Quant. Biol.* **78,** 313–324.

Wünning, I., and Lipps, H. J. (1983). *EMBO J.* **2,** 1753–1757.

Yao, M.-C., and Yao, C. H. (1981). *Proc. Natl. Acad. Sci. U.S.A.* **78,** 7436–7439.

Yao, M.-C., Blackburn, E. H., and Gall, J. G. (1981). *J. Cell Biol.* **90,** 515–520.

Yao, M.-C., Choi, J., Yokoyama, S., Austerberry, C. F., and Yao, C. H. (1984). *Cell* **36,** 433–440.

INTERNATIONAL REVIEW OF CYTOLOGY, VOL. 99

Structure and Formation of Telomeres in Holotrichous Ciliates

Elizabeth H. Blackburn

Department of Molecular Biology, University of California, Berkeley, California

I. Introduction

Telomeres are the ends of linear eukaryotic chromosomes. Many cytogenetic observations indicate that they must have highly specialized properties (reviewed in Blackburn and Szostak, 1984). One functional aspect of chromosomal telomeres is apparent when their behavior *in vivo* is contrasted with that of newly broken chromosome ends, produced by either mechanical rupture or X-irradiation. Such freshly broken ends generally will undergo fusion with other broken ends or are highly recombinogenic and in some situations are subject to degradation. A manifestation of these properties at a molecular level is seen when linear DNA molecules with restriction enzyme-cut ends are introduced into yeast cells: they are generally highly recombinogenic or are degraded (Orr-Weaver *et al.,* 1981). In contrast, normal telomeres do not show this

29

reactivity, suggesting that one role of a telomere is to protect and stabilize the end of the chromosome.

Another question which needs to be addressed when the structure of telomeres is considered is that of their replication. All known DNA polymerases require a primer (either DNA or RNA) bearing a 3'-hydroxyl group to initiate DNA synthesis. In the case of an RNA primer, it is removed once synthesis has been primed. The removal of the primer would result in a 5'-terminal gap being left in one strand at each end of a replicated linear DNA molecule. Although linear DNA viruses have solved this problem in a variety of ways, there was until recently no direct evidence on how the ends of linear eukaryotic chromosomes are able to be maintained indefinitely through repeated rounds of replication. Various models were proposed to explain telomere replication (reviewed in Blackburn and Szostak, 1984). However, structural analysis of telomeres was necessary before such models could be evaluated.

II. Ciliated Protozoa as Systems for Studying Telomeres

Ciliated protozoa have been uniquely useful for studying telomeres, largely because of the structure of their macronuclear genomes. Each single-celled organism contains two types of nuclei: the diploid, germline micronucleus and the polygenomic somatic macronucleus, which is the site of gene expression and which divides amitotically in vegetative growth. The micronucleus divides mitotically during vegetative growth and carries out meiosis in the sexual process, conjugation. After conjugation each cell receives a new macronucleus differentiated from a mitotic sister of the micronucleus, and the old macronucleus is destroyed. Several studies have shown that a profound reorganization of the genome occurs in macronuclear differentiation in many ciliates. Although the quantitative and cytological properties of this genomic reorganization process are very different among the different classes of ciliates, examination of their resulting macronuclear genomes has revealed major common features even between organisms as apparently unlike in their macronuclear development as *Tetrahymena*, a holotrichous ciliate, and *Oxytricha*, a hypotrichous ciliate (see Kraut *et al.*, this volume, and Brunk, this volume).

In both hypotrichous and holotrichous ciliates, this reorganization involves the fragmentation of the micronuclear chromosomes into subchromosomal linear DNA molecules. Thus, as described in Kraut *et al.* (this volume) and Brunk (this volume), the macronuclear genomes of many ciliates consist of relatively short, subchromosomal, multicopy linear DNA molecules.

The lengths of eukaryotic chromosomes commonly exceed thousands of kilobase pairs (kb), making direct analyses of the telomeres of such chromosomes impractical until, as described below, recent methods for cloning at least some telomeres in yeast were developed. Therefore, molecular and biochemical analyses of chromosomal termini were confined to systems, such as ciliates, with naturally occurring, short, linear nuclear DNAs. In ciliates each macronuclear DNA molecule is replicated autonomously in the course of vegetative cell divisions. Indeed, as will be discussed in this article, it appears that the macronuclear DNA molecules of ciliates terminate in structures which can accurately be described as telomeres by the criteria outlined above, i.e., their ability to stabilize and allow indefinite replication of the end of a linear DNA or chromosome. Furthermore, detailed molecular analyses have shown that the DNA sequences and structures at the molecular ends of macronuclear DNAs in ciliates are remarkably similar to those of the telomeres of a number of other eukaryotic species (reviewed in Blackburn, 1984).

III. Macronuclear DNA Molecules in Holotrichous Ciliates

The macronuclear DNAs in ciliates have average lengths of ~2–3 kb in hypotrichs, such as *Oxytrichia*, *Stylonychia*, and *Euplotes* (see Kraut *et al.*, this volume) and ~10^2–10^3 kb in holotrichous ciliates such as *Tetrahymena*, *Glaucoma*, and *Paramecium* (Preer and Preer, 1979; Katzen *et al.*, 1981; see Brunk, this volume). Although the average length of macronuclear DNA molecules in these holotrichs is about two orders of magnitude greater than that of hypotrichs, *Tetrahymena* and *Glaucoma* also have amplified ribosomal RNA genes (rDNA) in their macronuclei whose telomeres have been amenable to direct structural analysis.

In *Tetrahymena thermophila*, rearrangement and amplification of the single, chromosomally integrated micronuclear copy of the ribosomal RNA gene (Yao and Gall, 1977; Pan *et al.*, 1982) results in its conversion to several thousand linear rDNA molecules in the macronucleus. These relatively short (21-kb) molecules are in the form of palindromic dimers in the mature macronucleus, each consisting of two rRNA genes in head-to-head arrangement (reviewed in Blackburn, 1982). The rDNA molecules are present at an estimated ~9000 copies per macronucleus in *T. thermophila* (Pearlman *et al.*, 1979), comprising 1–2% of the total macronuclear DNA. Their small size and relative abundance allowed direct analysis of their telomeric sequences and structures (Blackburn and

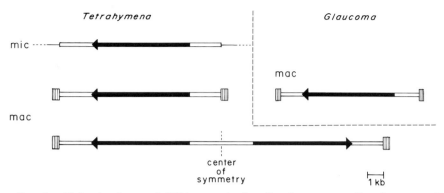

FIG. 1. Molecular forms of rRNA genes in the ciliated protozoans *Tetrahymena* and *Glaucoma*. A single chromosomally integrated rRNA gene, indicated by the bar, is found in the micronuclear genome of *T. thermophila* flanked by micronuclear-limited DNA sequences, as indicated by horizontal lines (King and Yao, 1982; Yao *et al.*, 1985). The comparable sequence in *Glaucoma* has not been analyzed. Solid bars and arrowheads indicate rRNA transcription units and $5' \rightarrow 3'$ polarity of transcripts; open bars are nontranscribed spacer regions. Rectangles with vertical bars indicate blocks of telomeric C_4A_2 repeats (Blackburn and Gall, 1978; Katzen *et al.*, 1981; Pan and Blackburn, 1981; Challoner and Blackburn, 1985). Adapted from Blackburn *et al.* (1984).

Gall, 1978). Early in macronuclear development free rRNA genes are also found in the form of linear, multicopy, self-replicating 11-kb molecules, each carrying a single gene (Pan and Blackburn, 1981), although these are eventually lost from mature macronuclei. A similar molecular form of amplified rDNA, ~9 kb in length, is found in the mature macronucleus of the related tetrahymenid *Glaucoma chattoni* (Katzen *et al.*, 1981). The different molecular forms of the rDNA in these ciliates are diagrammed in Fig. 1.

Many non-rDNA molecules in the *G. chattoni* macronuclear genome are less than 20 kb in length and have also been amenable to analysis (Katzen *et al.*, 1981). Even *T. thermophila* non-rDNA macronuclear DNAs (average length ~600 kb) represent a source from which telomeric sequences have been directly analyzed by cloning and sequencing (Blackburn *et al.*, 1984; T. Ryan, E. Spangler, and E. Blackburn, unpublished data). The macronuclear gene for immobilization antigen A in *Paramecium tetraurelia* and for antigen G in *Paramecium primaurelia* has in both cases been located by restriction mapping to within a few kilobase pairs of the end of a macronuclear genomic DNA molecule (Epstein and Forney, 1984). As described below, this finding has led to some promising genetic approaches to analyzing the process of telomere formation in *P. tetraurelia*.

A. STRUCTURE OF TELOMERES OF THE RIBOSOMAL RNA GENES OF *Tetrahymena*

The DNA sequence and structure at the ends of the macronuclear palindromic, linear rDNA in *T. thermophila* have been determined (Blackburn and Gall, 1978; Blackburn *et al.*, 1983). No significant differences in these end structures and their DNA sequences have been found between strains of *T. thermophila* or between different *Tetrahymena* species.

With the exception of the termini, the population of purified rDNA molecules from a clone of any given inbred *Tetrahymena* species is homogeneous with respect to length and DNA sequence. Thus, upon digestion of the rDNA with any restriction endonuclease, internal fragments containing sequences out to about 400 bp (base pairs) from the ends are homogeneous. However, the population of restriction fragments containing the ends of the palindromic rDNA molecules always shows length heterogeneity. For example, in *T. thermophila*, the restriction endonuclease *Alu*I cleaves the rDNA into several fragments, and the heterogeneity of the population of small terminal fragments is clearly seen (Blackburn and Gall, 1978). After gel electrophoresis, they comprise a heterodisperse band made up of fragments whose lengths range from about 400–550 bp. Similar heterogeneity has been found at the rDNA termini of all *Tetrahymena* species (e.g., Wild and Gall, 1979). As might be expected from this observation, the structure at the termini of these molecules was found to be complex (Blackburn and Gall, 1978; Blackburn *et al.*, 1983). Each end of the rDNA molecule contains tandem repeats of the hexanucleotide:

$$5'\text{-CCCCAA-}3'$$
$$3'\text{-GGGGTT-}5'$$

Several single-strand, one-nucleotide gaps were found on the CCC-CAA(C_4A_2) strand at specific positions within the cluster of repeats. On the GGGGTT (G_4T_2) strand, at least one single-strand gap was found at position inferred to be internal to the C_4A_2 strand breaks. More complete analysis of the rDNA ends in later experiments allowed a more accurate estimate of the number of C_4A_2 repeats and the positions of the specific single-stranded breaks to be made. Over 50 tandem repeats of the C_4T_2 were found in the majority, if not all, of the rDNA molecules. The single-strand breaks were confined to the terminal (distal) ~100-bp region of the block of repeats (Blackburn *et al.*, 1983). The structure proposed for the rDNA telomere is shown in Fig. 2.

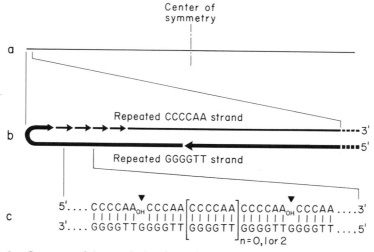

FIG. 2.   Structure of the terminal regions of the palindromic macronuclear rDNA molecules of *T. thermophila* (Blackburn and Gall, 1978; Blackburn *et al.*, 1983). (a) The palindromic rDNA molecule. (b) The terminal few hundred base pairs. Positions of specific
single-strand breaks on the repeated G_4T_2 strand (thick line) and repeated C_4A_2 strand (thin
line) are shown as arrowheads. The innermost break is $\sim$100 bp from the extreme end of the
molecule, which is shown here as a putative fold-back structure formed from the G_4T_2 strand
(see text). Arrows indicate $5' \rightarrow 3'$ polarity of each strand. (c) DNA sequence of the region
encompassing two adjacent single-strand breaks ($\blacktriangledown$) on the C_4A_2 strand. Adapted from
Blackburn (1982).

    The evidence for this structure was obtained by a combination of exper-
iments involving *in vitro* labeling of purified rDNA, fractionation of the
labeled polynucleotides produced, and their analysis by DNA-sequencing
methods (Blackburn and Gall, 1978; Blackburn *et al.*, 1983). Purified
native rDNA molecules can act as a template-primer system for nick-
translation synthesis by *Escherichia coli* DNA polymerase I, which
preferentially incorporates labeled [α-^{32}P]deoxynucleoside triphosphates
into the terminal restriction fragments of the rDNA by template-directed
synthesis of C_4A_2 repeats and/or G_4T_2 repeats. This synthesis is specifically initiated at the $3'$-OH groups of the single-strand breaks shown in
Fig. 2.
    As described below, the extreme termini of the rDNA molecules were
not readily accessible to a variety of end-labeling techniques. Based on
this observation together with studies on the presence and behavior of
end-to-end associated structures and on theoretical considerations, it was
postulated that the rDNA molecules are terminated by a fold-back structure (Blackburn and Gall, 1978). The possible nature of this structure is

further discussed below in Section VI. Another possibility to explain the inaccessibility of the rDNA ends was that they were tightly or covalently bound to protein, as is the case for the termini of the linear adenovirus or phage ϕ29 DNAs (reviewed in Blackburn and Szostak, 1984). Evidence for packaging of the C_4A_2 repeats in a nonnucleosomal form in macronuclei had been found in nuclease protection experiments (Blackburn and Chiou, 1981). However, a search for such an rDNA-bound protein strongly suggested that there are no tightly or covalently attached proteins at the rDNA ends (Blackburn et al., 1983). Because of their complex structure, cloning of the rDNA ends (and, as described below, the ends of other macronuclear DNAs of holotrichs) requires special procedures. Ligation of such telomeric ends to vectors or linkers must be preceded by converting them to blunt ends by treatment with S1 or Bal31 nuclease, which results in removal of structures distal to the single-strand breaks. The remaining internal portions of the molecule are generally stably maintained (Challoner, 1984).

The ends of the 11-kb single, free rRNA genes synthesized in conjugating and recently conjugated T. thermophila have been analyzed (Pan and Blackburn, 1981; Blackburn et al., 1985). Both termini of this rDNA species consist of a block of repeated C_4A_2 sequences (Blackburn et al., 1985) and are heterogeneous in length. The end of the 11-kb rDNA molecules which corresponds to the position of the center of the palindromic rDNA (see Fig. 1) has a length heterogeneity of ± 50 bp in the population of molecules. An array of single-strand breaks within this sequence, similar to that identified in the 21-kb molecules, is also present in the termini of the 11-kb rDNA. The 11-kb rDNA molecules are therefore structurally similar to the single rRNA genes in G. chattoni, which, as described below, also contain repeated C_4A_2 sequences at both ends.

B. TELOMERES OF OTHER MACRONUCLEAR DNAs: *Tetrahymena, Glaucoma,* AND *Paramecium*

Brief digestion of macronuclear DNA of T. thermophila by the combined exonuclease and endonuclease Bal31, which shortens linear DNA molecules from their ends, resulted in shortening of all the macronuclear DNA restriction fragments which hybridized to a C_4A_2 repeat probe. The kinetics of removal of the hybridization signal to the C_4A_2 repeat probe suggested that the C_4A_2 repeats were all within ~400–500 bp of the ends of macronuclear DNAs (Yao and Yao, 1981). It was therefore inferred that, like the rDNA, the other macronuclear DNA molecules of T. thermophila also terminate in C_4A_2 repeats. DNA sequencing of the telomeric regions of three non-rDNA macronuclear DNAs cloned in pBR322 has confirmed this result (Blackburn *et al.*, 1985; T. Ryan, M. Cherry, E. Spangler, and E. Blackburn, unpublished data).

Other species of *Tetrahymena* and the holotrichous ciliates *Paramecium* (Yao and Yao, 1981) and *Glaucoma* (Katzen *et al.*, 1981) also have C_4A_2 repeats at the ends of their macronuclear DNA molecules. When purified repeated C_4A_2 sequence was hybridized to *G. chattoni* DNA, all the subchromosomal macronuclear DNA fragments resolvable by gel electrophoresis, including the rDNA, hybridized strongly (Katzen *et al.*, 1981). Repeated C_4A_2 sequence was found to hybridize to the restriction fragments at both ends of the rDNA of *G. chattoni*. These fragments were cloned, and DNA sequence analysis confirmed the presence of a block of 30–40 C_4A_2 repeats at each end of the rDNA, each block in inverted orientation to the other (Challoner *et al.*, 1985). Total macronuclear DNA from *G. chattoni* was also analyzed for the presence of terminal inverted repeat sequences analogous to those first identified in *Oxytricha* macronuclear DNA, as described in Kraut *et al.* (this volume). As with *Oxytricha* macronuclear DNA, electron microscopy showed that 80% of total *G. chattoni* macronuclear DNA molecules, which had been denatured and briefly renatured to allow only intramolecular base pairing of complementary DNA sequences, consisted of circular, single-stranded molecules (Katzen *et al.*, 1981). This finding strongly suggested the presence of terminal inverted sequences, which was confirmed by direct sequence analysis. To determine whether these terminal sequences were common to all the macronuclear DNA molecules of *G. chattoni*, denatured macronuclear DNA was labeled with 5'-polynucleotide kinase using [γ-^{32}P]rATP. This labeled denatured DNA was then size fractionated to remove the smaller polynucleotides that dissociated from the ends of these molecules, owing to the presence of single-strand breaks within the repeated C_4A_2 block like those in *Tetrahymena* rDNA (see Fig. 2). The mixture of the larger internal bodies of the macronuclear DNA molecules was then subjected to Maxam–Gilbert DNA sequence determination. The common terminal sequence of such purified, denatured *G. chattoni* DNA was determined to be 5'-CAA(CCCCAA)$\geq_{38}$ . . . 3'. The same 5'-terminal DNA sequence was identified in a fraction of macronuclear DNA, greater than 50 kb in length, that comprises the bulk of the macronuclear DNA and contained no detectable rDNA (Katzen *et al.*, 1981).

The presence of an array of single-strand breaks in the terminal regions of the repeated C_4A_2 sequence was demonstrated by subjecting the macronuclear DNA of *G. chattoni* to the same *in vitro* end labeling and DNA polymerase I reactions described above for *Tetrahymena* rDNA. The same patterns of labeling and therefore of single-strand breaks in the termini were found. Similar results were obtained with either rDNA or a high-molecular-weight (>50-kb) fraction of macronuclear DNA which contained no detectable rDNA.

In summary, the common terminal DNA sequence of the macronuclear DNAs in *Tetrahymena* and *Glaucoma* consists of C_4A_2 repeats. The arrangement of single-strand breaks appears to be similar in all these DNA species.

IV. Functioning of *Tetrahymena* Telomeres in Yeast

A direct assay for the functioning and structural requirements of a telomere in ciliates would involve its reintroduction into a ciliate cell. However, because DNA-mediated transformation techniques have not been established for any ciliate, it has been necessary to use heterologous systems.

A favorable system for direct analysis of telomere structure and function has been the yeast *Saccharomyces cerevisiae* because this organism can be transformed by exogenously added DNA. To test the ability of *Tetrahymena* DNA termini to function in yeast, a linear yeast plasmid was constructed *in vitro* by ligating the purified terminal regions of *Tetrahymena* rDNA to a yeast plasmid vector (Szostak and Blackburn, 1982). This vector in its original circular form had been capable of autonomous replication and maintenance in yeast cells as a free plasmid. The plasmid vector was linearized *in vitro* by a single cut with a restriction enzyme, and one purified rDNA terminal restriction fragment was ligated in correct orientation onto each of its ends. When the resulting linear molecule was transformed into yeast, it was found to be replicated and maintained intact in linear form in yeast, whereas the linearized vector by itself was not. Repeated C_4A_2 sequences characteristic of *Tetrahymena* rDNA termini are not detectable by hybridization to the genome of untransformed yeast, but were still present in the linear plasmid which was replicating in yeast. Experiments to test other DNA sequences showed that it was the C_4A_2 repeats at its ends which conferred on the plasmid the ability to establish itself as a linear self-replicating molecule in yeast (Szostak and Blackburn, 1982; Blackburn and Szostak, 1984; A. Murray, T. Claus, and J. Szostak, personal communication).

V. Evidence for Addition of Telomeric Repeats to the Ends of Replicating Chromosomes

As described above, specific single-strand breaks were found in the telomeric sequences in the majority of the DNA molecules in preparations of the rDNAs and other macronuclear non-rDNA molecules of *Tetrahymena* and *Glaucoma*. Similar observations have been made for linear nuclear DNAs in the slime mold *Physarum* (Johnson, 1980), yeast (Szostak and Blackburn, 1982), and the flagellate *Trypanosoma brucei* (Blackburn and Challoner, 1984). However, this observation is not predicted by previous models for the replication of telomeres (reviewed in Blackburn and Szostak, 1984). In these models, such interruptions in the telomeric DNA regions were expected to occur only transitorily.

Further evidence that previous ideas on the replication of telomeres were insufficient to explain the observed properties of telomeres has come from findings on the variable lengths of telomeric regions. Data from several sources strongly suggest that telomeric DNA sequences are normally added to the ends of replicating chromosomal DNAs, a result not predicted by any earlier model for telomere replication. First, as described above for *Tetrahymena* rDNA, telomeric restriction fragments purified from holotrichous ciliates, slime molds, yeast, and trypanosomes all show length heterogeneity (reviewed in Blackburn *et al.*, 1983; Blackburn, 1984; Shampay *et al.*, 1984). The size variability commonly exceeds ± 100 bp for a given telomeric fragment purified from an unsynchronized cell population. The mixture of lengths in the populations of telomeric fragments obtained from asynchronously growing cells, from which telomeric DNAs are usually isolated, reflects different numbers of telomeric repeat sequences on the ends of chromosomes. Second, telomeric restriction fragments from *T. thermophila* rDNA have been analyzed in the growing progeny of newly conjugated cells. The mean size of the population of terminal rDNA fragments increased as the cells continued to divide for over 100 cell generations following conjugation. Growth rates of up to 6–10 bp per cell generation, continuing over 150 generations of log phase growth, have been observed (D. Larson and E. Blackburn, unpublished data). The same observation was previously made for telomeres in *T. brucei* (Bernards *et al.*, 1983). Thus, replication of the telomeres in these systems appears to be accompanied by regular additions of DNA.

When the terminal repeated sequences of *Tetrahymena* rDNAs function as telomeres in yeast, in every case examined they acquire, after replication in yeast, yeast-characteristic telomeric repeats. Thus, sequence analysis showed that *Tetrahymena* rDNA-derived telomeres on linear plasmids replicating in yeast have yeast telomeric sequences, which consist of irregular repeats of general formula $C_{1-3}A$, added onto the distal end of the stretch of *Tetrahymena* C_4A_2 repeats (Shampay *et al.*, 1984).

A model for telomere replication has been proposed to take into account all these observations (Shampay *et al.*, 1984; reviewed in Blackburn, 1984). In this model, it was suggested that cells have a mechanism to add, by nontemplated synthesis, telomeric repeat sequences to the distal ends of their linear DNAs. Such a terminal transferase-like activity was proposed to act by adding telomeric repeats to the 3' end of the G-rich strand at each end of the linear molecule. The extended 3' single-stranded region would then act as a template for synthesis of the complementary C-rich strand by the combined actions of primase and DNA polymerase. The addition but incomplete ligation of telomeric repeats might then account, at least in part, for the otherwise unexplained single-strand breaks found near the ends of many telomeric DNAs.

VI. Accessibility of Telomeres to End-Labeling Reactions

The presence of single-strand breaks in the distal portion of the block of telomeric repeats appears widespread, the possible exception being the macronuclear telomeres of hypotrichous ciliates (see Kraut et al., this volume). However, another shared feature of telomeres from diverse organisms is the relative inaccessibility of their extreme ends to certain enzymatic end-labeling reactions (Blackburn and Gall, 1978; Emery and Weiner, 1981). To characterize this property further, the native telomeres of the palindromic *Tetrahymena* rDNA molecules shown in Fig. 2 and a linear yeast plasmid were analyzed by end labeling (E. Blackburn, unpublished results). The linear yeast plasmid carried a yeast chromosomal telomere plus associated sequences at one end. The terminal ~500 bp of this telomeric region was made up entirely of yeast telomeric $C_{1-3}A$ repeats (Shampay et al., 1984). The other end consisted of the rDNA telomeric region of *Tetrahymena* with a block of 52 C_4A_2 repeats to whose distal side >228 bp of $C_{1-3}A$ repeat sequence were joined directly, forming the other telomere. Two end-labeling reactions were tested (1) the addition of the chain-terminating deoxynucleotide analog cordycepin to the 3' end of a DNA by terminal deoxynucleotidyltransferase, and (2) exchange of a 5' phosphate of a DNA strand with the γ-^{32}P of rATP, catalyzed by T_4 polynucleotide kinase.

Terminal transferase acts most efficiently on single-stranded DNA or single-stranded protruding 3' extensions of a double-stranded DNA (Bollum, 1974). Divalent cation is required for the reaction: Co^{2+} stimulates addition of nucleotides to a 3'-hydroxyl end by terminal transferase compared with Mg^{2+} for both single-stranded 3' extensions and recessed 3' ends of DNA (Roychoudhury et al., 1976). In contrast, polynucleotide kinase acts more efficiently on a protruding 5' terminus of a DNA strand than a recessed 5' end (Lillehaug et al., 1976).

The accessibility of native telomeric ends to these reactions was tested by measuring the amount of label incorporated at telomeric ends compared with various types of restriction-cut ends. *Tetrahymena* rDNA and the linear yeast plasmid were purified, respectively, from *Tetrahymena* and yeast cells. They were cut with restriction enzymes which leave 3′- or 5′-overhanging ends: *Hha*I (2 base 3′ overhang), *Pst*I (4 base 3′ overhang), *Sac*I (4 base 3′ overhang), *Cla*I (2 base 5′ overhang), and *Bam*H1 (4 base 5′ overhang). The various restriction fragments, terminating with the telomere and one of these restriction-cut ends, were labeled *in vitro* with [α-^{32}P]cordycepin triphosphate by terminal transferase. This reaction adds a single [α-^{32}P]cordycepin residue to the 3′-hydroxyl end of a DNA substrate. The effect of using Mg^{2+} or Co^{2+} as the divalent cation was examined by performing parallel reactions in the presence of one or other of these cations. In a similar series of experiments, the restriction fragments were labeled with [γ-^{32}P]rATP by polynucleotide kinase in the phosphate exchange reaction with the 5′ end of the DNA substrate (Berkner and Folk, 1977).

To analyze the labeling of the telomeres in these reactions and compare the extents of labeling of telomeres with the restriction enzyme-cut end, end-labeled fragments were cut with a second restriction enzyme to separate the labeled telomere from the labeled restriction-cut end. The labeled fragments were separated by gel electrophoresis, the gels dried and autoradiographed, and the autoradiograms were scanned with a densitometer. The amount of radioactivity in each gel band was quantitated by measuring the area under each peak of the densitometric scans.

These experiments showed two interesting properties of the telomeric ends. First, in the presence of Co^{2+} as the divalent cation in the terminal transferase reactions, the telomeric end was always labeled to a lesser extent than a protruding single-stranded 3′ end. However, the labeling of the rDNA telomere by polynucleotide kinase was also less efficient than the labeling of a protruding single-stranded 5′ end. The low extent of telomere labeling is all the more striking given the fact that there is more than one DNA end, owing to the presence of single-strand breaks, in both the *Tetrahymena* and yeast telomere (Fig. 2; E. H. Blackburn, unpublished work). Similar inaccessibility to end labeling by polynucleotide kinase has been seen with *Dictyostelium* rDNA telomeres (Emery and Weiner, 1981). Second, the effect of Mg^{2+} on the terminal transferase labeling of telomeres was unexpected; unlike the restriction-cut ends, whose labeling in the presence of Mg^{2+} was decreased relative to the labeling in Co^{2+} as expected (Roychoudury *et al.*, 1976), the labeling of telomeres was enhanced. This result suggests that the inaccessibility of the telomere to end labeling is not the result of its being engaged in a conventional Watson–Crick duplex structure, since such a DNA duplex would be more stabilized by Mg^{2+} than Co^{2+}, and hence less accessible to 3'-end labeling.

To account for these observations, it is noteworthy that the 3'-protruding strand of the *Tetrahymena* or yeast telomere would consist of a G-rich sequence. Dugaiczyk *et al.* (1980) have postulated that poly(dG) sequences can form two- and three-strand structures mediated by non-Watson–Crick base pairings. Other multistrand structures involving poly(G) homopolymers have been described (reviewed in Saenger, 1983). The terminal sequences of protein-free hypotrich macronuclear DNAs appear to be responsible in part for intermolecular aggregations (Lipps *et al.*, 1982). Taken together, these observations suggest that such non-Watson–Crick pairing interactions could play a role in the function of a telomere in protecting the end of a chromosomal DNA. For example, the protruding 3' end of such a G-rich strand, added to the end of a replicating linear DNA as described above, could fold back on itself, as depicted in Fig. 2 for *Tetrahymena* rDNA. The proposed addition of repeats in the course of telomere replication would therefore be expected to involve enzymes which could invade or destabilize such a fold-back or multistrand structure so that repeats could be added to its end.

VII. Formation of New Telomeres in Macronuclear Development

In the developing macronucleus, micronuclear chromosomal DNA becomes rearranged as the result of at least two kinds of processes: DNA recombinations or rejoinings, which juxtapose previously separated DNA segments, and DNA fragmentation (see Brunk, this volume). As an (albeit sepcialized) example of the latter process, in *T. thermophila*, rDNA amplification involves excision of the single copy of the rRNA gene in the micronuclear genome from flanking chromosomal DNA sequences (Yao and Yao, 1981; Pan and Blackburn, 1981; Yao *et al.*, 1985). This and other rearrangement events take place in a relatively short time period in macronuclear development: in a culture of *T. thermophila* cells conjugating at 30°C, the process is initiated in the period about 11–13 hours after cells of two different mating types are mixed to initiate the conjugation process, and the process appears complete within a few hours (Pan and Blackburn, 1981).

A. Source of Macronuclear Telomeres

The molecules generated in macronuclear development are able to replicate as free, linear DNAs; therefore, as discussed above, they must have telomeric sequences at their termini. Are these sequences already present at the ends of the micronuclear DNA segments destined to become the terminal regions of macronuclear DNA molecules? That is, are telomeric sequences already positioned in correct orientation, one at each end of each macronuclear-destined DNA sequence, so that micronuclear chromosomes are fragmented just outside telomeric sequences? Alternatively, are telomeric sequences added to the ends produced by the excision or cutting of macronuclear sequences away from flanking DNA? Data from both holotrichs and hypotrichs show that the second alternative is true for several instances of, and possibly all, macronuclear DNA molecules.

The first alternative for macronuclear telomere formation predicts the occurrence of telomeric sequences in chromosome-internal positions in the micronuclear genome. Southern blotting of purified micronuclear and macronuclear DNAs of $T.$ $thermophila$ digested with the same restriction enzyme and probed with a C_4A_2 repeat probe had shown that there was roughly the same amount of hybridization to each genome (Yao et $al.,$ 1979, 1981). The pattern of hybridizing bands for micronuclear DNA was different, regardless of the restriction enzyme used, from that of macronuclear DNA. This result appeared to argue that the C_4A_2 repeats in the micronuclear genome could be largely conserved, but rearranged in macronuclear development. Furthermore, apparently consistent with the first alternative described above, bands in the macronuclear DNA hybridizing to C_4A_2 repeats were sensitive to shortening and removal by a short digestion with $Bal31$ nuclease, showing they were telomeric in macronuclear DNA; in contrast, in micronuclear DNA, the C_4A_2-repeat hybridizing bands were not $Bal31$ sensitive (Yao and Yao, 1981). The sequences of the telomeres of $T.$ $thermophila$ micronuclear chromosomes are not yet known, but their total maximum number (10, since there are five chromosomes per haploid genome) is far exceeded by the $\sim10^2$ internally located micronuclear C_4A_2 repeat blocks.

Despite the above observations, evidence that the micronuclear C_4A_2 repeats of $T.$ $thermophila$ are not located at the ends of DNA segments destined to become macronuclear DNA molecules has come from two sources. First, the micronuclear precursor to macronuclear rDNA has been analyzed directly. The single chromosomally integrated rDNA gene of $T.$ $thermophila$ has no C_4A_2 sequence anywhere within a 2-kb region spanning one end of the integrated gene (Yao et $al.,$ 1985). This end is known to give rise to a telomere carrying C_4A_2 repeats in the course of macronuclear development to form the 11-kb rDNA molecules (Blackburn et $al.,$ 1985; Challoner and Blackburn, 1985). At the other end of the micronuclear rRNA gene, giving rise to the other 11-kb rDNA telomere, which is the same as the telomeres of the stable palindromic rDNA molecules (Fig. 1), there is a single C_4A_2 repeat unit. However, it is located $\sim$5 bp distal to the point at which the C_4A_2 repeats become joined to the adjacent retained sequence in the macronuclear rDNA in this strain (King and Yao, 1982).

The second type of evidence arguing that telomeric repeats are added to the ends of newly forming macronuclear DNA molecules in the course of telomere formation in macronuclear development has come from analysis of several cloned examples of micronuclear C_4A_2 repeats. Middle repetitive DNA sequences in the regions flanking several separately cloned C_4A_2 repeat clusters from the *T. thermophila* micronuclear genome were shown by Southern blotting to be eliminated from the macronuclear genome (Brunk *et al.*, 1982; Yao, 1982; Cherry and Blackburn, 1985). Even DNA sequences immediately abutting a cloned micronuclear C_4A_2 repeat sequence from the *T. thermophila* micronuclear genome are eliminated from the macronuclear genome (Blackburn *et al.*, 1984; Cherry and Blackburn, 1985). Determination of the DNA sequences of six randomly cloned micronuclear C_4A_2 repeat blocks and their flanking DNA also revealed that, first, the blocks themselves do not resemble macronuclear telomeric C_4A_2 repeats. Three out of the six micronuclear C_4A_2 repeat blocks had <20 repeats in the block, and single base transitions away from the canonical CCCCAA sequence were frequent. In contrast, macronuclear C_4A_2 repeats have been seen in only one cloned example to differ from the sequence C_4A_2 (the single divergent repeat was C_5A_2). Second, on the side of the micronuclear C_4A_2 repeats which would correspond, in the macronuclear DNA, to the adjoining, telomere-adjacent region of the molecule, a conserved 30-bp sequence was found immediately abutting all six cloned examples of micronuclear C_4A_2 repeats. This sequence is generally eliminated from the macronuclear genome, as judged by Southern blotting with a synthetic oligonucleotide probe consisting of this same 30-bp sequence (J. M. Cherry, unpublished results). Furthermore, none of five different cloned examples of *T. thermophila* macronuclear telomere regions contains this sequence or any of the other micronuclear sequences flanking the six sequenced micronuclear C_4A_2 repeat regions (Blackburn *et al.*, 1985; E. Spangler, J. M. Cherry, and E. Blackburn, unpublished data).

All the above observations are consistent with the notion that C_4A_2 repeats at the ends of macronuclear DNAs are not derived from repeats already present at the ends of excised macronuclear sequences. If they are added to the ends of macronuclear DNAs, are preexisting micronuclear C_4A_2 repeat sequences transposed to these ends? Although this cannot be rigorously excluded, the available data suggest such a mechanism would have to be complex; indeed, the existence of micronuclear

C_4A_2 repeats is the only evidence for such a mechanism. As described above, first, not all micronuclear C_4A_2 repeats are long enough or perfect enough clusters to be transferred directly to make complete macronuclear telomeres. Second, there is no trace of the abutting conserved micronuclear 30-bp sequence in the five sequenced macronuclear telomere-associated regions, arguing against recombination in the region close to C_4A_2 repeats. Instead, a mechanism like that proposed to occur during replication of telomeres, namely, the addition of telomeric repeats, can be considered.

As discussed above, many observations on telomeres in both *Tetrahymena* and other eukaryotes argue that there is a mechanism which adds telomeric repeats to preexisting telomeric repeated sequences at the ends of linear DNAs in the course of their replication. A model for the formation of macronuclear telomeres concomitant with chromosome fragmentation can be proposed in which the same mechanism is used. However, as described in Section IV, the presence of at least some telomeric sequences at the end of the DNA seems, at least in yeast, to be necessary for such addition. Therefore, in this model for *de novo* telomere formation in macronuclear development, the initial requirement for preexisting telomeric repeats at the newly produced end would have to be relaxed. The addition of repeats, in this model, could be uncoupled from overall DNA replication, although in *Tetrahymena,* formation of macronuclear DNA molecules is followed by their rapid replication up to an apparently regulated final copy number. In this species, the average number of telomeric repeats on newly formed non-rDNA macronuclear DNAs is about 40–80 (E. Spangler and E. Blackburn, unpublished data), and the gene copy number achieved after polyploidization of the macronucleus is about 45. The rDNA molecules are amplified to a final copy number of $\sim 10^4$ per macronucleus; their telomeres initially are very heterodisperse in size, with many telomeres with over 90 repeats in the population. The initial appearance of overly long telomeres and the rapid replication of new macronuclear DNA perhaps, therefore, are related. However, the temporal relationships between the rounds of replication, which bring up the DNA copy number, and the lengthening of the telomeres are not yet clearly worked out in *T. thermophila.*

B. Specificity of Telomere Formation

The available evidence indicates that telomere formation in ciliate macronuclear development is a highly specific and reproducible process. By Southern blot analysis of telomeric macronuclear restriction fragments fractionated by agarose gel electrophoresis, it has been shown that different clones of cells with independently developed macronuclei generate the same-sized telomeric fragments (Pan *et al.*, 1982; Blackburn *et al.*, 1983). Whether this is always the case to the nucleotide level within a particular cell strain is not known. Comparison of the sequences adjoining the C_4A_2 repeats in the macronuclear rDNAs of two different strains of *T. thermophila* and two strains of *Tetrahymena pigmentosa* showed small differences (4–14 bp) in the exact position of the junction of the C_4A_2 repeats with the internally adjacent rDNA sequence (Blackburn *et al.*, 1985). However, in this region a small number of single base-pair differences was found between the two strains that were compared in each species. This raises the possibility that these sequence differences influenced the precise point of excision or DNA processing.

This leads to the following question: at which step of the DNA processing is the specificity of telomere addition defined? The sequences adjacent to the internal edge of the C_4A_2 repeats of several sequenced macronuclear telomeres of *T. thermophila* and *G. chattoni* reveal no obvious common sequences or structural details, although there is a frequent occurrence of short, inverted repeats close to the C_4A_2 repeats (Blackburn *et al.*, 1985; Challoner *et al.*, 1985). This lack of homology of the sequences to which C_4A_2 repeats must become joined suggests that the specificity for telomere formation may reside instead at the step in which macronuclear ends are separated or cut from their flanking micronuclear sequences. The recognition sequences for cutting may therefore be in these flanking micronuclear sequences. Consistent with this possibility, inspection of the sequences flanking both ends of the micronuclear rRNA gene of *T. thermophila* suggests that invert repeats on each side of the excised rDNA have a role in its excision (Yao *et al.*, 1985). If the requirement for addition of C_4A_2 repeats in macronuclear development is simply for a free DNA end, then any end produced at this stage would have telomeric repeats added to it, by this model.

Certain mutants of *P. tetraurelia* with altered expression of the gene for immobilization antigen A are interesting in this context (Epstein and Forney, 1984). The wild-type gene in the macronuclear genome is situated 5–10 kb from a macronuclear telomere which lies 3′ to the gene. In one mutant, although a complete gene is present in the micronuclear genome, the process of genomic reorganization in macronuclear development appears to lead to a telomere being formed in the wrong position, namely, close to the 5′ side of the surface antigen *A* gene, removing the macronuclear gene. The inference is that C_4A_2 repeats are added to the misprocessed macronuclear region, since the resulting macronuclear DNA molecule apparently replicates and is maintained normally. This system may be very useful for identifying cis-acting sequences and other factors involved in the recognition process for telomere formation.

VIII. Is Telomere Formation in Ciliates Related to Processes Occurring in Other Eukaryotes?

The formation of the telomeres of macronuclear DNA is at least superficially similar to processes in which new telomeres are formed as a regular part of development in certain ascarid worms (see White, 1973). As in ciliates, chromosome fragmentation occurs early in the development of the fertilized zygote of the worm *Parascaris* (reviewed in Blackburn and Szostak, 1984). Similarly, developmentally determined chromosomal rearrangements, notably the DNA sequence eliminations of *Ascaris* and some copepods, also involve remodeling of chromosomal DNAs in a manner suggesting that new telomere formation is involved (White, 1973; Beerman, 1977).

The process of "chromosome healing" has been observed to occur in maize and insects (reviewed in Blackburn and Szostak, 1984). In these systems, the ability to heal a broken chromosome end is confined to a specific developmental stage, soon after fertilization. The healed end so formed acquires the stability of a normal telomeric end. By implication, it acquires telomeric DNA sequences. Whether chromosome healing shares features in common with the developmentally programmed formation of new telomeres seen in ciliates or nematodes is a question awaiting answers at a molecular level.

REFERENCES

Beerman, S. (1977). *Chromosoma* **60**, 297–344.
Berkner, K., and Folk, B. (1977). *J. Biol. Chem.* **252**, 3176–3184.
Bernards, A., Michels, P. A. M., Lincke, C. R., and Borst, P. (1983). *Nature (London)* **303**, 592–597.
Blackburn, E. H. (1982). *In* "The Cell Nucleus" (H. Busch and L. Rothblum, eds.), Vol. X, Part A, pp. 145–170. Academic Press, New York.
Blackburn, E. H. (1984). *Cell* **37**, 7–8.
Blackburn, E. H., and Challoner, P. B. (1984). *Cell* **36**, 447–457.
Blackburn, E. H., and Chiou, S. S. (1981). *Proc. Natl. Acad. Sci. U.S.A.* **78**, 2263–2267.
Blackburn, E. H., and Gall, J. G. (1978). *J. Mol. Biol.* **120**, 33–53.
Blackburn, E. H., and Szostak, J. W. (1984). *Annu. Rev. Biochem.* **53**, 163–194.
Blackburn, E. H., Budarf, M. L., Challoner, P. B., Cherry, J. M., Howard, E. A., Katzen, A. L., Pan, W.-C., and Ryan, T. (1983). *Cold Spring Harbor Symp. Quant. Biol.* **47**, 1195–1207.
Blackburn, E., Challoner, P., Cherry, M., Howard, E., Ryan, T., and Spangler, E. (1985). *In* "Genome Rearrangement" (I. Herskowitz and M. Simon, eds.), Vol. 20, pp. 191–203, New Series. Liss, New York.
Bollum, F. J. (1974). *In* "The Enzymes" (P. D. Boyer, ed.), Vol. X, 3rd Ed., pp. 145–171. Academic Press, New York.

Brunk, C. F., Tsao, S. G. S., Diamond, C. H., Oshashi, P. S., Tsao, N. N. G., and Pearlman, R. E. (1982). *Can. J. Biochem.* **60**, 847–853.

Challoner, P. B. (1984). Ph.D. thesis, University of California, Berkeley.

Challoner, P. B., and Blackburn, E. H. (1985). In preparation.

Challoner, P. B., Amin, A. A., Pearlman, R. E., and Blackburn, E. H. (1985). *Nucleic Acids Res.* **13**, 2661–2680.

Cherry, J. M., and Blackburn, E. H. (1985). *Cell,* in press.

Dugaiczyk, A., Robberson, D. L., and Ullrich, A. (1980). *Biochemistry* **19**, 5869–5873.

Emery, H. S., and Weiner, A. M. (1981). *Cell* **26**, 411–419.

Epstein, L. M., and Forney, J. D. (1984). *Mol. Cell. Biol.* **4**, 1583–1590.

Johnson, E. M. (1980). *Cell* **22**, 875–886.

Katzen, A. L., Cann, G. M., and Blackburn, E. H. (1981). *Cell* **24**, 313–320.

King, B. O., and Yao, M.-C. (1982). *Cell* **31**, 177–182.

Lillehaug, J. R., Kleppe, R. K., and Kleppe, K. (1976). *Biochemistry* **15**, 1858–1865.

Lipps, H. J., Gruissem, W., and Prescott, D. M. (1982). *Proc. Natl. Acad. Sci. U.S.A.* **79**, 2495–2499.

Orr-Weaver, T. L., Szostak, J. W., and Rothstein, R. J. (1981). *Proc. Natl. Acad. Sci. U.S.A.* **78**, 6354–6358.

Pan, W.-C., and Blackburn, E. H. (1981). *Cell* **23**, 459–466.

Pan, W.-C., Orias, E., Flacks, M., and Blackburn, E. H. (1982). *Cell* **28**, 595–604.

Pearlman, R. E., Andersson, P., Engberg, J., and Nilsson, J. R. (1979). *In* "Alfred Benzon Symposium: Specific Eukaryotic Genes" (J. Engberg, H. Klenow, and V. Leick, eds.), Vol. 13, pp. 351–368. Munksgaard, Copenhagen.

Preer, J. R., and Preer, L. B. (1979). *J. Protozool.* **26**, 14–18.

Roychoudhury, R., Jay, E., and Wu, R. (1976). *Nucleic Acids. Res.* **3**, 101–116.

Saenger, W. (1983). "Principles of Nucleic Acid Structure," Chap. 13. Springer-Verlag, Berlin and New York.

Shampay, J., Szostak, J. W., and Blackburn, E. H. (1984). *Nature (London)* **310**, 154–157.

Szostak, J. W., and Blackburn, E. H. (1982). *Cell* **29**, 245–255.

White, M. J. D. (1973). "Animal Cytology and Evolution," 3rd Ed., pp. 54–56. Cambridge Univ. Press, London and New York.

Wild, M. A., and Gall, J. G. (1979). *Cell* **16**, 565–573.

Yao, M.-C. (1982). *J. Cell Biol.* **92**, 783–789.

Yao, M.-C., and Gall, J. G. (1977). *Cell* **12**, 121–132.

Yao, M-C., and Yao, C. H. (1981). *Proc. Natl. Acad. Sci. U.S.A.* **78**, 7436–7439.

Yao, M.-C., Blackburn, E. H., and Gall, J. G. (1979). *Cold Spring Harbor Symp. Quant. Biol.* **43**, 1293–1296.

Yao, M.-C., Blackburn, E. H., and Gall, J. G. (1981). *J. Cell Biol.* **90**, 515–520.

Yao, M.-C., Zhu, S. G., and Yao, C.-H. (1985). *Mol. Cell. Biol.* **5**, 1260–1267.

INTERNATIONAL REVIEW OF CYTOLOGY, VOL. 99

Genome Reorganization in *Tetrahymena*

Clifford F. Brunk

Biology Department and Molecular Biology Institute, University of California, Los Angeles, California

I. Introduction

Our ideas about the organization and stability of the eukaryotic genome have fundamentally changed during the past decade, returning closer to a view prevalent early in this century. The concept of reorganization and

49

elimination of DNA sequences as a developmental mechanism for producing stable changes in cell lineages was a prominent idea during the early part of this century (reviewed by Wilson, 1925). There are a number of well-documented cases of DNA sequence elimination during the development of somatic cell lines; prominent among these are chromatin diminution in *Ascaris* (Tobler *et al.,* 1972; Streeck *et al.,* 1982) and chromosome elimination in Diptera (Bantock, 1970; Crouse *et al.,* 1971). Actually, DNA sequence loss and rearrangement are quite common in developing systems (Beams and Kessel, 1974; White, 1973). In spite of these examples, "Not many years ago geneticists and molecular biologists had come to accept as axiomatic the principle . . . that the total genome present in embryonic cells remains unchanged as somatic cells differentiate to form various tissues . . ." (Leder, 1982).

A major reason for the belief that the genome remained virtually unchanged during development was due to an interpretation of nuclear transplant experiments. Working with amphibians, King and Briggs (1956) and Gurdon (1962, 1974) were able to transplant nuclei from various tissues into enucleated oocytes and achieve varying degrees of development. Although these authors differed in their interpretation of these experiments, many investigators inferred from this work that nuclei in differentiated cells remained totipotent and in particular had not suffered irreversible reorganization or loss of DNA sequences.

The suggestion that somatic recombination is a central event in producing the diversity observed in the immunoglobins (Dreyer and Bennett, 1965) was at variance with the existing concept of genome stability. Recent molecular characterization of immunoglobin genes has confirmed the general proposal of somatic genome reorganization among portions of the immunoglobin genes (Leder *et al.,* 1980; Davis *et al.,* 1980; Tonegawa, 1983). Following this work, a number of examples in which genome reorganization plays a role in development and differentiation have been reported. The switching of mating type in yeast involves DNA sequence rearrangement (Naysmyth, 1982; Abraham *et al.,* 1982). The variation of surface antigens in trypanosomes is also mediated by somatic genome reorganization (Borst *et al.,* 1980; Van der Ploeg *et al.,* 1984). In bacteria, the alternation of flagellar antigenic type is accomplished by a DNA rearrangement (Silverman and Simon, 1980; Silverman *et al.,* 1980).

Genome reorganization as a developmental process is now well established, but it is not a major feature of most developmental programs. Among the ciliates, however, genome reorganization is a way of life. These organisms have a dual nuclear system. Each cell possesses one or more micronuclei and one or more macronuclei. The macronuclei are responsible for gene expression during the vegetative growth of the cell and the micronuclei are functional during the sexual phase of its life cycle. For some time it has been known that the hypotrichs, exemplified by *Stylonychia* and *Oxytricha,* go through an extensive reduction in genome complexity and massive physical reorganization in the process of producing new macronuclei during conjugation (Ammermann, 1965; Prescott, 1983). During the stage of conjugation that follows the formation of a new zygotic nucleus, there is a major polyploidization of the genome, followed by a subsequent loss of 90% or more of the DNA sequences (Kloetzel, 1970; Bostock and Prescott, 1972; Lauth *et al.,* 1976; Spear and Lauth, 1976). This loss is followed by a severalfold amplification of the remaining sequences. As a net result the macronucleus, which is responsible for vegetative gene expression, has only a small fraction of the sequences present in the germline micronucleus.

The physical state of the macronuclear DNA, in hypotrichs, is also remarkably different from the DNA of the micronucleus. The macronuclear genome consists of literally millions of linear DNA molecules or fragments. Each of these fragments contains one (or very few) gene(s) and each fragment is represented by several thousand copies. The macronuclear DNA fragments are relatively short, ranging in length from 400 bp (base pairs) to 10 kb (kilobase pairs), and averaging 2200 bp (Prescott *et al.,* 1971; Prescott and Murti, 1974; Lawn *et al.,* 1978; Rae and Spear, 1978; Swanton *et al.,* 1980). In contrast, the micronuclear DNA is very large, probably greater than 1000 kb.

The use of the term "fragment" with regard to macronuclear DNA does not imply that these molecules are created randomly. In fact, each macronuclear fragment is derived from a unique micronuclear DNA sequence which has been reorganized in a precise manner during the development of the macronucleus. Each macronuclear fragment is linear, with specialized telomeric regions at each end (Oka *et al.,* 1980; Yao and Yao, 1981; Klobutcher *et al.,* 1981). Although these macronuclear fragments might be referred to as macronuclear chromosomes, since they are composed of chromatin, they do not contain functional centromeres and do not participate in mitosis. Thus, the term "macronuclear fragment" seems more appropriate than the term "macronuclear chromosome," so long as it is understood that each of these fragments contains a unique DNA sequence generated by specific molecular processes.

Among the ciliates, *Paramecium* and *Tetrahymena* have been the most thoroughly investigated. Dating from the classic studies of Sonneborn and his students, *Paramecium* has been well characterized, particularly at the genetic level. In recent years, *Tetrahymena* has in some respects replaced *Paramecium* as the organism of choice. In the last several years, dramatic new findings have emerged from molecular studies on *Tetrahymena,* which are of interest to biologists in general. These include areas as diverse as telomere function and self-splicing RNA (Szostak and Blackburn, 1982; Cech *et al.*, 1982). In many respects *Tetrahymena* is an ideal organism for molecular biological investigations.

As pointed out by Gorovsky, *Tetrahymena* affords a unique opportunity to investigate the functions and interactions of germline and somatic nuclei (Gorovsky, 1973, 1980). Given the dominant role of genome reorganization in the life cycle of the ciliates, *Tetrahymena* is a logical choice for an in-depth study of this process. Five years ago Gorovsky reviewed the organization and reorganization of the genome of *Tetrahymena* (Gorovsky, 1980). At that time it was apparent that *Tetrahymena*, like the hypotrichs, undergoes substantial genome reorganization during conjugation. In the 5 years since Gorovsky's review a remarkable amount has been learned about the reorganization process.

II. Background Studies on the Macronuclear Genome

A. Loss of Sequences from the Macronucleus

The complexity of macronuclear DNA has been reported from several independent studies. The complexity is measured by comparing the rate of reassociation of denatured DNA pieces with that of a reference DNA of known complexity, usually *Escherichia coli* DNA. An early report by Flavell and Jones (1970) indicated that the macronuclear complexity was about 60,000 kb. Several later reports, however, indicate a macronuclear complexity in the range of 220,000 kb (Allen and Gibson, 1972; Yao and Gorovsky, 1974; Borchsenius *et al.*, 1978). These later estimates of the macronuclear complexity are very close to the haploid DNA content of micronuclei determined by microspectrophotometry (Woodard *et al.*, 1972; Gibson and Martin, 1971; Seyfert, 1979; Seyfert and Preparata, 1979; Doerder and DeBault, 1975). It is impossible, using this technique, to detect any signficant difference in the complexity of the macronucleus and the micronuclear DNA.

A direct comparison of macronuclear and micronuclear DNA complexity by hybridization analysis in which one DNA is present as a radioactive tracer and the other DNA provides the bulk of the DNA affords a better estimate of any loss in complexity in the macronucleus. This requires separate isolation of the two different nuclei in relatively pure form. An analysis of this type performed by Yao and Gorovsky (1974) indicates that 10–20% of the micronuclear complexity is lost (or grossly underrepresented) in the macronucleus. Moderately repetitive sequences are predominant among the sequences lost. A more recent report by Iwamura *et al.* (1979) also indicates that 80% of the sequences present in the micronucleus are present in similar relative amounts in the macronucleus. There is a suggestion in this latter report that the underrepresented sequences in the macronucleus are present in limited amounts rather than being totally lost.

There is clearly not the wholesale loss of micronuclear sequences in *Tetrahymena* that occurs in the hypotrichs. Reports suggest that there is a loss of genome complexity in the macronucleus relative to the micronucleus, but this loss is nominally at the level of about 20%. In fact, the technique of renaturation kinetics is not sufficiently accurate to indicate more specifically the amount of loss or to distinguish between complete loss and gross underrepresentation.

Investigation of the loss of specific sequences during macronuclear development can be approached in a direct manner by using recombinant DNA technology. Cloned micronuclear sequences have been used as probes in Southern analysis (Southern, 1975) of macronuclear and micronuclear DNA. These studies show that some of the micronuclear sequences are completely absent in the mature macronuclear genome (Brunk *et al.*, 1982; Yao, 1982a; Karrer, 1983). The cloned micronuclear sequences examined indicate that both repetitive and single-copy sequences are deleted from the macronuclear genome. A number of families of repeated sequences predominate among the eliminated sequences. An estimate of the degree of elimination of micronuclear sequences, based on examination of randomly selected micronuclear clones, suggests that 20–25% of the micronuclear genome is eliminated from the mature macronucleus (Yao, 1982a; Karrer, 1983). Thus, two quite different approaches indicate a loss of about 20% of the micronuclear complexity during anlagen development.

B. Size of Macronuclear DNA

The micronuclei of *Tetrahymena* contain five chromosomes, which can be observed during meiosis (Ray, 1956). Cytologically, the chromosomes all appear to be in the same size range. Assuming that the micronuclear chromosomes are uninemic and of roughly equal size, each chromosome would contain about 44,000 kb. The physical state of macronuclear DNA in *Tetrahymena* has been unclear. Early measurements of the size of macronuclear DNA produced conflicting estimates. Williams *et al.* (1978), using viscoelastometric measurements, estimated the majority of macronuclear DNA molecules to be in the range of 30,000–45,000 kb. This would imply little or no fragmentation of most micronuclear DNA molecules during the development of macronuclei (anlagen). In contrast, studies employing velocity sedimentation centrifugation suggest macronuclear DNA molecules in the range of 600 kb (Merkulova and Borchsenius, 1976; Preer and Preer, 1979). Actually, neither of these techniques is highly reliable for DNA in this size range. The viscoelastic technique tends to give an estimate biased by the largest molecules in the preparation, and velocity sedimentation centrifugation has potential artifacts related to local DNA concentrations when used for very large DNA molecules.

Another ciliate, *Glaucoma chattoni,* presents an interesting intermediate between the macronuclear organization found in *Tetrahymena* and that found in the hypotrichs. The macronuclear DNA of *G. chattoni* contains many molecules that are relatively small and can be characterized accurately by velocity sedimentation centrifugation (Katzen *et al.,* 1981). The larger molecules range in size up to well above the resolution limits of velocity sedimentation centrifugation. The indication is that macronuclear DNA fragments can exist in a wide range from very small molecules only a few kilobase pairs in length up to molecules several thousand kilobase pairs in length.

All reports indicate that the majority of macronuclear DNA molecules in *Tetrahymena* are very large relative to that of the hypotrichs, but just how large is unclear. The range could be from 500 up to 40,000 kb. Unfortunately, DNA molecules of this size are difficult to handle without shearing. Conventional isolation and characterization techniques are simply not effective for DNA molecules in this range. As will be discussed later, a new technique, alternating orthogonal field (AOF) gel electrophoresis, is capable of resolving DNA molecules up to several million base pairs in length. Using this technique, it is clear that the macronuclear DNA of *Tetrahymena* exists as a collection of fragments ranging from less than 100 kb to at least several thousand kilobase pairs.

C. C$_4$A$_2$ Repeats in Macronuclear and Micronuclear DNA

Early examinations of macronuclear ribosomal RNA (rRNA) genes in *Tetrahymena* revealed an interesting example of genome reorganization. The rRNA genes in the macronucleus are found on linear palindromic molecules 21 kb in length (Gall, 1974; Engberg *et al.,* 1976; Karrer and Gall, 1976). These rDNA molecules are highly amplified, with about 9000 copies present per macronucleus (Engberg and Pearlman, 1972; Yao *et al.,* 1974). This makes the rDNA the most prevalent sequence in the macronucleus. A careful examination of the micronuclear genome indicates that the rDNA is represented by a single copy per haploid genome in the micronucleus (Yao and Gall, 1977). This work also indicates that there are no copies of the rRNA genes integrated into high-molecular-weight DNA in the macronuclear genome. Thus, during anlagen development the rDNA sequences are removed from an integrated location in high-molecular-weight DNA, reorganized into an autonomously replicating molecule (fragment), and extensively amplified.

Blackburn and Gall (1978) found that the macronuclear rDNA molecules have on each end a sequence consisting of 20–70 repeats of the hexanucleotide CCCCAA (C_4A_2). A large family of repeated sequences in the micronucleus capable of hybridization with the C_4A_2 from the rDNA terminus was recognized by Yao *et al.* (1981), who found roughly equivalent amounts of C_4A_2 homologous sequences in the micronuclear and macronuclear genomes. When C_4A_2 was used as a probe to examine micronuclear and macronuclear DNA digested with a restriction endonuclease, it was apparent that the organization of these sequences in the two nuclei is very different (Yao and Gall, 1979). Other sequences used as probes for similar examinations of restricted DNA appeared to have virtually identical patterns for macronuclear and micronuclear DNA. These results suggested that the C_4A_2 sequences may have a role in reorganization. It appeared reasonable to assume that C_4A_2 repeats might be acting as a signal sequence at which the macronuclear DNA reorganization occurs. During anlagen development cleavage within or near the C_4A_2 repeats could be responsible for producing macronuclear fragments. This would dramatically alter the organization of C_4A_2 sequences in the macronucleus relative to that in the micronucleus. Consistent with this, Yao and Yao (1981) and Brunk *et al.* (1982) have demonstrated that the majority of C_4A_2 sequences found in the macronucleus are located at the termini of macronuclear fragments.

Several lines of evidence now make the idea that C_4A_2 repeats are signals for reorganization a less attractive suggestion. First, a number of micronuclear sequences, which are eliminated in the macronucleus, do not appear to be adjacent to C_4A_2 sequences (Brunk *et al.*, 1982). Thus, not all sequences eliminated from the developing anlagen contain C_4A_2. Apparently, C_4A_2 is not the sole reorganization signal sequence. Further, the role of C_4A_2 sequences has been investigated, and a convincing case can be made for their function as telomeric sequences (Szostak and Blackburn, 1982). In addition, the micronuclear C_4A_2 homologous sequences are not all conserved as exact C_4A_2 repeats. Some of the sequences have drifted, which may indicate that these sequences are not under functional constraint and, thus, are not signals for reorganization (Blackburn *et al.*, 1982).

III. The Formation of Macronuclear DNA Molecules

A. Ribosomal DNA Molecules

The best characterized case of genome reorganization in *Tetrahymena* is the removal of the rRNA gene from an integrated site in the micronuclear chromosome and its rearrangement into a 21-kb palindrome in the macronucleus. *Tetrahymena* is remarkable with regard to the rRNA genes in both its micronucleus and its macronucleus. Hybridization and genetic evidence indicates that there is a single copy of the rRNA gene per haploid genome of the micronucleus (Yao and Gall, 1977; Pan *et al.*, 1982). Southern analysis of micronuclear DNA digested with restriction endonucleases and using the rDNA sequence as a probe indicates that there are two unique flanking sequences for the integrated copy of the rRNA gene in the micronuclear genome. This arrangement precludes multiple integrated copies either in a tandem array (which would give a minimum of three flanking sequences) or dispersed integrated copies (which would give a minimum of four flanking regions). Hybridization of an rDNA probe to various nullisomic strains of *Tetrahymena* indicates that the integrated copy of the rRNA gene in the micronucleus is on the left arm of chromosome 2 (Bruns and Brussard, 1981; Bruns, 1984). *Tetrahymena* is most unusual in having just a single rRNA gene in its germline. Usually, there is a family of rRNA genes in the germline, often in tandem array (Tobler, 1975).

The organization of the rRNA genes in the macronucleus of *Tetrahymena thermophila* is also unusual. The macronuclear rRNA genes are on a linear 21-kb DNA molecule, which is almost a perfect palindrome. Only the central 24–30 nucleotides are nonpalindromic (Kiss and Pearlman, 1981; Engberg, 1983). The rRNA genes are transcribed toward the termini, with the 17 S rRNA center proximal and the 26 S rRNA center distal (Cech *et al.*, 1982). There are ~9000 palindromic molecules or 18,000 rRNA genes in the macronucleus (Yao, 1982b; Pearlman *et al.*, 1979). Although amplification of rRNA genes as extrachromosomal molecules is not uncommon, particularly in oocytes, palindromic rDNA molecules are much less common (Gall, 1968, 1974).

The micronuclear rDNA sequence, including the right end (telomeric end of the mature palindromic rDNA) and 5.8 kb of the flanking sequence, was cloned and characterized (Yao, 1981). Using the flanking region as a probe for Southern analysis of macronuclear DNA digested with restriction of endonucleases, it is clear that a 2.8-kb region of DNA immediately flanking the right end of the integrated rRNA gene in the micronucleus is missing in the macronuclear genome. This 2.8-kb sequence, which is a single-copy sequence in the micronucleus, is neither in the palindromic rDNA molecule nor in high-molecular-weight macronuclear DNA. Apparently, the removal of the rDNA sequence from the macronuclear DNA and the loss of the 2.8-kb region has created a new terminus within the flanking sequence. This was directly demonstrated by exonuclease *Bal*31 digestion of the flanking region in macronuclear DNA (Yao, 1981). When macronuclear DNA is digested for varying times with *Bal*31, the telomeres and adjacent regions are progressively removed. If the *Bal*31 digested DNA is now digested with a restriction endonuclease, telomeric regions will be shortened while internal regions remain unaltered. Using the micronuclear flank of the rDNA as a probe for Southern analysis of *Bal*31-digested macronuclear DNA, it is clear that this region is a telomere in macronuclear DNA.

The micronuclear clone containing the right end of the rRNA gene and the flanking region has been sequenced at the junction between the rDNA and the flank (King and Yao, 1982). There is a single C_4A_2 hexanucleotide at the end of the rRNA sequence. The existence of only a single C_4A_2 decreases the probability that C_4A_2 alone is acting as a signal sequence for reorganization. Apparently, the 20–70 C_4A_2 found in the mature macronuclear rDNA molecules are added to the molecule after excision from the micronuclear DNA.

A similar characterization has been performed for the flanking sequence at the left end of the integrated rRNA gene, the portion of the sequence at the center of the palindrome (Yao *et al.*, 1985). Again, a portion of the flanking sequence is absent from the macronuclear DNA, and the flanking sequence that remains in the macronucleus exists as a telomere. In this case, less than 1 kb is eliminated during the reorganization. Unlike the right end, however, no C_4A_2 sequence is found at the junction between the rDNA and the flanking micronuclear sequence. This is interesting because this sequence will be the central region of the palindrome in the mature rDNA molecule, and the central region of the mature molecule has no C_4A_2 sequence.

B. Type I Reorganization

Reorganization of the genome during anlagen development in which there is fragmentation of a chromosome with the creation of new telomeres will be called type I reorganization. Type I reorganization is diagrammed in Fig. 1. Clearly, type I reorganization occurs in the left arm of chromosome 2 with the removal of the rRNA gene. The excision of the rRNA gene is the only well-characterized example of type I reorganization. In this case, there is a loss of micronuclear sequences accompanying

Micronuclear DNA

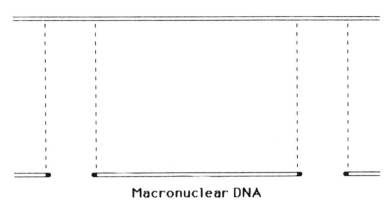

Macronuclear DNA

Teleomeres ➤

FIG. 1.   A diagram of type I reorganization. The micronuclear DNA is fragmented into smaller molecules by cleavage at specific sites. DNA regions between the macronuclear fragments are eliminated in this process. New telomeres are created at the ends of the macronuclear fragments by the addition of C_4A_2 regions.

the rearrangement. However, it is easy to conceive of type I reorganization in which a loss of DNA sequence does not occur.

The fragmentation of the macronuclear DNA is a direct measure of type I reorganization. The genome of *Tetrahymena* is not reduced to fragments the size of those found in the hypotrichs, but there is a good deal of circumstantial evidence indicating that the macronuclear DNA of *Tetrahymena* exists in fragments much smaller than the length of micronuclear chromosomes. A direct measure of the size and the number of macronuclear DNA molecules would indicate how prevalent type I reorganization is in the development of macronuclei.

C. MACRONUCLEAR DNA FRAGMENTS

Recently, a new type of gel electrophoresis capable of resolving DNA molecules with lengths up to several million base pairs has been described (Schwartz *et al.*, 1982; Schwartz and Cantor, 1984; Carle and Olson, 1984). A convergent nonuniform electric field, generated between a single pole and an array of electrodes, is applied across the gel. The electric field is applied for a short period (10–100 seconds) in one direction followed by a second, orthogonal (essentially at right angles) nonuniform electric field applied for the same period. This alternation of fields is continued for about 1000 minutes. The net result of AOF gel electrophoresis is that the DNA molecules migrate along the median between the two fields (generally at 45° to the applied fields). The alternation of the fields prevents the large DNA molecules from aligning with the electric field and migrating more rapidly than would be expected for their molecular weight. Even very large DNA molecules have migration rates proportional to their lengths in AOF gel electrophoresis. This technique has been used to resolve the chromosomes from yeast (Carle and Olson, 1985). The largest yeast DNA molecules that are regularly resolved are about 1800 kb in length (Schwartz and Cantor, 1984).

We have employed this technique to determine the size of the macronuclear DNA molecules in *T. thermophila* (Conover and Brunk, 1985). The cells are starved to reduce the cellular mass and are cast as thin slabs into low-melting-temperature agarose. The slabs are incubated at 50° overnight in buffer containing 0.5 *M* EDTA, pronase (2 mg/ml), and detergent (1% Sarkosyl). This incubation lyses the cells and frees much of the DNA without shearing or substantial cleavage by endogenous nucleases. The slabs are then loaded on a 1.5% agarose gel and subjected to AOF gel electrophoresis. Figure 2 shows a typical ethidium bromide stained AOF gel used to separate *T. thermophila* macronuclear DNA fragments. Four inbred strains of *T. thermophila* are shown alone with intact yeast chromosomes as molecular-weight markers. The wells are at the top. The first prominent band below the well (puddle) is produced by DNA molecules that are small enough to enter the gel, but too large to be resolved. In the yeast lane there is little DNA in the puddle. The largest yeast band (faint) is well below the puddle. The *T. thermophila* lanes have a band just below the puddle, considerably larger than the largest yeast band. The macronuclear fragments range in size from 21 kb up to several million base pairs.

Clearly, the *Tetrahymena* DNA displayed on the AOF gel is in fragments, but are these fragments the result of the isolation procedure or do they reflect the *in situ* state of the macronuclear DNA? If the fragments represent the native state of macronuclear DNA, then each single-copy macronuclear DNA sequence should be found on a single fragment of unique size. Alternatively, if the fragments are an artifact of isolation, then each macronuclear sequence would be expected to be distributed among a number of fragments of different sizes. To verify that the fragments we observe by AOF gel electrophoresis are the native macronuclear fragments rather than an artifact of isolation, an AOF gel was dried and probed with different unique macronuclear sequences (Tsao *et al.*, 1983). A number of different unique DNA sequence probes each hybridized to a different band. We also verified that the fragments observed on

CLIFFORD F. BRUNK

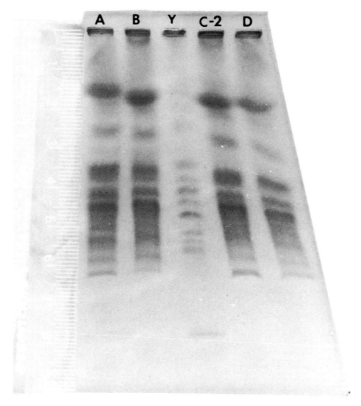

Fig. 2. An ethidium bromide stained AOF gel of *T. thermophila* macronuclear DNA. Four inbred strains, A, B, C-2, and D, are displayed. The center lane is yeast (Y) used as a molecular-weight marker. The first two yeast bands are not well resolved; bands 5, 7, and 9 in this strain are more intense (two or more chromosomes per band). The band below the well (puddle) is larger than the largest yeast band (faint); thus, DNA molecules up to about 2000 kb are resolved on this gel. The gel was run for 1000 minutes at 300 V and a switching period of 220 seconds.

the AOF gels were of macronuclear origin by probing the dried gels with a micronuclear-specific DNA sequence (Brunk *et al.*, 1982). In this case, hybridization occurred primarily within the well and to a slight degree within the puddle. Apparently the micronuclear DNA is too large to enter the gel to a significant degree. The micronuclear DNA in the puddle probably results from random breakage.

The smallest macronuclear fragments are the rDNA molecules (21 kb). The mitochondrial DNA is another relatively small DNA molecule, slightly larger than the rDNA. Both of these species have been identified by hybridization. The rDNA and mitochondrial bands can be resolved on

AOF gels using a short alternation period (10–15 seconds). In Fig. 2, however, the rDNA and mitochondrial bands combine to form the intense band at the bottom of the gel. We observe several macronuclear fragments with a size of 100 kb or less. About one-half of the mass of macronuclear fragments that enter the gel have a size smaller than 750 kb. Thus, the majority of macronuclear fragments are smaller than 750 kb. Figure 3 shows a microdensitometer scan of an AOF gel similar to the one shown in Fig. 2. There are a number of bands of macronuclear DNA ranging in size up to several million base pairs. Some of the macronuclear DNA fragments are larger than the largest resolved yeast chromosome. It is possible that some of the macronuclear fragments are too large to enter the gel and, like the micronuclear DNA, are confined to the well. A few of the larger bands may represent single macronuclear fragments, while others are composed of several fragments. Based on the size of the genome and the average fragment size, there should be about 300–500 fragments. This is consistent with the distributions of DNA observed on the AOF gels.

Clearly, the DNA of the developing anlagen is fragmented into several hundred pieces during the conjugation process. Thus, type I genome reor-

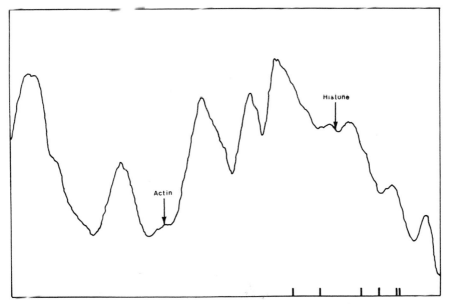

FIG. 3. A microdensitometer scan of an AOF gel like that in Fig. 2. The marks along the bottom show the relative position of yeast bands. Right to left: the approximate size of the yeast bands are 260, 290, 460, 580, and 700 kb (Carle and Olson, 1984). The arrows indicate the relative positions at which *T. thermophila* actin and histone H4-I hybridize to the gel.

ganization is a signficant process during anlagen development. Although only one example of this type of reorganization has been investigated in detail, it is probable that a similar process is operative for production of all the fragments.

IV. Elimination of Sequences without the Creation of New Telomeres

A. Type II Reorganizations

In addition to type I reorganizations, which create new telomeres, developing macronuclei of *T. thermophila* also make internal deletions. Sequences are eliminated and the retained regions rejoined so that no new telomeres are created. We will refer to reorganizations which eliminate sequences without the creation of new telomeres as type II reorganizations. Type II reorganization is shown in Fig. 4. This process of breakage and rejoining can be readily detected by comparing the restriction maps of micronuclear and macronuclear DNA. Yao *et al.* (1984) have characterized a micronuclear and macronuclear set of cloned sequences that contain homologous regions. Within a particular 9.5-kb micronuclear DNA sequence analyzed, there were three regions deleted from the macronuclear DNA. Each deleted region averages about 1–2 kb in length, with a total of about 3–5 kb of micronuclear DNA lost from this 9.5-kb region. This second type of rearrangement, like type I reorganizations, is a well-

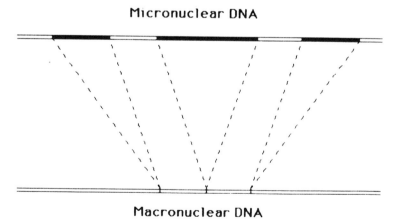

Micronuclear DNA

Macronuclear DNA

FIG. 4.   A diagram of type II reorganization. Regions of the micronuclear DNA, shown as black bars, are deleted and the flanking sequences are rejoined. No new telomeres are created.

regulated developmental event. Different strains of *T. thermophila* or the cell lines derived from different conjugations all display very similar reorganization patterns.

When the reorganization of this specific 9.5-kb micronuclear region was examined in detail following conjugation, two alternatives for the rejoining of the middle region were observed (Austerberry *et al.*, 1984). The middle region of the macronuclear DNA clone in the parental strains has a deletion of 600 bp relative to the micronuclear DNA. Some of the progeny have this 600-bp region deleted in their macronuclear DNA, while others have an extended region of 900 bp eliminated. Apparently, both types of deletions occur in most anlagen and then assort during vegetative propagation (phenotypic assortment). Although most of the reorganizations are highly specific, there are alternative patterns at some sites. Both of the alternative patterns in this 9.5-kb micronuclear region are stable and undergo phenotypic assortment.

The 5 S rRNA genes occur as a family and provide good markers for following genome reorganization. There are about 150 5 S rRNA sequences per haploid genome, arranged into some 30 clusters in both the micronuclear and macronuclear DNA (Pederson *et al.*, 1984). The 5 S rRNA genes occur on all micronuclear chromosomes except chromosome 2 (Allen *et al.*, 1984). Genome reorganization occurs in the vicinity of some of the 5 S rRNA genes, producing differences between the micronuclear and macronuclear patterns (Pederson *et al.*, 1984). Alternative patterns of 5 S rRNA sequence reorganization have been observed during anlagen development for some 5 S genes (Allen *et al.*, 1985). The pattern observed immediately after conjugation changes during vegetative growth, apparently by phenotypic assortment. Although the majority of the rearrangements occur during anlagen development, occasional changes in the 5 S rRNA pattern in addition to those produced by phenotypic assortment are observed during vegetative growth. These vegetative changes are observed most frequently in strains other than strain B.

The complexity of the macronucleus is 10–20% less than that of the micronucleus. The comparison of homologous micronuclear and macronuclear sequences indicates that type II reorganization could account for a good deal of this reduction in complexity. Twenty randomly selected macronuclear clones were used as probes in a Southern analysis of restriction-endonuclease-digested micronuclear and macronuclear DNA. In a number of these cases, one band in the micronuclear DNA is replaced by one or more different bands in the macronucleus, indicating type II reorganization. It was found that 9 of the 20 probes revealed differences between the 2 nuclei indicative of type II rearrangements (Yao *et al.*, 1984). These cloned regions are about 15 kb in length. If they are representative of the micronuclear genome as a whole, then there are over 5000 sites of type II reorganization. This frequency of internal reorganizations suggests that there is a great deal of type II reorganization during anlagen development.

The micronuclear genome consists of about 220,000 kb of DNA (Gorovsky, 1980). About 20% of this DNA, 44,000 kb, is eliminated during anlagen development. There are insufficient data to accurately estimate the amount of DNA eliminated via type I reorganizations relative to type II reorganizations. Estimates of the size and number of macronuclear fragments on AOF gels suggest there are about 300–500 macronuclear fragments. Each fragment is generated by a type I reorganization; thus, there are 300–500 type I reorganizations. The occurrence of internal deletions suggests over 5000 type II reorganizations. A characterization of internal eliminated micronuclear sequences, type II reorganizations, suggests that up to 15% of the micronuclear DNA is eliminated by this mode (Howard and Blackburn, 1985). This is consistent with the vast majority of sequence elimination occurring via type II reorganizations. If we assume that each reorganization leads to a similar amount of elimination without regard to type, then an average of about 8 kb is eliminated during each reorganization. This estimate of the average amount of DNA eliminated in each reorganization implies that many of the eliminations are substantially larger than the ones which have been well characterized thus far. This is consistent with the result obtained when a number of clones containing relatively long inserts (10–12 kb) were selected from a bank of micronuclear DNA sequences using the telomeric C_4A_2 as a probe. In most cases the entire cloned sequence was absent from the macronuclear genome (Brunk et al., 1982).

B. Loss of Repetitive Sequences

The sequences eliminated during anlagen development are predominantly repetitive in the micronucleus. This was initially observed during the hybridization kinetics analysis of Yao and Gorovsky (1974). Characterization of cloned sequences eliminated from developing anlagen also shows that the majority of these sequences are repetitive in the micronucleus (Brunk et al., 1982; Yao, 1982a). A prominent class of repetitive micronuclear sequences hybridize with the C_4A_2 sequence, although they are internal sequences (not telomeric). These internal micronuclear C_4A_2 homologous regions often include deviations from the simple C_4A_2 sequence found at the telomeres (Blackburn et al., 1982). It is interesting to note that virtually all of these internal C_4A_2 sequences are removed during anlagen development. The vast majority of C_4A_2 homologous sequences in macronuclear DNA are found at telomeres (Yao and Yao, 1981). When a library was prepared from macronuclear DNA, very few clones were found to have homology with C_4A_2. Of course, telomeric C_4A_2 sequences would not be represented in such a library, since they have only one restriction site. Some of clones showing C_4A_2 homology in this macronuclear library were characterized. In all cases, these clones contained micronuclear specific sequences, indicating that they were the result of micronuclear DNA contamination (Brunk et al., 1982).

A second family of micronuclear specific repetitive sequences was characterized by Brunk et al. (1982). The first representative of this family was found on a micronuclear clone containing a C_4A_2 homologous sequence. A 450-bp sequence between an *Xba*I and *Hind*III (X–H) restriction site was used as a probe to characterize this family. Southern analysis of the micronuclear and macronuclear genomes indicates that there are several hundred representatives of this family in the micronucleus and none in the macronucleus (Brunk et al., 1982; Brunk and Conover, 1985). A number of additional micronuclear clones homologous to the X–H probe were characterized. Interestingly, none of these additional representatives of the X–H family has a C_4A_2 homologous sequence within the same clone. Apparently, these two micronuclear repetitive families are not generally distributed together; both families are eliminated during anlagen development.

Most of micronuclear-specific cloned sequences are quite long, averaging over 10 kb. Those that have been characterized appear to be a collection of various combinations of repetitive families. In addition to the micronuclear repetitive sequences, a few micronuclear unique sequences have been cloned (Brunk et al., 1982; Yao, 1982a). These unique sequences were found adjacent to members of repetitive families. It is quite possible that short unique sequences occur frequently among the repetitive sequences. If there is a repetitive element in the probe, the sequence appears to be repetitive even though it might contain a unique region. Thus, our analysis would underestimate the unique sequences.

V. Timing of Reorganization during Anlagen Development

A. Timing of Type I Reorganizations

During conjugation, micronuclei give rise to gametic nuclei which are exchanged between paired cells and fuse to become the source of all the nuclei of the exconjugate cells (Elliott, 1973; Doerder and Debault, 1975). Cells of different mating types pair, the micronuclei go through meiosis, and three of the meiotic products are destroyed. The remaining haploid micronucleus goes through mitosis to yield two identical pronuclei. One of these nuclei is exchanged with the other cell in the pair. Following exchange, the haploid nuclei fuse to form a zygotic diploid nucleus which undergoes two mitotic divisions. Two of the nuclei migrate to the anterior portion of the cell and develop into new macronuclei (anlagen), while the other two nuclei migrate to the posterior portion of the cell and become the new micronuclei. During anlagen development, the old macronucleus degenerates and is lost. Thus, we have both the macronucleus and the micronucleus developing from a common diploid fertilization nucleus. At some stage in this development the physical organization of the anlagen DNA must be altered from the very long chromosomal DNA molecules (probably uninemic) in the micronucleus to the relatively short fragments found in the macronucleus.

Gene expression in *T. thermophila* is interesting. There is no evidence of gene expression from the micronucleus during vegetative growth (Mayo and Orias, 1981). Gene expression from the developing anlagen occurs only if the cells are refed following conjugation (Mayo and Orias, 1979). When the cells are refed, gene expression begins 12–14 hours following the initiation of conjugation. Autoradiographic evidence suggests that there is some transcription from micronuclei early in conjugations (Sugai and Hiwatashi, 1974). In general, there is little gene expression during conjugation and no expression from the developing anlagen until the cells have been refed.

The conversion of the rDNA sequence from a single integrated copy in chromosomal DNA to an amplified extrachromosomal fragment has been investigated in detail (Pan and Blackburn, 1981; Pan et al., 1982). Early during anlagen development, the rDNA sequence is found as an 11-kb fragment containing a single rDNA gene. This 11-kb rDNA molecule is unusual, as it is only one-half of the mature rDNA palindrome and has a C_4A_2 telomere at each end. Using hetrokaryons with a restriction polymorphism in the rDNA gene, it can be shown that this unusual 11-kb rDNA is derived from the developing anlagen and is probably an intermediate in the production of the mature 21-kb palindrome. Labeling experiments indicate that the 11-kb rDNA is first detected at 12 hours following conjugation. It is interesting to note that both 11- and 21-kb (palindromes) are detected at 12 hours; thus, there is a rapid production of palindromes from the 11-kb structures if they are indeed intermediates. The production of extrachromosomal rDNA molecules is a clear example of type I reorganization, and thus, this type of reorganization can occur early in anlagen development.

During conjugation, the micronuclei and developing anlagen replicate their DNA, but the old macronuclei do not undergo extensive DNA replication. Thus, if DNA is labeled during conjugation, the label will occur primarily in the micronuclei and the anlagen. Immediately after mixing the cells for conjugation, there is a stimulation of micronuclear division in some cells accompanied by micronuclear DNA synthesis. There is also some mitrochondrial DNA synthesis and a little rDNA synthesis at this time. If radioactivity is introduced 3 hours following mixing, only micronuclei and anlagen are labeled.

The recent development of AOF gel electrophoresis has made it feasible to examine type I reorganizations in detail. The course of type I reorganizations can be followed by introducing label 3 hours after the initiation of conjugation and then isolating DNA for AOF gel analysis at various times during conjugation. Prior to 12 hours after the initiation of conjugation, all of the label is found in high-molecular-weight molecules characteristic of micronuclear DNA. Labeled macronuclear fragments begin to be observed between 12 and 16 hours. By 18 hours virtually all of the fragmentation is complete, and at this time the labeled fragments resemble the pattern found in mature macronuclei. This reorganization process is very rapid, requiring only a few hours from onset to completion.

B. TIMING OF TYPE II REORGANIZATIONS

The elimination of micronuclear-specific sequences has been investigated in developing anlagen. An autoradiographic approach was employed by Yokoyama and Yao (1982). A micronuclear-specific sequence was used as a probe for *in situ* hybridization to developing anlagen. The data indicate that the hybridization signal decreased to a significant extent prior to the first cell division. The elimination of micronuclear-specific sequences was also measured in isolated anlagen (Brunk and Conover, 1985). The developing anlagen progress through a series of endoreplications without nuclear division. This allows anlagen to be isolated at specific stages (e.g., 4, 8, or 16C). In this study DNA was prepared from anlagen at different stages of development, and this DNA was probed with a micronuclear-specific sequence on an analytical Southern blot. This analysis indicates that the 4C anlagen has essentially the same amount of homology with the micronuclear-specific sequence as micronuclei, while the 8C and 16C anlagen have lost virtually all of this sequence. Both of these studies indicate that micronuclear-specific sequences are lost early during anlagen development. The loss of these sequences is preciptous and occurs in the absence of nuclear division. These data rule out underreplication and dilution as a mechanism for elimination. Instead, the kinetics of elimination strongly suggest an active degradation of the micronuclear sequences between the 4 and 8C stages. The most plausible mechanism would involve recognition and rapid degradation of the regions to be eliminated by a specific nuclease.

Anlagen development progresses to the 8C stage in the absence of refeeding (Brunk and Bohman, 1985). It is evident that the loss of micronuclear-specific sequences does not require refeeding. There is no detectable gene expression from anlagen until refeeding. Thus, the gene products responsible for the elimination of micronuclear-specific sequences in the developing anlagen must be expressed from the old macronuclei or be products of the limited micronuclear transcription that apparently occurs early in conjugation.

The micronuclear-specific sequences used in these analyses have not been thoroughly characterized. However, in view of the large number of occurrences of these sequences in the micronuclear genome and preliminary restriction data, it appears that at least some of the probe sequences are eliminated by type II reorganizations. The anlagen development 12 hours after conjugation is predominantly at the 4C stage, while at 18 hours the 8C stage predominates (Brunk and Conover, 1985). Thus, the micronuclear-specific sequences are eliminated between 12 and 18 hours following the initiation of conjugation. This is similar to the time at which rDNA is excised from high-molecular-weight DNA. Apparently, both type I and II reorganizations occur early during conjugation.

The rearrangement of a well-characterized sequence that undergoes type II reorganization has also been monitored during anlagen development (Austerberry *et al.*, 1984). In this study, a probe was prepared from a cloned region of micronuclear DNA that contains three regions which are eliminated during anlagen development. All of the regions are removed between 12 and 14 hours after the initiation of conjugation. One of the regions is eliminated slightly earlier than the other two, generating a reorganization intermediate. In the middle region there are two alternative sizes for the eliminated sequence, which differ in length by 300 bp. Both alternative reorganizations can occur within a single developing anlagen and both are stable. The alternatives eventually assort in the progeny in a manner similar to any other allele. These details clearly indicate that the reorganization process is under precise molecular control both with regard to the sites of reorganization and the timing of reorganization.

In summary, these results suggest that both type I and II reorganizations occur within a short period of time during the early portion of anlagen development. The kinetics of elimination for these sequences strongly indicate that the process involves an active elimination, probably mediated by a specific nuclease, rather than a passive elimination by underreplication and dilution.

VI. The Relation between Reorganization and Gene Expression

The function of the DNA fragments in the macronucleus is not sufficiently well understood for the role of fragmentation to be definitively assigned. In an attempt to suggest a role for fragmentation, it seems reasonable to examine the differences in macronuclear and micronuclear processes that might account for this difference in physical organization. One of the most apparent differences in the two nuclei is the fact that virtually all gene expression occurs in the macronuclei (Gorovsky, 1973). There is extensive literature indicating that rearrangement of DNA can lead to activation of gene expression (Leder *et al.*, 1980; Tonegawa, 1983; Naysmyth, 1982; Borst *et al.*, 1980). As an initial hypothesis, it seems reasonable to assume that the reorganization may be involved in activation of gene expression in *Tetrahymena*. If reorganization is involved in gene activation, then transcriptionally active sequences in the macronucleus would be expected to be proximal to reorganization sites. A comparison of the arrangement of genes within the micronuclear and macronuclear genomes should reveal such reorganization sites.

A. Ribosomal RNA Genes

The ribosomal RNA genes are among the most actively expressed genes in the cell. They are reorganized into 21 kb DNA fragments during anlagen development. This would seem to be consistent with the hypothesis that reorganization plays a role in the activation of expression for these genes. The ribosomal genes may not be typical genes, however, since a number of organisms have their rDNA amplified, often as extrachromosomal units, without extensive general reorganization of the genome (Tobler, 1975). In *Tetrahymena,* the rDNA fragments are present in about 9000 copies in the macronucleus, while most fragments are represented by about 50 copies It is quite probable that the primary role of reorganization of the rDNA sequences in *Tetrahymena* is amplification rather than activation of gene expression.

B. α-Tubulin Genes

Several genes have been examined relative to their organization in the micronuclear and macronuclear DNA. Callahan *et al.* (1984) have examined the genomic organization of α-tubulin in micronuclear and macronuclear DNA. α-Tubulin is a highly expressed gene in *Tetrahymena*, as in most organisms. There is apparently a single α-tubulin gene in *T. thermophila*. A comparison of restriction fragments homologous to a heterologous α-tubulin probe indicates type II reorganizations in the immediate vicinity of the *T. thermophila* α-tubulin gene. The macronuclear DNA shows deletions and rejoinings within 5 kb or less on both the 3' and 5' sides of the α-tubulin gene. Again, this would fit well with the hypothesis that reorganization has a role in gene activation. The authors of this study do not feel that these reorganizations are necessarily involved in gene activation when considered in the context of the other examples presented below.

C. 5 S rRNA GENES

Restriction mapping of the 5 S rRNA gene clusters indicated that about one-fourth of the clusters were found on different-sized restriction fragments in macronuclear and micronuclear DNA (Pederson *et al.*, 1984). When these rearrangements were characterized to determine if they occurred preferentially on the 5' or 3' side of the 5 S gene clusters, the rearrangements were found to occur about equally on the 5' and 3' side. The pattern of 5 S gene clusters was essentially unaltered by digestion with exonuclease *Bal*31, indicating that the reorganizations did not create new termini in the immediate vicinity of the 5 S gene clusters. Thus, these rearrangements are type II reorganizations. The 5 S gene clusters are widely dispersed throughout the macronuclear genome, precluding a single large-scale reorganization event as the mechanism activating the genes in the macronucleus (Allen *et al.*, 1984). The random occurrence of the reorganizations in the vicinity of the macronuclear 5 S gene clusters suggests that these rearrangements may not play an important role in the transcriptional activation of the 5 S genes (Pederson *et al.*, 1984).

D. HISTONE GENES

One of the two histone *H4* genes, *H4-I*, of *T. thermophila* has been cloned and its organization in macronuclear and micronuclear DNA characterized (Bannon *et al.*, 1984). There are no detectable rearrangements within 5 kb to the left or 6 kb to the right of this gene. Using the histone *H4-I* gene as a probe, the organization of the T. thermophila histone *H4-II* gene was determined. Like *H4-I*, no rearrangements in the immediate vicinity of *H4-II* were observed. The absence of rearrangements proximal to these two highly transcribed genes makes it less likely that reorganization plays an essential role in activation of gene expression in the macronucleus.

T. thermophila can have any one of seven different mating types (Nanney and Caughey, 1953). Mating type is determined in a stochastic fashion during anlagen development and is extremely stable during vegetative growth. Orias (1981) has proposed that the mating type determination may involve DNA reorganization similar to that occurring in the immunoglobin system in mammals. There are genetic data consistent with this suggestion, but as yet no molecular results (Orias, 1981; Orias and Blum, 1984). If this proposal is correct, determination of mating type would be a clear example of gene expression which is activated by genome reorganization.

While it is not possible to rule out a role for genome reorganization in transcriptional activation, in view of the data available this hypothesis is less likely than one might assume a priori. Reorganizations, particularly type II rearrangements, occur relatively often, and thus, a reorganization may be found in the general vicinity of many DNA sequences strictly by chance. The observation of rearrangements near active genes in the macronucleus seems to be consistent with chance occurrence of these reorganizations. It is also possible that in certain cases such as mating type determination, genome reorganization may play an essential role in gene activation.

VII. The Genetic Consequences of Macronuclear DNA Fragmentation

The genetic behavior of *T. thermophila* has been studied in detail. Several years ago Nanney and Preparata (1979) summarized the genetic behavior associated with the macronucleus of *T. thermophila*, for which molecular explanations are required. These included phenotypic assortment, unequal allelic input for some loci, delayed assortment at some loci, linkage disruption, assortment depression, ploidy-related developmental differences in macronuclear primordia, and ploidy-independent macronuclear DNA content. Most of these genetic phenomena can be reasonably explained in terms of genomic reorganization occurring during anlagen development in combination with copy number control for the macronuclear fragments.

A. Phenotypic Assortment

One of the most remarkable features of *T. thermophila* genetics is a process called phenotypic assortment. When two strains of *T. thermophila* which are each homozygous for a different allele at a given locus are mated, the micronuclei and macronuclei in resulting exconjugates are initially heterozygous for the alleles at this locus. The macronucleus divides amitotically and, over a number of generations following conjugation, the alleles randomly assort to the progeny. With time, this random assortment gives rise to progeny with multiple copies of a single allele for the locus. At that time the macronuclear genotype and thus the phenotype are fixed. Thus, all loci drift and fix following conjugation, and each macronucleus becomes homozygous at each locus by random choice among the alleles present in the zygotic nucleus. This process, called phenotypic assortment, is a natural consequence of binary fission without a mechanism to ensure regular segregation (amitotic division).

While amitotic division readily explains phenotypic assortment, it presents an equally troublesome problem. If the fragments assort randomly so that the alleles drift and fix, why don't the loci also assort to produce aneuploidy? In order for the alleles to fix, fragments carrying the alternative alleles must be lost. In the absence of a regular segregatory mechanism the loss of a whole class of macronuclear fragments would be expected (Preer and Preer, 1979). Clearly, loci are not segregating, as the frequency of nonviable cells is much lower than the rate of phenotypic assortment (Nanney, 1959; Preer and Preer, 1979).

B. Copy Number Control of Macronuclear DNA Fragments

A mechanism that maintains the appropriate number of copies of each type of fragment during the DNA replication period will effectively prevent the loss of loci and allow the random assortment of alleles. A mechanism similar to that operative for the copy number control of bacterial plasmids (Prichard, 1979; Tomizawa, 1984) could maintain the appropriate copy number of each macronuclear DNA fragment. The macronuclear DNA replicates only during a well-defined portion of the cell cycle, thus a copy number control mechanism would therefore be under cell cycle control, as is the case with the 2-μm plasmid in yeast (Zakian et al., 1979). The macronuclear fragments would assort randomly to the daughter cells by binary fission of the macronucleus. During the next DNA synthetic period the appropriate number of each fragment would be restored. In this manner a full complement of each type of fragment would be present prior to the amitotic division of the macronucleus. In this arrangement there is a negligible probability that all the representatives of a given fragment would be lost at a given division. Although the appropriate number of each fragment is restored during each DNA synthetic period, the alleles do assort and become fixed because the copy number restoration operates on the fragment type and not on the alleles. Given the clear physical evidence for fragmentation of the macronuclear DNA, a copy number control mechanism which adjusts the abundance of each macronuclear fragment is essential to maintain genome balance, since even a moderate genic imbalance would be expected to lead to loss of fitness. A copy number control mechanism operating on the macronuclear fragment type coupled with the fragmentation process explains some of the unique aspects of *T. thermophila* genetics.

When the micronuclear chromosomes are fragmented during anlagen development, the control of DNA replication is shifted from the mechanism operative in the micronucleus with uninemic chromosomes to a copy number control system in which the abundance of each individual fragment type is independently controlled. A system similar to that of bacterial cells hosting a number of compatible plasmids (Novick and Hoppenstedt, 1978) is the type of mechanism that is postulated. In general outline, DNA replication is controlled by some substance which acts as an inhibitor of DNA replication initiation, either a protein or RNA. The level of this inhibitor is maintained proportional to the total number of copies of the replicating molecule (Prichard, 1979). Replication of the molecule is regulated by the concentration of the inhibitor. When the number of copies is low, inhibition is relieved and replication increases the number to the appropriate level. This negative feedback regulates the copy number of each type of macronuclear fragment independently. In this system different fragments may have different copy numbers. In fact, the copy number of individual fragments may be modulated as a part of the developmental program of the cell.

This type of control is operative for prokaryotic plasmids, but it has been hard to envision this sort of feedback mechanism working on a large ensemble of independently replicating DNA molecules. The degree of specificity required to distinguish among and independently control several hundred replication origins is difficult to imagine, particularly if the inhibitor is a protein. The coli E1 plasmid in *E. coli* is under a type of copy number control that makes an attractive model for the regulation of replication in macronuclear fragments. An RNA molecule acts as the inhibitor controlling copy number in this system (Tomizawa and Itoh, 1981; Som and Tomizawa, 1983). In this system a small transcript (RNA 1) from the region of the genome, including the origin of replication, acts as the inhibitor (Tomizawa et al., 1981). The specificity resides in the homology between this RNA and the DNA sequence adjacent to the origin of replication, which acts as the control region. A single base change in this region can change the specificity (compatibility) of the inhibitor (Lacatena and Cesareni, 1981). In this model a large number of independent inhibitors with a high degree of specificity for different types of macronuclear fragments are easily conceivable. A single base change in the control region can also alter the copy number of the plasmid (Muesing *et al.*, 1981). This mechanism can provide effective copy number control of a large number of macronuclear fragments. It can account for the degree of specificity required for independent control of several hundred fragments without requiring elaborate protein inhibitors. It also provides for different copy numbers among the macronuclear fragments.

While the prokaryotic plasmids are replicated throughout the cell cycle, the replication of macronuclear fragments is confined to a specific portion of the cell cycle, the S phase. Thus, there is a cell-cycle control superimposed upon the copy number control. The 2-μm plasmid of yeast suggests a model for this type of control. Although the copy number control of the 2-μm plasmid is not understood in the same detail as the prokaryotic plasmids, it seems to have a similar mechanism. The replication of the 2-μm plasmid in yeast is confined to the S phase of the cell cycle (Zakian *et al.*, 1979). Thus, some combination of the mechanisms operative in the coli E1 plasmid and yeast 2-μm plasmid provides a reasonable model for the replication control of the macronuclear fragments.

C. Ribosomal DNA under Copy Number Control

The rDNA can be viewed as one of the macronuclear DNA fragments. During anlagen development it is reorganized into a small macronuclear fragment under independent copy number control. In this case the copy number is about 9000 copies per macronucleus. The copy number of the rDNA also appears to vary with physiological conditions. During starvation there is apparently a decrease in the rDNA relative to the other macronuclear fragments of about 30% (Engberg and Pearlman, 1972). An rDNA with a restriction polymorphism has been found in inbred strain C-3 (Pan and Blackburn, 1981). This rDNA has an additional *Bam*HI site 1.5 kb from the termini and an additional *Mbo*I site 0.6-kb telomere proximal from the additional *Bam*HI site. The locus for the rDNA in *T. thermophila* has been designated *rdn* and the allele with the polymorphisms is named *rdnA1* (Pan *et al.*, 1982). This rDNA allele has an unusual phenotype; when strain C-3 is conjugated with another inbred strain, over 95% of the total macronuclear rDNA molecules of the progeny population are of the rDNA-A1 type (Pan *et al.*, 1982). This dominance is not the result of a selective advantage conferred on cells containing the rDNA-A1-type molecules during vegetative growth, such that these cells simply outgrow cells with wild-type rDNA$^+$ molecules. In cultures where rDNA-A1 and rDNA$^+$ cells were mixed in known ratios, there was no loss of the rDNA$^+$ cells during prolonged vegetative growth.

The domination of one specific rDNA allele was originally believed to occur during the initial amplification of the rDNA early in anlagen development (Pan *et al.*, 1982). More recent results suggest that the rdnA1-type rDNA may be preferentially replicated within each cell during vegetative growth (E. Orias *et al.*, personal communication). Only a slight alteration in the copy control of the *rdnA1* allele would be necessary to give this rDNA molecule a selective advantage and result in its dominance. Thus, the dominance of the *rdnA1* locus is probably a result of increased copy number for these rDNA molecules.

D. Genetic Effects of Fragmentation

The process of fragmentation and copy number control of the fragments also provides a basis for understanding a number of genetic phenomena in *T. thermophila*, including linkage disruption, unequal allelic input, delayed assortment, and ploidy-independent macronuclear DNA content. Linkage disruption is a direct result of the fragmentation of the macronuclear DNA. There are two studies that clearly indicate that loci closely linked by meiotic analysis coassort randomly (Allen, 1964; Doerder, 1973). Fragmentation of macronuclear DNA during anlagen development would be expected to split even closely linked markers to different macronuclear fragments. The clonal distribution of markers is strictly a function of their association on macronuclear fragments rather than their association in micronuclear chromosomes. As a corollary, clonal assortment of meiotically linked markers is often the result of macronuclear DNA fragmentation rather than somatic recombination.

The rate of phenotypic assortment and the time at which the assortment begins are a function of the number of copies of the macronuclear fragment assorting. The copy number of a macronuclear fragment may vary during anlagen development. It is also possible that there may be slight differences in the copy numbers of homologous fragments, in the same manner as there are different alleles at homologous loci on the fragments. The time at which assortment can first be detected is related to the total number of each allele present. This in turn is determined by the copy number of the fragment on which the allele is found. Early and late assortment may simply be the result of a lower or higher copy number for specific fragments present during the early stages of anlagen development. Doerder *et al.* (1977) have suggested that the early assortment is related to a biased input of alleles and that late assortment is the result of the higher DNA content of the macronucleus following conjugation. Essentially the same concept can be applied to differential copy number of macronuclear fragments. In effect, the copy number control of individual macronuclear fragments allows for uncoordinated replication of macronuclear replicons without causing genic imbalance.

The copy number control of individual macronuclear fragments provides a mechanism to account for early and late assortment. It also explains the ploidy-independent macronuclear DNA content. Macronuclei in clones derived from haploid, diploid, triploid, and aneuploid primordia tend to have similar DNA contents. During anlagen development, once the macronuclear fragments have come under copy number control, the DNA content is set by the collective copy numbers of the fragments rather than the ploidy of the zygotic nuclei. In essence, as soon as fragmentation occurs the copy number control of each fragment comes into play. This not only restores the abundance of each fragment preventing aneuploidy, it also becomes the dominant factor in determining the DNA content of the macronucleus.

E. Role of Macronuclear Fragments in Development

A copy number control mechanism regulating the abundance of macronuclear fragments may play a significant role in development and gene expression as well as providing a process by which aneuploidy may be avoided. *T. thermophila* are apparently unusual in having a single copy of the ribosomal gene in the germline nucleus (micronucleus). Generally, an organism has several hundred copies of the ribosomal genes, which are often augmented by amplification (Brown and Dawid, 1968; Gall, 1968). Multiple copies of the ribosomal genes present the problem of genetic drift among the various copies. A single copy simplifies this problem, possibly at the price of vulnerability to loss of that single copy. The process of fragmentation and amplification occurring during anlagen development provides an opportune time to expand the copies of highly transcribed genes.

The histones in *T. thermophila* may be another example of limited gene copies in the germline. There are apparently only two copies of histone H4 found in the micronucleus, and these are on different chromosomes (Bannon *et al.,* 1984). Most organisms have a whole family of histone genes arrayed in clusters (Old and Woodland, 1984). The fragmentation process and the shift to copy number control allow the appropriate amplification of genes in the somatic nucleus (macronucleus), while only a minimal set is required in the germline (micronucleus). The intensity of hybridization of a probe for the histone H4 sequence with the macronuclear fragment carrying the gene suggests an amplification of that macronuclear fragment and the histone *H4* gene.

Actin seems to present a similar situation. Using the yeast actin gene as a probe in a Southern analysis of genomic DNA, there appears to be a single actin gene in *T. thermophila* (C. Cupples and R. Pearlman, personal communication). When the *T. thermophila* sequence, homologous to the yeast actin, is used as a probe, the intensity of hybridization with a single macronuclear fragment suggests a high copy number for this fragment.

An investigation of the distribution of 5 S rRNA gene clusters in micronuclear and macronuclear DNA apparently shows different copy numbers for some of the clusters. The relative intensities (relative copy numbers) of 5 S genes in macronuclear and micronuclear restriction fragments of similar size indicate that many of the macronuclear clusters have the same abundance in both genomes. Some of the clusters are about twice as abundant in the macronuclear genome as the micronuclear genome and a few are four times as prevalent in the macronuclear genome (Allen *et al.*, 1984).

It is clear that macronuclei have the rDNA molecules present in a very high copy number. It is also likely that other macronuclear fragments are present in varying copy numbers. Thus, copy number control of the macronuclear fragments may provide not only a mechanism for preventing aneuploidy, it may also allow differential gene amplification.

VIII. DNA Replication in the Macronucleus and Micronucleus

One of the striking features of DNA replication in *T. thermophila* is that the macronuclear and micronuclear S phases are in different portions of the cell cycle. The micronuclear S phase starts immediately following mitosis; thus, there is virtually no G_1 period for the micronuclei (Prescott and Stone, 1967; Woodard *et al.*, 1972). The vast majority of micronuclei have a 4C (G_2 phase) complement of DNA. Alternatively, the macronucleus has a more conventional cell cycle, with G_1, S, and G_2 phases roughly equal in duration. Mitosis in the micronucleus and amitotic nuclear division in the macronucleus both occur just prior to cytokinesis.

It is unusual to have nuclei in a common cytoplasm replicate their DNA at different times. This is even more unusual because the micronucleus resides tucked into a pocket in the macronucleus during the majority of the cell cycle, only emerging during mitosis (and the micronuclear S phase). Thus, the micronucleus is actually in very close contact with the macronucleus during the macronuclear DNA replication period. In most cases substances in the cytoplasm of the cell induce DNA replication in the nuclei (Goldstein, 1963). In this case the micronuclei and macronuclei are not responding to the same inducers of DNA replication. It seems probable that the fragmentation of the macronuclear genome and the shift to a copy number control of DNA replication may be involved in the different responses of the macronucleus and the micronucleus.

The micronuclear chromosomes are assumed to be uninemic and are on the order of 40,000 kb. The micronuclear S phase is relatively short, 10 minutes in rapidly growing cells. The rate of DNA replication in most eukaryotes is about 3000–5000 bp/minute (Kornberg, 1974). If *T. thermophila* has a similar rate, then in 10 minutes a pair of bidirectional replication forks could replicate 60–100 kb of DNA. A micronuclear chromosome would require 100 or more tandemly arrayed bidirectional origins of replication in order to duplicate the genome during the S period. This is a common mode of DNA replication in eukaryotes (Hand, 1978). Possibly the majority of the origins of replication in the micronuclear genome are among the DNA sequences eliminated from the macronuclear DNA. It might also be expected that sequences involved in mitosis would also be eliminated from the macronuclear DNA.

The macronuclear DNA replication period is about 1 hour in rapidly growing *T. thermophila*. If the macronuclear DNA is replicated at a rate common for eukaryotes, then many of the macronuclear fragments could be replicated from a single centrally located origin of replication. A single (or very few) origin(s) of replication on a macronuclear fragment might be expected for a mechanism operating in the mode of the prokaryotic copy number control.

IX. Conclusions and Summary

In overview, the reorganization process in *T. thermophila* is diagrammed in Fig. 5. During anlagen development, type I reorganization leads to fragmentation of the DNA into molecules ranging in size from 21 to several thousand kilobase pairs. This process creates 300–500 new linear fragments, each terminated in a C_4A_2 telomere. In addition, over 5000 internal deletions are removed by type II reorganizations. Virtually all of this reorganization takes place during a relatively short period early in anlagen development. The net result of these reorganizations is a loss of 10–20% of the complexity found in the micronuclear genome.

Along with these eliminations, the control of DNA replication in the macronucleus is shifted to a copy number control mechanism which prevents aneuploidy and allows differential amplification of specific sequences. There is a general amplification of the macronuclear fragments so that on average each fragment is present in about 50 copies. The

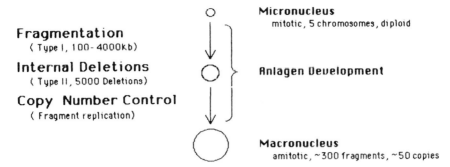

Fragmentation
(Type I, 100- 4000kb)

Internal Deletions
(Type II, 5000 Deletions)

Copy Number Control
(Fragment replication)

Micronucleus
mitotic, 5 chromosomes, diploid

Anlagen Development

Macronucleus
amitotic, ~300 fragments, ~50 copies

FIG. 5. An overview of the genome reorganization process in *Tetrahymena*. During anlagen development the DNA is fragmented into 300–500 molecules, ranging in size from less than 100 to as much as 4000 kb. Over 5000 internal eliminations also occur, reducing the complexity 10–20%. The replication of macronuclear fragments is under individual copy number control for each type of fragment. During this process the macronuclear DNA is amplified so that each fragment is represented by about 50 copies.

resulting macronuclei have a DNA replication period during a different part of the cell cycle than the micronuclei.

Reorganization of the genome into fragments and the resulting copy number control has consequences for the cells at the levels of phenotypic assortment, DNA replication, and differential gene amplification. This reorganization process, which is predominant in ciliates, occurs in specialized cases in many higher organisms and appears to be a general biological strategy.

ACKNOWLEDGMENTS

I am greatly indebted to Dr. E. Orias for many constructive discussions on the model of copy number control for macronuclear fragments and to a number of colleagues who assisted with constructive criticism of this manuscript and who communicated the results of their investigations prior to their publication. In particular, I wish to thank Drs. S. Allen, E. Blackburn, K. Conover, K. Jones, V. Merriam, E. Orias, R. Pearlman, and M.-C. Yao.

Experimental work in the author's laboratory was supported by research grants from the National Science Foundation (PCM-8215666) and the Academic Senate of the University of California (2455).

REFERENCES

Abraham, J., Feldman, J. R., Naysmyth, K. A., Strathern, J. N., Klar, A. J. S., Broach, J. R., and Hicks, J. B. (1982). *Cold Spring Harbor Symp. Quant. Biol.* **47,** 989–999.
Allen, S. L. (1964). *Genetics* **49,** 617–627.

Allen, S. L., and Gibson, I. (1972). *Biochem. Gen.* **6,** 293–313.

Allen, S. L., Ervin, P. R., McLaren, N. C., and Brand, R. E. (1984). *Mol. Gen. Genet.* **197,** 244–253.

Allen, S. L., Ervin, P. R., White, T. C., and McLaren, N. C. (1985). *Dev. Genet.* **5,** 181–200.

Ammermann, D. (1965). *Arch. Protistenkd.* **108,** 109–152.

Austerberry, C. F., Allis, C. D., and Yao, M.-C. (1984). *Proc. Natl. Acad. Sci. U.S.A.* **81,** 7383–7387.

Bannon, G. A., Bowen, J. K., Yao, M.-C., and Gorovsky, M. A. (1984). *Nucleic Acids Res.* **12,** 1961–1975.

Bantock, C. R. (1970). *J. Embryol. Exp. Morphol.* **24,** 257–286.

Beams, H. W., and Kessel, R. G. (1974). *Int. Rev. Cytol.* **39,** 413–479.

Blackburn, E. H., and Gall, J. G. (1978). *J. Mol. Biol.* **120,** 33–53.

Blackburn, E. H., Budarf, M. L., Challoner, P. B., Cherry, J. M., Howard, E. A., Katzen, A. L., Pan, W.-C., and Ryan, T. (1982). *Cold Spring Harbor Symp. Quant. Biol.* **47,** 1195–1207.

Borchsenius, S. N., Belozerskaya, N. A., Merkulova, N. A., Wolfson, V. G., and Vorob'ev, V. I. (1978). *Chromosoma* **69,** 275–289.

Borst, P., Frasch, A. C. C., Brenards, A., Van der Ploeg, L. H. J., Hoeijmakers, J. H. J., Arnberg, A. C., and Cross, G. A. M. (1980). *Cold Spring Harbor Symp. Quant. Biol.* **45,** 935–943.

Bostock, C. J., and Prescott, D. M. (1972). *Proc. Natl. Acad. Sci. U.S.A.* **69,** 139–142.

Boswell, R. E., Klobutcher, L. A., and Prescott, D. M. (1982). *Proc. Natl. Acad. Sci. U.S.A.* **69,** 139–142.

Brown, D. D., and Dawid, I. B. (1968). *Science* **160,** 272–280.

Brunk, C. F., and Bohman, R. E. (1985). *Exp. Cell Res.,* in press.

Brunk, C. F., and Conover, R. K. (1985). *Mol. Cell. Biol.* **5,** 93–98.

Brunk, C. F., Tsao, G. S. G., Diamond, C. H., Ohashi, P. S., Tsao, N. N. G., and Pearlman, R. E. (1982). *Can. J. Biochem.* **60,** 847–853.

Bruns, P. J. (1984). *In* "Genetic Maps" (S. O'Brien, ed.). Cold Spring Harbor Laboratory, Cold Spring Harbor, New York.

Bruns, P. J., and Brussard, T. E. B. (1981). *Science* **213,** 549–551.

Callahan, R. C., Shalke, G., and Gorovsky, M. A. (1984). *Cell* **36,** 441–445.

Carle, G. F., and Olson, M. V. (1984). *Nucleic Acids Res.* **12,** 5647–5664.

Carle, G. F., and Olson, M. V. (1985). *Proc. Natl. Acad. Sci. U.S.A.* **82,** 3756–3760.

Cech, T. R., Zaug, A. J., Grabowski, P. J., and Brehm, S. L. (1982). *In* "The Cell Nucleus" (H. Bush and L. Rothblum, eds.), Vol. 10, pp. 171–204. Academic Press, New York.

Conover, R. K., and Brunk, C. F. (1985). Submitted.

Crouse, H. V., Brown, A., and Mumford, B. C. (1971). *Chromosoma* **34,** 324–339.

Davis, M. M., Kim, S. K., and Hood, L. (1980). *Cell* **22,** 1–2, 374–382.

Doerder, F. P. (1973). *Genetics* **74,** 81–106.

Doerder, F. P., and DeBault, L. E. (1975). *J. Cell Sci.* **17,** 47–93.

Doerder, F. P., Lief, J. H., and DeBault, L. E. (1977). *Science* **198,** 946–948.

Dreyer, W. J., and Bennett, J. C. (1965). *Proc. Natl. Acad. Sci. U.S.A.* **54,** 864–869.

Elliot, A. M. (1973). *In* "Biology of Tetrahymena" (A. M. Elliot, ed.), pp. 259–288. Hutchinson & Ross, Stroudsburg, Pennsylvania.

Engberg, J. (1983). *Nucleic Acids Res.* **11,** 4939–4946.

Engberg, J., and Pearlman, R. E. (1972). *Eur. J. Biochem.* **26,** 393–400.

Engberg, J., Andersson, P., Leick, V., and Collins, J. (1976). *J. Mol. Biol.* **104,** 455–470.

Flavell, R. A., and Jones, I. G. (1970). *Biochem. J.* **116,** 155–157.

Gall, J. G. (1968). *Proc. Natl. Acad. Sci. U.S.A.* **60,** 553–560.

Gall, J. G. (1974). *Proc. Natl. Acad. Sci. U.S.A.* **71**, 3078–3081.
Gibson, I., and Martin, N. (1971). *Chromosoma* **35**, 374–382.
Goldstein, L. (1963). *In* "Cell Growth and Cell Division" (R. J. C. Harris, ed.), Vol. 2, pp. 129–145. Academic Press, New York.
Gorovsky, M. A. (1973). *J. Protozool.* **20**, 19–25.
Gorovsky, M. A. (1980). *Annu. Rev. Genet.* **14**, 203–239.
Gurdon, J. B. (1962). *J. Embryol. Exp. Morphol.* **10**, 622–640.
Gurdon, J. B. (1974). "The Control of Gene Expression in Animal Development." Harvard Univ. Press, Cambridge, Massachusetts.
Hand, R. (1978). *Cell* **15**, 317–325.
Howard, E. A., and Blackburn, E. H. (1985). *Mol. Cell. Biol.* **5**, 2039–2050.
Iwamura, Y., Sakai, M., Mita, T., and Muramatsu, M. (1979). *Biochemistry* **18**, 5289–5294.
Karrer, K. M. (1983). *Mol. Cell. Biol.* **3**, 1909–1919.
Karrer, K. M., and Gall, J. G. (1976). *J. Mol. Biol.* **104**, 421–453.
Katzen, A. L., Cann, G. M., and Blackburn, E. H. (1981). *Cell* **24**, 313–320.
King, B. O., and Yao, M.-C. (1982). *Cell* **31**, 177–182.
King, T. J., and Briggs, R. (1956). *Cold Spring Harbor Symp. Quant. Biol.* **21**, 271–289.
Kiss, G. B., and Pearlman, R. E. (1981). *Mol. Cell. Biol.* **1**, 535–543.
Klobutcher, L. A., Swanton, M. T., Donini, P., and Prescott, D. M. (1981). *Proc. Natl. Acad. Sci. U.S.A.* **78**, 3015–3019.
Kloetzel, J. A. (1970). *J. Cell Biol.* **47**, 395–407.
Kornberg, A. (1974). "DNA Replication." Freeman, San Francisco, California.
Lacatena, R. M., and Cesareni, G. (1981). *Nature (London)* **294**, 623–626.
Lauth, M. R., Spear, B. B., Heumann, J. M., and Prescott, D. M. (1976). *Cell* **7**, 67–74.
Lawn, R. M., Heumann, J. M., Herrick, G., and Prescott, D. M. (1978). *Cold Spring Harbor Symp. Quant. Biol.* **42**, 483–492.
Leder, P. (1982). *Sci. Am.* **246**, 102–115.
Leder, P., Max, E. E., Seidman, J. G., Kwan, S.-P., Scharff, M., Nau, M., and Norman, B. (1980). *Cold Spring Harbor Symp. Quant. Biol.* **45**, 859–865.
Mayo, K. A., and Orias, E. (1979). *J. Cell Biol.* **83**, 210a.
Mayo, K. A., and Orias, E. (1981). *Genetics* **98**, 747–762.
Merkulova, N. A., and Borchsenius, S. N. (1976). *Mol. Biol. USSR* **10**, 875–880.
Muesing, M., Tamm, J., Shepard, H. M., and Polisky, B. (1981). *Cell* **24**, 235–242.
Nanney, D. L. (1959). *J. Protozool.* **6**, 171–177.
Nanney, D. L., and Caughey, P. A. (1953). *Proc. Natl. Acad. Sci. U.S.A.* **39**, 1057–1063.
Nanney, D. M., and Preparata, R. M. (1979). *J. Protozool.* **26**, 2–9.
Naysmyth, K. A. (1982). *Annu. Rev. Genet.* **16**, 439–500.
Novick, R. P., and Hoppensteadt, F. C. (1978). *Plasmid* **1**, 421–434.
Oka, Y., Shiota, S., Nakai, S., Nishida, Y., and Okubo, S. (1980). *Gene* **10**, 301–306.
Old, R. W., and Woodland, H. R. (1984). *Cell* **38**, 624–626.
Orias, E. (1981). *Dev. Genet.* **2**, 185–202.
Orias, E., and Baum, M. P. (1984). *Dev. Genet.* **4**, 145–158.
Pan, W.-C., and Blackburn, E. H. (1981). *Cell* **23**, 459–466.
Pan, W.-C., Orias, E., Flacks, M., and Blackburn, E. H. (1982). *Cell* **28**, 595–604.
Pearlman, R. E., Andersson, P., Engberg, J., and Nilsson, J. R. (1979). *Exp. Cell Res.* **123**, 147–155.
Pederson, D. S., Yao, M.-C., Kimmel, A. R., and Gorovsky, M. A. (1984). *Nucleic Acids Res.* **12**, 3003–3021.
Preer, J. R., Jr., and Preer, L. B. (1979). *J. Protozool.* **26**, 14–18.

Prescott, D. M. (1983). *Mod. Cell Biol.* **2**, 329–352.

Prescott, D. M., and Murti, K. G. (1974). *Cold Spring Harbor Symp. Quant. Biol.* **38**, 609–618.

Prescott, D. M., and Stone, G. E. (1967). *In* "Research in Protozoology" (T. T. Chen, ed.), Vol. 2, pp. 119–146. Pergamon, Oxford.

Prescott, D. M., Bostock, C. J., Murti, K. G., Lauth, M. R., and Gamow, E. (1971). *Chromosoma* **34**, 355–366.

Prichard, R. E. (1979). *In* "Control of DNA Replication in Bacteria" (I. Molineur and M. Kohiyama, eds.), pp. 1–26. Plenum, New York.

Rae, P. M. M., and Spear, B. B. (1978). *Proc. Natl. Acad. Sci. U.S.A.* **75**, 4992–4996.

Ray, C., Jr. (1956). *J. Protozool.* **3**, 88–96.

Schwartz, D. C., and Cantor, C. R. (1984). *Cell* **37**, 67–75.

Schwartz, D. C., Saffran, W., Welsh, J., Hass, R., Goldenberg, M., and Cantor, C. R. (1982). *Cold Spring Harbor Symp. Quant. Biol.* **47**, 189–195.

Seyfert, H.-M. (1979). *J. Protozool.* **26**, 66–74.

Seyfert, H.-M., and Preparata, R. M. (1979). *J. Cell Sci.* **40**, 111–123.

Silverman, M., and Simon, M. (1980). *Cell* **19**, 845–854.

Silverman, M., Zieg, J., Mandel, G., and Simon, M. (1980). *Cold Spring Harbor Symp. Quant. Biol.* **45**, 17–26.

Som, T., and Tomizawa, J. (1983). *Proc. Natl. Acad. Sci. U.S.A.* **80**, 3232–3236.

Southern, E. (1975). *J. Mol. Biol.* **98**, 503–517.

Spear, B. B., and Lauth, M. R. (1976). *Chromosoma* **54**, 1–13.

Streeck, R. E., Moritz, K. B., and Beer, K. (1982). *Nucleic Acids Res.* **10**, 3495–3502.

Sugai, T., and Hiwatashi, K. (1974). *J. Protozool.* **21**, 542–548.

Swanton, M. T., Heumann, J. M., and Prescott, D. M. (1980). *Chromosoma* **77**, 217–227.

Szostak, J. W., and Blackburn, E. H. (1982). *Cell* **29**, 245–255.

Tobler, H. (1975). *In* "Biochemistry of Animal Development" (R. Weber, ed.), Vol. 3, pp. 91–143. Academic Press, New York.

Tobler, H., Smith, K. D., and Ursprung, H. (1972). *Dev. Biol.* **27**, 109–203.

Tomizawa, J.-I. (1984). *Cell* **38**, 861–870.

Tomizawa, J., and Itoh, T. (1981). *Proc. Natl. Acad. Sci. U.S.A.* **78**, 6096–6100.

Tomizawa, J.-I., Itoh, T., Selzer, G., and Som, T. (1981). *Proc. Natl. Acad. Sci. U.S.A.* **78**, 1421–1425.

Tonegawa, S. (1983). *Nature (London)* **302**, 575–581.

Tsao, S. G. S., Brunk, C. F., and Pearlman, R. E. (1983). *Anal. Biochem.* **131**, 365–372.

Van der Ploeg, L. H. T., Cornelissen, A. W. C. A., Michels, P. A. M., and Borst, P. (1984). *Cell* **39**, 213–221.

White, M. J. D. (1973). "Animal Cytology and Evolution," 3rd Ed. Cambridge Univ. Press, London and New York.

Williams, J. B., Fleck, D. W., Hellier, L. E., and Uhlenhopp, E. (1978). *Proc. Natl. Acad. Sci. U.S.A.* **75**, 5062–5065.

Wilson, E. B. (1925). "The Cell in Development and Heredity," 3rd Ed. Macmillan, New York.

Woodard, J., Kaneshiro, E., and Gorovsky, M. A. (1972). *Genetics* **70**, 251–260.

Yao, M.-C. (1981). *Cell* **24**, 765–774.

Yao, M.-C. (1982a). *J. Cell Biol.* **92**, 783–789.

Yao, M.-C. (1982b). *In* "The Cell Nucleus" (H. Bush and L. Rothblum, eds.), Vol. 12, pp. 127–153. Academic Press, New York.

Yao, M.-C., and Gall, J. G. (1977). *Cell* **12**, 121–132.

Yao, M.-C., and Gall, J. G. (1979). *J. Protozool.* **26**, 10–13.

Yao, M.-C., and Gorovsky, M. A. (1974). *Chromosoma* **48**, 1–18.

Yao, M.-C., and Yao, H.-C. (1981). *Proc. Natl. Acad. Sci. U.S.A.* **78**, 7436–7439.

Yao, M.-C., Kimmel, A. R., and Gorovsky, M. A. (1974). *Proc. Natl. Acad. Sci. U.S.A.* **71**, 3082–3086.

Yao, M.-C., Blackburn, E. H., and Gall, J. G. (1981). *J. Cell Biol.* **90**, 515–520.

Yao, M.-C., Chio, J., Yokoyama, S., Austerberry, C. F., and Yao, H.-C. (1984). *Cell* **36**, 433–440.

Yao, M.-C., Zhu, S.-G., and Yao, H.-C. (1985). *Mol. Cell. Biol.* **5**, 1260–1267.

Yokoyama, R. W., and Yao, M.-C. (1982). *Chromosoma* **85**, 11–22.

Zakian, V. A., Brewer, B. J., and Fangman, W. L. (1979). *Cell* **17**, 923–934.

The Molecular Biology of Antigenic Variation in Trypanosomes: Gene Rearrangements and Discontinuous Transcription

TITIA DE LANGE[1]

Division of Molecular Biology, Antoni van Leeuwenhoekhuis, The Netherlands Cancer Institute, Amsterdam, The Netherlands

I. Introduction

Protozoan parasites of mammals have developed various strategems to cope with their host's defense system (reviewed by Bloom, 1979). First, some invaders have managed to infiltrate the ultimate hiding place: the inside of a mammalian cell. Examples are *Leishmania*, surviving in mac-

---

[1] Present address: Department of Microbiology and Immunology, University of California, San Francisco, California.

rophages, and *Trypanosoma cruzi,* which shelters in various cell types. Second, many parasites have found ways to impair the immune system. A third strategy, cunningly employed by some trypanosomes, matches the immense collection of immunoglobulins with an equally impressive assortment of sequentially expressed antigens. The molecular biology of trypanosome antigenic variation will be discussed in detail here; other aspects of the biochemistry of trypanosomes have been discussed elsewhere (cf. Vickerman and Barry, 1982; Englund *et al.,* 1982; Opperdoes, 1984).

The bloodstream-dwelling African trypanosomes are covered by a homogeneous layer of coat protein, variant surface glycoprotein, (VSG),[2] of which antigenically different types can be synthesized during a chronic infection (cf. Cross, 1978; Vickerman, 1978). Trypanosomes have reserved ~10% of their genome to program antigenic variation; every trypanosome may have as many as 1000 genes that code for VSG (Van der Ploeg *et al.,* 1982a, 1984a).

The occasional switches in the type of coat protein sythesized are a consequence of the sequential activation of different VSG genes (cf. Borst and Cross, 1982). Among the mechanisms for gene activation, gene rearrangements feature prominently. Often, surface-antigen genes are activated by duplicative transposition to a dominant telomeric expression site, where they may procure a transcriptional promoter (cf. Bernards *et al.,* 1981; De Lange *et al.,* 1985). In addition, some telomeric VSG genes are activated without any apparent rearrangements or transposition to the dominant expression site (cf. Van der Ploeg *et al.,* 1984a). Hence, there must be more than one locus where surface-antigen genes can be expressed. In spite of the presence of multiple expression sites, only one gene is expressed at a time. The molecular basis of this exclusion is unknown.

The study of transcripts for VSG has uncovered a novel mode of mRNA synthesis that appears to be a general characteristic of the family of Trypanosomatidae. The first 35 nucleotides of many, if not all, mRNAs are the same and not encoded contiguously with the rest of the transcript (cf. De Lange *et al.,* 1984b; Parsons *et al.,* 1984a). The common leader sequence is provided in trans from highly repetitive genes that are probably not linked to protein-coding genes. These mini-exon genes are transcribed into a small precursor RNA which is thought to donate its 5′ end

---

[2] Abbreviations: BC, basic copy; bp, base pair(s); cDNA, complementary DNA; ELC, expression-linked extra copy; kb, kilobase pair(s); L-ELC, lingering expression-linked extra copy; Mb, megabase pair(s); mf, membrane form; mRNA, messenger RNA; PFG, pulsed filed gradient; VSG, variant surface glycoprotein.

sequence to mRNAs either by bimolecular splicing or by serving as a primer in the transcription of structural genes.

II. Antigenic Variation in African Trypanosomes

A. TRYPANOSOME BIOLOGY

All trypanosomes and their allies discussed here belong to the family of the Trypanosomatidae [see Fig. 1 for classification and Lumsden and Evans (1976, 1979) for a comprehensive treatise on trypanosome biology]. They are uniflagellated, parasitic protozoa with a DNA-containing kinetoplast. The trypanosomes that show antigenic variation belong to the section Salivaria of the genus *Trypanosoma*.

The life cycle of *Trypanosoma brucei*, a representative of the Salivarian trypanosomes, is shown in Fig. 2. The infection of a mammal usually results from the injection of a few hundred or thousand trypanosomes by an infected tsetse fly. The trypanosomes multiply in a chancre formed at the site of the fly bite, from which they drain into lymph vessels and enter the circulatory system.

In the mammal, trypanosomes hide behind a compact layer of glycoprotein that covers the parasite body and flagellum. This surface coat protects against the nonspecific host defense and, moreover, presents antigens that readily elicit a humoral immune response, eventually removing the trypanosomes from the circulation (Vickerman, 1969, 1978). A few trypanosomes, however, survive the massacre because they have switched to another antigenically distinct coat protein and initiate a new parasitemic wave. Antigenic switching occurs at an apparent frequency of 10^{-4}–10^{-6} per cell division (Doyle, 1977) and is indepenent of the immune response: *in vitro* cultured bloodstream trypanosomes are also capable of

Phylum	Protozoa				
Class	Mastigophora				
Order	Kinetoplastida				
Family	Trypanosomatidae				
Genera	Leptomonas Leishmania Trypanosoma Crithidia Phytomonas Herpetomonas				
Sections	Stercoraria	Salivaria			
Subgenera	Schizotrypanum	Duttonella Trypanozoon Nannomonas			
Species	T. cruzi	T. vivax	T. equiperdum T. congolense T. evansi T. brucei		

FIG. 1.   Classification of some trypanosomes (after Lumsden and Evans, 1976).

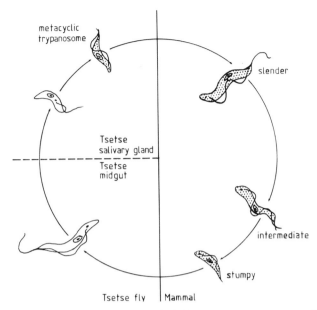

FIG. 2. The life cycle of *T. brucei*. The shaded trypanosomes carry a surface coat (see Section II,A for explanation).

coat exchange (Doyle *et al.*, 1980). Due to the large number of different coats available [e.g., more than 100 for *Trypanosoma equiperdum* (Capbern *et al.*, 1977)] and the economical use of this repertoire, trypanosomes can maintain a lengthy chronic infection with on average one parasitemic wave each week. The human disease associated with the infection is called sleeping sickness. The causative agents are *Trypanosoma brucei rhodesiense* and *Trypanosoma brucei gambiense*. *Trypanosoma brucei brucei* is not man infective due to lysis in human serum (Rifkin, 1978) and is, therefore, rather popular with biochemists. This subspecies, along with *Trypanosoma vivax* and *Trypanosoma congolense*, gives rise to diseases in cattle and other domestic mammals.

In nature, bloodstream trypanosomes are pleomorphic (see Fig. 2). Long slender forms, which divide by binary fission, are accompanied by nondividing short stumpy forms thought to be preadapted to life in the tsetse fly (Vickerman and Barry, 1982). Rodent-adapted virulent laboratory strains are usually monomorphic and often not transmissible by tsetse flies. Here, the fly is substituted by a syringe.

Upon entering the fly as part of its bloodmeal, the trypanosomes lose their surface coat and transform to the procyclic stage. A similar transformation occurs *in vitro* when bloodstream trypanosomes are incubated in the appropriate media at 25°C (Brown *et al.*, 1973; Overath *et al.*, 1983). These cultured procyclic trypanosomes are not infective to mammals: in the absence of their surface coat, trypanosomes are lysed by serum. The fly stage of the life cycle is completed in a few weeks. After a period of differentiation in the midgut, trypanosomes migrate to the salivary glands where they reacquire the surface coat and become infective. These metacyclic trypanosomes (see Fig. 2) show a limited heterogeneity with respect to their surface antigens (Barry *et al.*, 1979): often reexpression of the same set of probably less than 20 metacyclic coats occurs after cyclic transmission of a trypanosome strain (Crowe *et al.*, 1983). Variations in the metacyclic coat repertoire are, however, not rare (Barry *et al.*, 1983). Therefore, the prospects of vaccination against metacyclic trypanosomes are rather dim.

The occurrence of trypanosomiasis in tropical Africa is more or less restricted to the distribution of the insect vector, i.e., the tsetse belt. Salivarian trypanosome species found outside this area (e.g., in North Africa, South America, and Asia) have lost the need for transmission by the tsetse fly; e.g., *Trypanosoma evansi*, which is pathogenic to camels and horses and is mechanically transmitted by bloodsucking flies and infected vampire bats, and *T. equiperdum*, which is venereally transmitted between horses.

Trypanosomes are relatively easy organisms for molecular biologists to study. First, trypanosomes can be cloned and maintained in small laboratory rodents, such as mice and rats, or stored under liquid nitrogen (Van Meirvenne *et al.*, 1975a,b). The study of antigenic variation especially profits from the availability of fairly efficient cloning procedures. Due to the low switching frequency and the high growth rate (generation time about 6 hours), large batches of antigenically identical trypanosomes can be analyzed. Second, the genome of trypanosomes is small, about 4×10^4 kb per haploid genome for *T. brucei*, with ~70% single-copy sequences (Borst *et al.*, 1980, 1982). On the other hand, the absence of *in vitro* and *in vivo* genetics that can be manipulated (no DNA-mediated transformation system, no mutants, and no test-tube sex) hampers the study of trypanosome molecular biology.

Several lines of evidence indicate that trypanosomes are diploid organisms. Studies on isoenzyme patterns (Gibson *et al.*, 1980; Tait, 1980) and on the DNA content of the nucleus (Borst *et al.*, 1982) are compatible with this idea. Recently, more information on the ploidy of *T. brucei* has come from the analysis of restriction-site polymorphisms in and around three genes that code for glycolytic enzymes (Gibson *et al.*, 1985). Like the isoenzyme patterns, the polymorphisms reveal the presence of allelic copies of each gene, which appear to segregate independently. This supports the suggestion that *T. brucei* has a diploid genome and indulges in genetic exchange. Preliminary experiments suggest that recombination of genomes may occur in the fly (L. Jenni, personal communication).

B. The Variant Surface Glycoprotein

The surface coat of trypanosomes consists of a uniform layer of about 10^7 closely packed variant surface glycoprotein (VSG) molecules (Cross, 1975, 1978). Antigenically distinct *T. brucei* clones each yield a unique VSG of about 60,000 Da that differs from other VSGs in isoelectric point, peptide map, amino acid sequence, and carbohydrate content (Cross, 1975, 1978). The protein can be subdivided into an exposed N-terminal domain and a membrane-oriented C-terminal part, which is glycosylated. The N-terminal amino acid sequence is variable and probably determines the antigenic characteristics of the VSG (Bridgen *et al.*, 1976; Miller *et al.*, 1984). In contrast, the last 110 amino acids at the C-terminus show some conservation (Rice-Ficht *et al.*, 1981; Majumder *et al.*, 1981; Matthyssens *et al.*, 1981; Boothroyd *et al.*, 1981). This may reflect a limitation on the structure nearest the plasma membrane, imposed by the necessity to interact with other VSGs and/or by the membrane proximity.

In spite of the presence of sugar side chains, the C-terminal part is not immunogenic *in situ,* probably because the antigenic determinants are shielded by the rest of the protein. Indeed, one trypanosome variant does not bind concanavalin A unless the N-terminal domain is removed by trypsin (Cross and Johnson, 1976). Carbohydrate makes up 7–17% (w/w) of the soluble VSG that is released during mild detergent lysis and consists of mannose, galactose, and glucosamine residues (no sialic acid) (Cross, 1975, 1977; Johnson and Cross, 1977). The high-mannose oligosaccharides are added in a tunicamycin-sensitive manner using the Asn-X-Ser/Thr recognition site (Rovis and Dube, 1981).

The soluble VSG, which was initially analyzed, has recently turned out to be incomplete. The complete membrane form (mf) VSG can be obtained by boiling trypanosomes in detergent, thereby inhibiting a membrane-bound enzyme that under milder conditions clips off the membrane anchorage site (Cardoso de Almeida and Turner, 1983). Although the physiological role of this enzyme(s) has not been studied, it is tempting to speculate about a "stripte-ase" function for rapid coat shedding when trypanosomes enter the fly, or a role in coat turnover. In mf VSG the C-terminal aspartic acid or serine residue is covalently linked through an ethanolamine residue (Holder, 1983) to a phosphoglycolipid, which contains myristic acid (Ferguson and Cross, 1984; Ferguson *et al.*, 1985). This saturated fatty acid is thought to anchor VSGs in the plasma membrane. Acylation has also been reported for the VSGs of *T. equiperdum* (Duvillier *et al.*, 1983). The addition of the lipid-containing carbohydrate group happens early in the synthesis of the coat protein after removal of a short C-terminal extension peptide and the N-terminal signal peptide (Boothroyd *et al.*, 1980, 1981; Boothroyd and Cross, 1982; McConnel *et al.*, 1981, 1983).

III. The Structure and Organization of VSG Genes

A. GENE STRUCTURE

VSG mRNA is ~2 kb, polyadenylated, and very abundant (Williams *et al.*, 1978; Hoeijmakers *et al.*, 1980). Not surprisingly, therefore, large parts of VSG mRNA sequences could readily be obtained from cDNA libraries (Williams *et al.*, 1979; Hoeijmakers *et al.*, 1980; Pays *et al.*, 1980). The use of these cDNAs as VSG-specific probes in the analysis of DNA and RNA showed that each VSG is encoded by a separate gene and that each VSG gene is transcribed only when the corresponding VSG is synthesized; i.e., VSG synthesis is regulated at the level of transcription (Hoeijmakers *et al.*, 1980).

To date the structure of over 30 VSG genes has been investigated in detail, and (part of) the nucleotide sequence of many of these has been determined (cf. Majumder *et al.*, 1981; Boothroyd *et al.*, 1980, 1981; Matthyssens *et al.*, 1981; Rice-Ficht *et al.*, 1981, 1982; Bernards *et al.*, 1981; Boothroyd and Cross, 1982; Donelson *et al.*, 1982; Liu *et al.*, 1983; Young *et al.*, 1983; Pays *et al.*, 1983b–d; De Lange *et al.*, 1983a, Campbell *et al.*, 1984a; Bernards *et al.*, 1985). The scheme in Fig. 3 summarizes their most salient general features. As may be expected for a gene in a lower eukaryote, VSG-coding regions lack intervening sequences that commonly interrupt protein-coding genes in higher eukaryotes. At both extremes VSG genes encode untranslated RNA segments. The leader sequence is variable in size and sequence (Boothroyd and Cross, 1982; Liu *et al.*, 1983; Dorfman and Donelson, 1984); in contrast, the 3'-untranslated trailer (UT) is highly conserved with notable motifs like the 14-mer (5'-UGAUAUAUUUUAAC-3'), which is present within 15 nucleotides upstream of the poly(A) tail in nearly every VSG mRNA of *T. brucei* (Rice-Ficht *et al.*, 1981; Borst and Cross, 1982; Michels *et al.*, 1983; Liu *et al.*, 1983). Using the 14-mer sequence as priming site for reverse transcriptase catalyzed cDNA synthesis, the 3' end sequence of a score of VSG mRNAs has been determined directly on the RNA template, thus omitting the laborious cDNA cloning step (Michels *et al.*, 1983; Merrit *et*

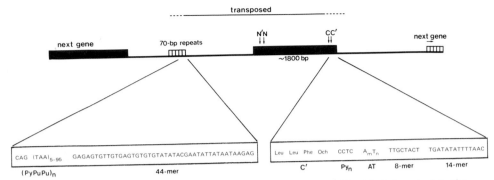

FIG. 3.   The structure of chromosome-internal VSG genes. The black box depicts the region that codes for the mature VSG (between N and C), the N-terminal signal peptide (N′), the C-terminal extension (C′), and the untranslated termini of VSG mRNA. Two VSG gene-specific repeat sequences are shown in the blocks below the scheme: the left-hand block shows the consensus sequence of the 70-bp repeats upstream of the gene, the right-hand block shows the conserved features at the 3′ end of VSG genes. The extent of the gene segment that is transposed upon activation is indicated, as is the close linkage between VSG genes.

al., 1983). From this analysis and data obtained with cloned cDNAs it is evident that the homology at the 3′ end of VSG genes extends into the C-terminus of the mature protein (Rice-Ficht et al., 1981; Borst and Cross, 1982). A homology gradient from 5′ to 3′ in the gene was already predicted on the basis of hybridization patterns of nuclear DNAs with cDNA probes (Borst et al., 1981). Whereas 5′-half gene probes usually detect one or two restriction fragments in nuclear DNA, 3′-half gene probes often hybridize to a multitude of bands, so much so that they have proved useless for mapping analyses in some cases. The unique 5′ halves of VSG genes coincide with a functional domain: the N-terminus of the protein that carries the variable antigenic determinants. Likewise, the conservation of the 3′ ends of VSG genes probably in part reflects a structural constraint on the amino acid sequence of the membrane-oriented C-terminus. In addition, the conserved sequence at the 3′ ends plays a role in the activation VSG genes (see Section IV,A).

Contrary to what is seen for housekeeping genes (see Section II,A), the copy numbers of many individual VSG genes are not in agreement with the presumed diploidy of trypanosomes: genes for surface antigens are present at one, two, or three copies per nucleus (Frasch et al., 1982; De Lange et al., 1983a). How trypanosomes manage a partly haploid genome is not clear (see Gibson et al., 1985, for discussion of this problem).

B. Genomic Context

VSG genes are found in two radically different genomic contexts: at chromosome-internal sites and at chromosome ends (telomeres) (Williams *et al.*, 1982; De Lange and Borst, 1982). Until recently, the contribution of telomeric VSG genes in the gene repertoire could not be fully appreciated; no reliable information on chromosome number had been obtained because trypanosomes do not have discrete condensed chromatin bodies during any phase of the cell cycle. The application of the pulsed field-gradient (PFG) gel electrophoresis technique, by which chromosome-sized DNA molecules can be separated, has recently uncovered the interesting structural details of the trypanosome genome (Schwartz and Cantor, 1984; Van der Ploeg *et al.*, 1984a). A representative PFG gel pattern of *T. brucei* chromosomes is shown in Fig. 4. Roughly four fractions are separated (I) large chromosomes that hardly move away from the origin, (II) several chromosomes that migrate in the megabase-pair size range and cannot be accurately sized due to a compression effect, (III) several intermediately sized chromosomes, and (IV) an impressive number of small (50–150 kb) chromosomes (Van der Ploeg *et al.*, 1984a; A. Bernards, personal communication). Obviously, there is ample room for VSG genes at telomeres. Indeed, many (if not all) small chromosomes may carry VSG genes at their ends, since probes for the conserved 3' ends of VSG genes hybridize strongly to this DNA fraction (Van der Ploeg *et al.*, 1984a).

The majority of VSG genes, however, has a chromosome-internal position. Unlike telomeric VSG genes, chromosome-internal genes and their surrounding sequences are usually well represented in genomic libraries made in *Escherichia coli*. Extensive analysis of cosmid clones containing the gene for VSG 118 in *T. brucei* strain 427 has shown that this gene is closely linked to other VSG genes, detected through the conserved 3' end sequence. Less than 3 kb separates the 118 gene from its nearest neighbors, and a similar VSG gene clustering was found around the gene for VSG 117 (Van der Ploeg *et al.*, 1982a).

Another point to emerge from the analysis of cloned genes is the presence of a second conserved feature of VSG genes, the 70-bp repeats (Van der Ploeg *et al.*, 1982a; Liu *et al.*, 1983; Campbell *et al.*, 1984a). This conserved sequence is found a few kilobase pairs upstream of the start codon and is made up of direct imperfect repeats, usually 70–80 bp long. Their consensus sequence (Campbell *et al.*, 1984a) is given in Fig. 3; the length variation is mainly due to the number of PyPuPu (typically TAA) motifs that precede the more conserved 44-bp 3' half of the repeats. Using

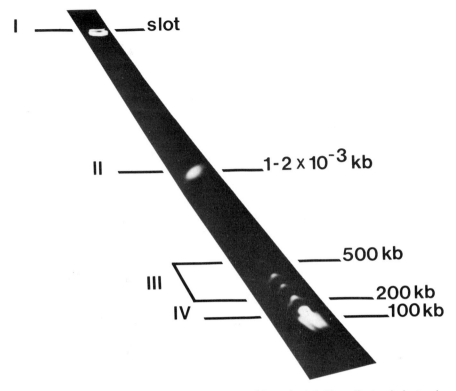

FIG. 4. Chromosomes of *T. brucei*, separated by pulsed field-gradient gel electrophoresis, stained with ethidium, and photographed over ultraviolet light. Four classes of chromosomes are visualized, I, large chromosomes that hardly leave the slot; II, a few chromosomes in the megabase-pair size range, which cannot be sized accurately due to a compression in this area; III, intermediate-sized chromosomes; IV, minichromosomes. The tentative molecular weights are derived from comigration with phage λ (50-kb) oligomers (A. Bernards, personal communication). For details on the electrophoresis technique, see Schwartz and Cantor (1984) and Van der Ploeg *et al.* (1984a).

this sequence to detect VSG genes in a cosmid library, Van der Ploeg *et al.* (1982a) estimated that 9% of the colonies contain VSG gene clusters, together representing an estimated total of 10^3 potential chromosome-internal VSG genes in the genome of *T. brucei*.

The only telomeric VSG gene that has been cloned together with a large stretch upstream codes for VSG 221. Like the chromosome-internal genes, the area upstream of gene 221 contains 70-bp repeats (Bernards *et al.*, 1984a, 1985) (see Fig. 5). A difference between telomeric and chromosome-internal VSG genes may be the number of 70-bp repeats: whereas the genes for VSG 117 and 118 have, respectively, four and five copies, the telomeric 221 gene has about 50. Although this has not been verified for other telomeric VSG genes, indirect evidence suggests this difference may be the rule. First, Van der Ploeg *et al.* (1982a) have shown that satellite-like long stretches of 70-bp repeats are present in the genome. Second, restriction maps of telomeric VSG genes invariably show a region of up to 40 kb that is scarce in restriction enzyme cutting sites (cf. Young *et al.*, 1983a; Pays *et al.*, 1983b; Laurent *et al.*, 1983; Bernards *et al.*, 1984a). These "barren" regions could very well contain large arrays of 70-bp repeats because these repeats lack sites for most restriction endonucleases (with notable exceptions such as *Acc*I, *Dde*I, *Hph*I, and *Mbo*I).

C. Telomere Structure and Instability

Originally some VSG genes were suspected to reside close to a duplex discontinuity in the DNA on the basis of their position upstream of a site where all tested restriction endonucleases appeared to cut (Van der Ploeg *et al.*, 1982b; Michels *et al.*, 1982). This supposition was verified by showing that the 3'-flanking sequence of these genes can be progressively shortened by digestion of intact nuclear DNA with exonuclease *Bal*31 (Williams *et al.*, 1982; De Lange and Borst, 1982).

Currently, there are additional criteria that set telomeric VSG genes apart from their chromosome-internal counterparts. Trypanosome telomeres have proved to be unusually unstable DNA segments that frequently show extensive length alterations. As a result, the distance between the gene and the end of the chromosome varies in different trypanosome clones, and now this feature is usually the first indication for the telomeric location of a newly analyzed VSG gene.

Fig. 5. Common sequence elements around telomeric VSG genes. A speculative scheme based on data from Van der Ploeg *et al.* (1984b), Blackburn and Challoner (1984), Bernards *et al.* (1985), and De Lange *et al.* (1983a). The repeats indicated are the 70-bp repeats (see Fig. 3 for consensus sequence), a T-rich element (T_n), a stretch of direct repeats of 29 bp [(29 mer)$_n$] with 5'-TTAGGG-3' motifs, and a tandem array of 5'-TTAGGG-3' hexamers (note that this region reads CCCTAA from telomere to centromere). The arrows symbolize the presence of single-stranded (ss) breaks in trypanosome telomeres. (See Fig. 3 for explanation of the structure of the gene and Sections III,B and C for further details.)

At least two processes are responsible for this curious phenomenon. First, telomeres in trypanosomes gradually increase in length by 6–7 bp per generation (28 bp per day) (Bernards *et al.*, 1983; Pays *et al.*, 1983a; Van der Ploeg *et al.*, 1984b). Second, this growth is balanced by occasional deletions that become apparent when trypanosomes are carried through cloning procedures or heat-shock treatment (Bernards *et al.*, 1983). Telomeres harboring actively transcribed VSG genes may be slightly different, as their growth rate is significantly higher (36–40 bp per day) and they appear more prone to deletions (Pays *et al.*, 1983a). A model explaining the growth of trypanosome telomeres has been presented by Van der Ploeg *et al.* (1984b).

The precise structure of the chromosome end flanking telomeric VSG genes is not known because no such structure has been cloned in its entirety. Nonetheless, a fairly accurate picture, schematically drawn in Fig. 5, can be extracted from the separate pieces of information now available. Van der Ploeg *et al.* (1984b) succeeded in cloning a DNA segment that is, in all likelihood, derived from a telomere. The most striking feature of this DNA is the presence of over 50 tandemly repeated hexamers, which read 5′-CCCTAA-3′ from telomere to centromere. As nuclear DNA fragments that contain 5′-$(CCCTAA)_n$-3′ behave like chromosome ends and are abundant, it is probable that this sequence is the common denominator of trypanosome telomeres; the sequence at the very end remains unknown, however. The telomeric sequence 5′-$(CCCTAA)_n$-3′ is common to all trypanosomatid genera and concurs nicely with the consensus sequence for protozoan telomeres, 5′-$[CCCP_y(A_n)]_n$-3′ (Van der Ploeg *et al.*, 1984b,c; Blackburn and Challoner, 1984). Furthermore, trypanosome telomeres contain single-stranded breaks in both strands, an additional feature shared with telomeres of other protozoa (Blackburn and Challoner, 1984). Two other more or less telomere-specific repeats are found more internal in the chromosome analyzed by Van der Ploeg *et al.*, a 29-bp 5′-$(CCCTAA)_n$-3′-derived repeat and a T-rich element (dT_n stretches are encountered, when reading 5′ to 3′ from centromere to telomere, as shown in Fig. 5). The latter had been found previously downstream of a telomeric VSG gene, showing the linkage between this telomeric repeat and VSG genes at chromosome ends (De Lange *et al.*, 1983a).

IV. Activation of VSG Genes

VSG genes can be activated in various ways. On the basis of observations at the DNA level, three different pathways can be discerned (see

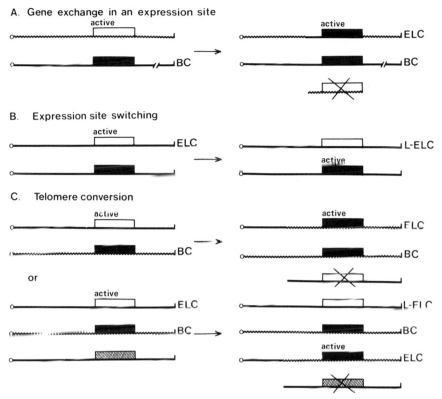

FIG. 6.   Three modes of VSG gene activation. Schemes A, B, and C are discussed separately in Sections IV,A, B, and C, respectively. BC, Basic copy; ELC, expression-linked extra copy; L-ELC, lingering expression-linked extra copy. A gene with a cross symbolizes that this gene copy is destroyed during the switching event.

Fig. 6). Gene switching can be achieved by an exchange of genes in an expression site (scheme A), by a switch from one expression site to another (scheme B), and by the duplication of a telomeric VSG gene and its surroundings (scheme C). The three pathways are discussed separately for reasons of clarity.

A. EXCHANGE OF VSG GENES IN A TELOMERIC EXPRESSION SITE

Chromosome-internal VSG are invariably activated by a duplicative transposition to a telomeric expression site. The expression-linked extra copy (ELC) of the gene thus formed is transcribed (Pays *et al.*, 1981a; Bernards *et al.*, 1981); the original basic copy (BC) of the gene is silent.

The transcription of ELC genes starts in the expression site (Bernards *et al.*, 1985; De Lange *et al.*, 1985), i.e., the duplicative transposition activates VSG genes by addition of a promoter (see Section VI,D). Here, the mechanism of duplicative transposition is discussed.

1. The Structure of the Expression Site

We have studied four separately generated ELC genes coding for VSG 118 and two for VSG 117 (Michels *et al.*, 1983). Their genomic environment, drawn in Fig. 7, is rather similar and characterized by three typical features. First, the ELC genes are proximal to a telomere, Second, upstream of the gene there are long barren regions of variable length. Third, 5' of the barren region there is a constant set of restriction enzyme cutting sites, suggesting that all six ELCs occupy the same expression site. In support, the 118 and 117 ELC genes invariably reside on a chromosome that migrates in the mega-base-pair range (fraction II in Fig. 4), whereas the corresponding BC genes are housed in larger chromosomes that do not enter the gel (Van der Ploeg *et al.*, 1984a).

Some of these characteristics of the dominant acceptor site for ELC genes are typical for *T. brucei* strain 427. Other strains have different expression sites. For example, the expression site in *T. brucei* strain 1125 has a different restriction map and resides on a small (about 300-kb) chromosome (Pays *et al.*, 1983b; M. Guyaux, E. Pays, A. W. C. A. Cornelissen, and P. Borst, unpublished). However, all expression sites are telomeric and contain a barren region 5' of the ELC gene.

2. The Transposed Gene Segment

The duplicative transposition of chromosome-internal VSG genes is probably a gene conversion of a previously expressed ELC by the incoming gene (Pays *et al.*, 1981b, 1983c; Borst and Cross, 1982; Michels *et al.*, 1982, 1983). The data on the boundaries of the transposed segment suggest as much, although other possibilities (see Borst, 1983) have not been rigorously excluded. The 3' end of the transposed gene segment has been determined in many cases, often by direct sequence analysis of the 3' end of VSG mRNA using the conserved 14-mer (see Fig. 3) as priming site. Michels *et al.* (1983) inferred that the recombination can occur anywhere in and around the conserved 3' end of the gene (see Fig. 3), from 16 bp into the mature C-terminus (140 bp upstream of the 14-mer) to downstream of the 14-mer. For instance, the point of sequence divergence between the 118 BC gene and its ELCs has been found from 42 bp upstream of the stop codon (86 bp upstream of the 14-mer) to 237 bp downstream of the 14-mer (Michels *et al.*, 1983; T. De Lange and M. Timmers, unpublished).

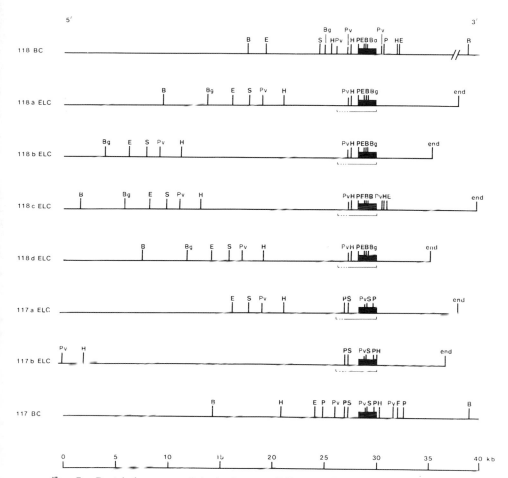

FIG. 7. Restriction maps of the basic copy (BC) genes for VSG 117 and 118 and six expression-linked extra copy (ELC) genes in six trypanosome clones that produce VSG 117 or 118 (see Michels *et al.*, 1983, and Section IV,A). The transposed gene segments are indicated beneath the maps. "End" denotes the end of the chromosome. B, *Bam*HI; Bg, *Bgl*II; E, *Eco*RI; H, *Hin*dIII; P, *Pst*I; Pv, *Pvu*II; S, *Sal*I.

A second region of homology that may play a role in the recombination in the expression site is located downstream of some VSG genes (see also Section V,A). Initially, the limited data on 3'-flanking regions indicated that each VSG gene has a unique sequence downstream. However, the analysis of additional genes now suggests that there are several classes of VSG genes with substantial sequence homology downstream. For example, the sequence downstream of the 118 BC gene is reiterated as judged

from the number of restriction fragments that are detected by a probe
from this region under relaxed hybridization conditions (P. Leegwater,
and A. Bernards, personal communication). By DNA sequence analysis
we have found that one of these homologous sequences indeed flanks the
3' end of a VSG gene, called 1.1006 (De Lange et al., 1983a). Another
example is the 3'-flanking sequence of the BC gene for VSG 117 that is
homologous to the sequence downstream of a pseudo-VSG gene (1.1010),
but not to the 118 BC gene area (Bernards et al., 1985).

The 5' end of the transposed gene segment resides some 1–2 kb upstream of the start codon in the region that contains the 70-bp repeats (see
Fig. 3) (Van der Ploeg et al., 1982b; Michels et al., 1983; Liu et al., 1983).
For this reason we proposed that these repeats are instrumental in the
gene activation process (Van der Ploeg et al., 1982b; Liu et al., 1983).
Verification of this proposal has been delayed, largely due to the fact that
the 5' barren region that should contain the 5' breakpoint has repeatedly
eluded cloning in E. coli. Recently, however, Boothroyd and co-workers
succeeded in cloning a small segment of the 5' barren region upstream of
the 117a ELC gene and determined the point where the sequence diverges
from the BC gene (Campbell et al., 1984a). As predicted, the 5' breakpoint falls in a 70-bp repeat, the first in front of the BC gene. We have
similarly determined the 5' breakpoint in the 117b ELC gene (De Lange et
al., 1985). In this case, the first sequence divergence is in the unusual
third repeat that lacks the 44-bp 3' repeat half. The presence of this rare
telltale repeat in the ELC strongly suggests that this sequence is cotransposed and therefore that the 5' breakpoints are different for these independent activations of the same VSG gene.

Furthermore, the ELC clones allow a glimpse of the structure of the 5'
barren region. In both cases the cloned expression site segments are made
up of 70-bp repeats. Whether other repeats inhabit the barren region
remains to be seen. The information now available suffices to explain the
length alterations of this DNA segment (see Fig. 7). Provided that the 70-
bp repeats in which the crossover takes place can be selected randomly
both in the expression site and in the incoming gene, as suggested by our
results with the 117 gene, contraction and expansion can be understood as
a consequence of gene replacement (Campbell et al., 1984a; De Lange et
al., 1985). A switch to a gene carrying many 70-bp repeats (e.g., gene 221,
see Section III,C) could increase the length of the barren region considerably and any switch could shorten it.

In the transposition of at least some telomeric VSG genes to the dominant expression site, the 5' breakpoint of the transposed segment is probably also located within the 70-bp repeats a few kilobase pairs upstream of
the start codon (Bernards et al., 1984a). In contrast, the 3' end of the

transposed segment in the two cases examined in detail resides at least 0.8 kb downstream of the gene (De Lange *et al.*, 1983a; Bernards *et al.*, 1984a). It is quite possible that here the complete telomere is transposed to the expression site by a telomere conversion. For lack of landmarks (e.g., restriction sites) in the last part of telomeres, this proposal is not easily verified.

The rules for gene conversion sketched here, although seemingly loose enough, are not always obeyed. A notable exception is an ELC of gene 1.1 in *T. brucei* strain 1125 that is accompanied by a second (earlier?) transposed element at the 5' side, appropriately referred to as companion (Pays *et al.*, 1983b). The boundary between the companion and the 1.1 ELC does not carry 70-bp repeats. Possibly these repeats are present more upstream at the companion–expression site junction. Another aberrant gene conversion event occurred during switch-off of the 1.10 ELC gene in the same strain (Pays *et al.*, 1983d). Instead of being completely replaced by an unrelated VSG gene, the 1.10 ELC underwent a partial gene conversion by the related 1.1 gene. This gene conversion involved the 5' end of the coding region, but not the cotransposed segment, and ended more than 630 bp upstream of the 3' end, i.e., much further into the protein-coding region than observed in the six switches of the 117 and 118 genes in strain 427. I note that this aberration may be a consequence of the high-sequence homology of the interacting genes.

B. Switching of Expression Sites

Obviously, the exchange of genes in a single expression site as discussed above would elegantly explain the mutually exclusive expression of VSG genes. However, this attractive model does not apply in its simplest form.

Early observations by the group of Williams and others indicated that telomeric genes can be activated without duplication or any other DNA rearrangement within the range of view of restriction maps (Williams *et al.*, 1980, 1982; Young *et al.*, 1982, 1983a,b; Majiwa *et al.*, 1982; Donelson *et al.*, 1982; Penncavage *et al.*, 1983; Laurent *et al.*, 1984; Bernards *et al.*, 1984a; Myler *et al.*, 1984). These data could only be reconciled with the single expression site hypothesis by invoking a reciprocal translocation of the expression site telomere and the telomere with the activated gene. If this recombination would occur far upstream or within highly homologous regions, it could have gone unnoticed, especially because the chromosomal location of these genes had not been studied.

This hypothesis was recently disproved by a study on the gene that codes for VSG 221. When not expressed, this gene is present at one copy per nucleus on a chromosome that remains close to the slot in PFG gels (Bernards *et al.*, 1984a; Van der Ploeg *et al.*, 1984a). When the gene is activated, it either follows the duplicative transposition route to the dominant expression site on a chromosomes of $1-2 \times 10^3$ kb (in fraction II in Fig. 4), or no significant change occurs within the gene or its surroundings. The latter situation is observed in trypanosome clone 221a, which expresses VSG 221. In this clone, the active 221 gene apparently occupies exactly the same telomere as the inactive gene, i.e., the gene is activated *in situ* without concomitant duplication (see Fig. 6B). The detailed restriction map reaching up to 45 kb 5' of the coding region indicates that no gross rearrangement has accompanied the activation of the gene, but does not exclude that the gene is activated in the dominant expression site: a reciprocal translocation with the expression site telomere could fully account for the results. This was ruled out by Van der Ploeg *et al.* (1984a), who showed that the active single copy of the 221 gene in 221a trypanosomes still resides on a large chromosome that hardly leaves the slot of PFG gels, implying that the genome of *T. brucei* strain 427 contains at least two separate loci where VSG genes can be expressed. The elucidation of the activation mechanism of the 221 gene in its own telomere awaits the location of the start of transcription of this gene (see Section VI,D).

There is much additional circumstantial evidence in favor of multiple expression sites. For instance, in some cases the switch-off of an ELC is not accompanied by its destruction by the next gene in line. Rather, the old ELC persists in an inactive form, called a lingering ELC (see Fig. 6) (Michels *et al.*, 1984; Buck *et al.*, 1984; Laurent *et al.*, 1984). As far as can be judged from restriction analysis and PFG gel electrophoresis, the lingering ELC remains in the dominant expression site. Obviously, a gene in another expression site has been activated in these trypanosomes.

The most challenging question raised by the use of multiple expression sites concerns the coordination of the activity of these telomeres. The simplest models retain the idea that only one expression site is active at a time. These models invoke a unique mobile element that confers telomere activity in cis (Borst *et al.*, 1983) or a unique site at the nuclear matrix where a telomere must be bound to be active (Van der Ploeg and Cornelissen, 1984). Other models do not incorporate a unique activating entity, but instead invoke a stochastic activation mechanism. For instance, in analogy with the phase variation system of *Salmonella* (reviewed by Silverman and Simon, 1983), every expression site could contain an invertible element that activates the telomere in one orientation. Bernards *et al.* (1984b) have argued that a stochastic on and off switch of expression sites could be mediated by DNA modification. All but one expression site would be repressed due to modification, and every expression site would have a low probability of losing its modification during replication. The attraction of this model resides in the fact that there is indirect experimental evidence for modified nucleotides in inactive telomeres (Bernards *et al.*, 1984b; Pays *et al.*, 1984). The restriction enzymes *Pst*I and *Pvu*II are incapable of cutting some sites in inactive telomeric VSG genes to completion. Most likely this is due to a partial modification of the recognition site. The level of modification is highest proximal to the telomere and strictly linked to the absence of gene expression.

In stochastic activation models, two types of aberrant trypanosomes are predicted, "zebras" that have two active expression sites and "naked" trypanosomes that have none. Neither have been observed. This is as expected because coatless trypanosomes are lysed in serum, and trypanosomes that do not cease production of the previous coat will probably be removed by the immune response (cf. Bernards *et al.*, 1984b). However, we have derived an exceptional trypanosome clone from variant 221 in which two expression sites appear to be simultaneously active (A. W. C. A. Cornelissen, P. Johnson, and P. Borst, unpublished). In this variant, the previously expressed 221 gene has been inactivated by an insertion of 30-kb DNA at a site about 3 kb upstream of start codon. The insertion prevents expression of the 221 gene from the VSG gene promoter far upstream (see Section VI,D), probably by providing a termination signal, since the region upstream of the insertion is transcribed, whereas the downstream area is silent. Hence, in this variant the expression site for VSG 221 is still active, although it yields no VSG mRNA. As the surface antigen of this variant (VSG 1.1) is produced from a VSG gene on another telomere, it appears that two expression sites are simultaneously active in this trypanosome clone. This argues against a mobile promoter or any other unique entity controlling VSG gene transcription and in favor of a stochastic mechanism of expression site activation/ inactivation.

C. TELOMERE CONVERSION

Telomeric VSG genes are often switched on without concomitant duplication; in some cases, however, the activation is accompanied by the duplication of the gene and a large area upstream (see Fig. 6C) (Pays *et al.*, 1983d; Michels *et al.*, 1984; Laurent *et al.*, 1984; Van der Ploeg *et al.*, 1984d). Usually, the extent of the duplication has remained undetermined because it exceeds the limits of the available restriction maps (about 30 kb). The duplication of one telomere most likely occurs at the expense of another, i.e., by an extensive telomere conversion instead of by duplication of the whole chromosome.

Three observations argue in favor of telomere conversion. First, the genesis of a new active copy of the telomere that harbors the gene for VSG 1.8 in our stock was followed using PFG analysis of the chromosomal location of this gene. The ELC is found on a new chromosome of about 425 kb, whose appearance accompanies the loss of a smaller chromosome (about 375 kb) (Van der Ploeg et al., 1984d). The simplest explanation for this set of events is a telomere conversion of the 375-kb chromosome by the 1.8 telomere, thereby changing its migration rate in PFG gels considerably. The second observation supporting telomere conversion was made on the antigenic switch of a trypanosome from clone 221a to expression of the telomeric gene for VSG 224 (J. M. Kooter, R. Moberts, A. Bernards, and P. Borst, unpublished). The 224 telomere in this case is duplicated over an area of less than 45 kb, as judged from the analysis of large restriction fragments on PFG gels. At the same time the previously active telomeric 221 gene is lost, together with at least 15 kb of upstream sequence. Although rigorous proof is still wanting, it is very likely that the 224 telomere has converted the 221 telomere during the switch. In fact, this rather drastic way of switching off a telomeric VSG gene seems to be a hallmark of gene 221. In five out of six inactivation events, the gene and its upstream region have been lost (Bernards et al., 1984a; Liu et al., 1984). Finally, Pays and co-workers (1983d) have reported a switch from expression of the gene for VSG 1.1C to expression of the telomeric gene for VSG 1.3B. In this switch the (telomeric) 1.1C gene is lost and the 1.3B gene together with at least 30 kb of upstream sequences are duplicated. Again, a conversion of the 1.1C telomere by the 1.3B telomere is the simplest explanation for these observations.

V. Changes in VSG Repertoire

A. The Order of VSG Gene Expression

Early serological data argue that the expression sequence of VSGs is nonrandom: some VSGs are predominant early and others usually appear later in a chronic infection (Van Meirvenne et al., 1975a,b; Capbern et al., 1977; Kosinski, 1980; Miller and Turner, 1981). More recent analyses support the hypothesis that the predominant early VSGs are encoded by genes with a permanent telomeric location and that these genes are switched on preferentially (Laurent et al., 1983; Michels et al., 1984; Myler et al., 1984).

This is clearly seen in single relapse experiments. For instance, when trypanosome clone 118a (expressing VSG 118a) is treated with antibodies against VSG 118, the few surviving trypanosomes that had apparently switched prior to the experiment were found to express either the telomeric gene for VSG 1.8 or the telomeric gene for VSG 1.1 (Michels *et al.*, 1984). An extreme case is the telomeric VSG 1 gene in *T. equiperdum* which is preferentially switched on in single relapse experiments and always dominates the early phase of chronic infections (Capbern *et al.*, 1977; Longacre *et al.*, 1983; Raibaud *et al.*, 1983). In contrast, chromosome-internal VSG genes are expressed late. Examples are the genes for VSG 118 and 117 that are activated reproducibly in the third week of chronic infections in rabbits (Michels *et al.*, 1983). Michels *et al.* (1984) elegantly showed that the genomic environment of the 118 gene determines the frequency with which it is activated. When the gene has a chromosome-internal position, it is never switched on in single relapse experiments. In contrast, when the 118 ELC is still present as a (telomeric) lingering ELC, it is switched on with high preference in such experiments. Similar results have been obtained by Laurent *et al.* (1984) with *T. brucei* strain 1125.

It is not clear whether there is an order of activation of the chromosome-internal genes later in infection. Capbern *et al.* (1977) found that in different chronic infections certain VSGs repeatedly come up at the same time, suggesting some order in the expression of VSGs. It is not excluded, however, that this order is the consequence of factors [growth rate, competition, host factors (see Kosinski, 1980)] other than a fixed sequence in the activation of VSG genes. Aside from these uncertainties, it could be argued that extensive homology downstream of VSG genes (see Section IV,A,2) may influence the sequence of gene conversion events in the expression site. This hypothesis has not been tested extensively yet, mainly because the sequences downstream of ELC genes are usually not present in genomic libraries in *E. coli*. The single exception to this rule is the downstream area of the 118c ELC that aberrantly has several restriction enzyme cutting sites (see Fig. 7). Analysis of this DNA indicated that it is derived from another (telomeric) VSG gene (1.1006) and, moreover, that the 3'-flanking sequences of the 1.1006 gene and the 118 BC gene are rather alike (De Lange *et al.*, 1983a). In this case, then, the recombination between these two genes may have been aided by the nucleotide sequence homology downstream of the 14-mer. More data are needed to assess whether such homologies indeed influence the order of VSG gene expression later in chronic infections.

Of course, an expression sequence based on nucleotide sequence homology would lead to many retrograde switches, yielding trypanosomes that probably would be destroyed by the host. However, this may be no intolerable burden for the trypanosome population. In view of the preferred switch-on of telomeric genes, one may expect that late in a chronic infection trypanosomes often switch to a coat protein that has already been used.

B. EVOLUTION OF VSG GENES

Given the high frequency of gene conversion at telomeres and the strong selection for such changes, it is not surprising that telomeric VSG genes are particularly unstable. This instability is illustrated by a survey of different trypanosome isolates for the presence of three telomeric genes (221, 1.1006, and 1.1005). These genes are only present in strain 427 (13, 6, and 5 strains analyzed, respectively) (Frasch *et al.*, 1982; De Lange *et al.*, 1983a; Borst *et al.*, 1984). It has been argued that the enormous number of small chromosomes in *T. brucei* has evolved to increase the number of early telomeric genes that can be changed rapidly (Van der Ploeg *et al.*, 1984a). However, the presence of small chromosomes is not essential for antigenic variation: *T. equiperdum* and *T. b. gambiense* manage without them (Van der Ploeg *et al.*, 1984c; W. Gibson and P. Borst, unpublished).

Some chromosome-internal VSG genes also show a remarkable instability. We have compared about 40 kb surrounding the 118 BC gene in strains 1125 and 427 and found a deletion of one neighboring VSG gene in strain 1125 and a quadruplication of another in strain 427 (Borst *et al.*, 1984; W. Gibson, A. Bernards, and L. H. T. Van der Ploeg, unpublished). The high frequency of recombination in and around chromosome-internal VSG genes is also indicated by the loss of the 118 gene in 5 out of 13 trypanosome strains (Frasch *et al.*, 1982). Another chromosome-internal gene, the 117 gene, is much more stable (Pays *et al.*, 1982; Frasch *et al.*, 1982).

VI. Discontinuous Transcription

The analysis of transcripts coding for VSGs in *T. brucei* has revealed a novel mode of mRNA synthesis that may be a general feature of kineto-plastid flagellates. The first 35 nucleotides of many (all?) mRNAs are identical and come from highly reiterated genes that are not linked to structural genes. Transcription therefore appears to be a discontinuous process.

A. Many Trypanosomal mRNAs Share a Common 5' Terminal Sequence

The common 5' end sequence of trypanosomal mRNAs was first found on the transcripts coding for VSG 118 in *T. brucei*. As noted, the protein-coding part of VSG BC genes does not contain introns. Still, VSG mRNAs are transcribed from two separate exons. The first observation that pointed in this direction was reported by Bernards *et al.* (1981), who characterized a VSG cDNA that contains 9 bp at the 5' end that are not encoded by the corresponding BC gene. Subsequent sequence analysis of the 5' end of the mRNA for VSG 118 and the transposed 118 gene segment showed that the first 35 nucleotides of the messenger are absent from the transposed segment. Van der Ploeg *et al.* (1982c) therefore proposed that these 35 nucleotides are encoded by a mini-exon in the expression site. In apparent agreement with this supposition, Boothroyd and Cross (1982) found the same mini-exon sequence at the 5' end of the transcript of the 117a ELC gene, which resides in the same expression site. A search for the mini-exon in this expression site yielded negative results: as judged from restriction analysis and quantitative hybridization experiments, the mini-exon is not present within 10 kb upstream of the 118d ELC gene (De Lange *et al.*, 1983b). In contrast, many mini-exons are present elsewhere in the genome. A 22-nucleotide probe for the mini-exon was found to hybridize to large arrays of a tandemly linked 1.35-kb repeat (De Lange *et al.*, 1983b; Nelson *et al.*, 1983). Sequence analysis showed that the complete mini-exon is present in these repeats, flanked by the sequence TTG/GTAGT that resembles the consensus sequence (CAG/GTPuAGT (see Mount, 1982) for splice donor sites in protein-coding genes (De Lange *et al.*, 1983b). We speculated that the expression site may contain one of these arrays of mini-exons and that this reiteration of initiation sites somehow would enhance the transcription of VSG genes. More recent evidence argues strongly against this model, however.

From the analysis of transcripts that do not code for surface antigens, it has become clear that the mini-exon sequence is not exclusive to the VSG messenger. Although VSG mRNA is very abundant, it accounts for at most 10% of the polyadenylated transcripts that hybridize to a 22-mer probe for the mini-exon (De Lange *et al.*, 1984b). The other RNAs are heterogeneous in size, also have the mini-exon sequence at a 5'-terminal position, but are not encoded in toto by the mini-exon repeat (De Lange *et al.*, 1984b). We have investigated 13 of these transcripts; another six were analyzed by Parsons *et al.* (1984a). Although the function of none of these RNAs is known, they appear to be part of the pool of messenger RNAs:

they are discretely sized transcripts of variable length and abundance that are encoded by single or low copy number genes whose expression is in some cases dependent on the differentiation stage of the trypanosomes. The simplest interpretation of these results is that the bulk of mRNAs in *T. brucei* has the same mini-exon-derived 5' end. We note, however, that the existence of a class of protein-coding genes that yield mRNAs without the common 5' end sequence cannot yet be excluded. The transcripts that were analyzed were selected for the presence of the mini-exon sequence. A systematic analysis of randomly picked mRNAs has not been carried out. Apart from VSG mRNAs, only two transcripts of known function thus far were available for study. These code for the α- and β-tubulins, and both have the mini-exon sequence, as shown by sandwich hybridization (T. De Lange, unpublished).

B. Evidence for Discontinuous Transcription

On theoretical grounds it seems unlikely that each structural gene has its own mini-exon upstream. Assuming that more than 1000 genes yield a transcript with the common 5' end sequence, there are simply too few mini-exons (about 200 per nucleus, De Lange *et al.*, 1983b) to serve each gene. Even if mini-exons are used by considerably fewer genes, the strong clustering of the repeats that contain mini-exons would probably demand that several genes share one repeat array. Therefore, we and others raised the possibility that mini-exons and structural genes are not linked, i.e., that transcription is a discontinuous process (cf. Borst *et al.*, 1983; Campbell *et al.*, 1984a; De Lange *et al.*, 1984b; Parsons *et al.*, 1984a). Indeed, Parsons *et al.* (1984a) did not find mini-exons within 8 kb upstream of four genes that yield a transcript with a mini-exon-derived 5' end. Another case in point is the genomic environment of the α- and β-tubulin genes. These two genes are closely linked on a 3.7-kb element that is repeated about 10 times per haploid genome in one cluster (Thomashow *et al.*, 1983; Seebeck *et al.*, 1983; T. De Lange, unpublished). The mini-exon is neither present on the 3.7-kb element nor within 13.5 kb upstream of the cluster (T. De Lange, unpublished).

A second line of evidence evolved from the analysis of the transcription of mini-exon repeats. Both Kooter *et al.* (1984) and Campbell *et al.* (1984b) have shown that these repeats are transcribed into an RNA of about 140 nucleotides that starts with the mini-exon. Kooter *et al.* also showed that this molecule is not a processing product: the rest of the mini-exon repeat is transcribed 750-fold less in isolated nuclei. Preliminary data indicate that the mini-exon repeats and VSG genes are not transcribed by the same polymerase (Kooter and Borst, 1984). Whereas transcription of mini-exon genes in isolated nuclei is 80% inhibited by the addition of 200 μg of α-amanitin per milliliter, VSG gene transcription is not affected by α-amanitin at all. Unlike other protein-coding genes whose transcription is blocked by 20 μg of α-amanitin per milliliter, genes for surface antigens still yield nascent RNA in the presence of 1000 μg of α-amanitin per milliliter. Whether this remarkable α-amanitin resistance also exists *in vivo* remains to be seen. Why VSG genes are transcribed by a (partly) different polymerase than other protein-coding genes is also not known.

A third line of evidence in favor of discontinuous transcription has recently come from the analysis of an expressed VSG gene in *T. brucei* strain 1125 (M. Guyeaux, E. Pays, A. W. C. A. Cornelissen, and P. Borst, unpublished). In these trypanosomes the ELC gene for VSG 1.1A is located on a chromosome of about 300 kb that does not contain a mini-exon. Nevertheless, the mRNA for VSG 1.1A has the mini-exon sequence. Thus, unless the mini-exon in this case is split and therefore undetectable with the available probes, transcription of this ELC gene is a discontinuous process. A similar observation was made by Van der Ploeg et al. (1984d).

Currently, two models for discontinuous transcription prevail (De Lange *et al.*, 1984a; Parsons *et al.*, 1984a; Campbell *et al.*, 1984b). One of these proposes that the transcripts of mini-exons and structural genes are synthesized independently and that the mini-exon sequence is transferred posttranscriptionally from the 140-nucleotide transcript onto mRNAs by a splicing event that involved two rather than one molecule. The other model says that (part of) the 140-nucleotide mini-exon transcript functions as a primer for the initiation of transcription in front of structural genes. A conventional splicing step would subsequently link the two parts of the mature mRNA.

The more fundamental question of why discontinuous transcription has evolved has also not been answered satisfactorily. Is the trypanosome RNA polymerase II crippled in that it needs a primer to start transcription or in that it does not start at a specific site and the spliced on mini exon has to create a defined 5' end? Does the mini-exon sequence have a function in the RNA, e.g., in stability, transport, or translation? If so, why is this sequence not encoded in each gene? Perhaps the clustering of mini-exon genes in tandem arrays helps to keep this sequence homogeneous. Elucidation of the mechanism of discontinuous transcription may shed some light on these questions.

C. Structure of Mini-Exon Genes

The repeat elements that carry mini-exons have been investigated in several trypanosome species. In general, the elements are small (1.35 kb or less) and tandemly linked in large clusters: tandem arrays are found in all *Trypanozoon* species and in *T. vivax, T. congolense,* and *T. cruzi* (De Lange *et al.,* 1983b, 1984b; Nelson *et al.,* 1983, 1984). A probe for the mini-exon of *T. brucei* furthermore detects putatively repeated elements in other trypanosomatid genera, such as *Crithidia, Leptomonas, Herpetomonas, Phytomonas,* and *Leishmania,* suggesting that discontinuous transcription is a general characteristic of trypanosomatids and may be used by other kinetoplastid flagellates (Nelson *et al.,* 1984; De Lange *et al.,* 1984b). The number of mini-exon repeat clusters in *T. brucei* is probably not much larger than 20 and many repeats are linked to more than 10 others (De Lange *et al.,* 1983b, 1984b). In addition, a few elements are linked in smaller clusters and others occur singly dispersed (Nelson *et al.,* 1983; Parsons *et al.,* 1984b).

The repeat units from *T. brucei, T. vivax,* and *T. cruzi* have been cloned and subjected to nucleotide sequence analysis (De Lange *et al.,* 1984b). From the comparison of these repeats, it has become clear that very little is conserved outside the mini-exon and its direct vicinity. The most prominent conserved features are (1) a 10-mer around the first nucleotide of the mini-exon, (2) the 3' part of the mini-exon and the putative 5' splice donor site, and (3) a T-rich stretch downstream (see Fig. 8). The T-rich stretch may play a role in defining the 3' end of the mini-exon transcripts because the 3' end of the mini-exon repeat transcript in *T. brucei* maps just in front of it (De Lange *et al.,* 1984b; Kooter *et al.,* 1984; Campbell *et al.,* 1984b). The significance and function of the other two conserved blocks are unclear. The conservation within the mini-exon could reflect a role in the RNA or in transcription of the genes. The same holds for the apparent conservation at the 3' border of the mini-exon. It is tempting to speculate that this sequence functions in the splicing event that links mini-exons to the rest of messenger RNAs. This idea is supported by the fact that VSG BC genes are flanked by a sequence that remotely resembles the 3' splice acceptor site in protein-coding genes in other eukaryotes, e.g., the AG rule for intron ends is obeyed (Van der Ploeg *et al.,* 1982c; Boothroyd and Cross, 1982; Liu *et al.,* 1983), but other possibilities are not ruled out. Information on which polymerase transcribes mini-exon genes and which nucleotides are necessary for transcription should clarify this matter.

Thus far, conflicting results as to the identity of the polymerase making the mini-exon transcript have been obtained. The effect of α-amantin on the transcription of these genes in isolated nuclei, i.e., inhibition between 20 and 200 μg/ml (Kooter and Borst, 1984), suggests that a type III poly-

mini-exon repeat array

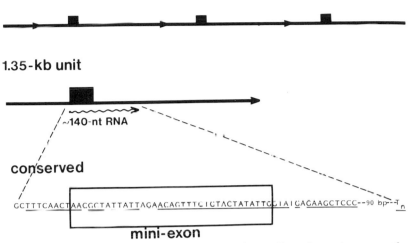

1.35-kb unit

~140-nt RNA

conserved

GCTTTCAACTAACGCTATTATTAGAACAGTTTCTCTACTATATTCGTATCACAAGCTCCC--90 bp--T$_n$

mini-exon

FIG. 8.   The structure of mini-exon genes in *T. brucei*. Part of a tandem array of mini-exon repeats is drawn schematically, and below (part of) one of the 1.35-kb repeat units is shown with the 140-nucleotide transcript that is encoded by mini-exon genes. The lower part shows the sequence of the mini-exon and its immediate flanking sequence in *T. brucei*. The conserved elements (also found in the mini-exon repeats of *T. vivax* and *T. cruzi*) are underlined. The position of the T-rich stretch (T_n) varies in different trypanosome species. (See also Section VI,C.)

merase transcribes these genes (cf. Chambon, 1975). However, recent experiments suggest that the 140-nucleotide mini-exon transcript contains a cap structure (P. Laird and P. Borst, unpublished). This is atypical for a RNA polymerase III product, and trypanosomes apparently obey this rule, since their 5 S RNA does not contain a cap. Finally, the conserved sequence blocks in the mini-exon repeats do not resemble the consensus sequence for RNA polymerase III promoters (RGYNNRRYNGG- 30–60 bp -GATTCRANNC; see Ciliberto *et al.*, 1983). Rather, the presence of a conserved 10-mer around the initiation site is reminiscent of an RNA polymerase I transcription unit (Bach *et al.*, 1981; Financzek *et al.*, 1982; Verbeet *et al.*, 1984).

D. TRANSCRIPTION OF VSG GENES

Notwithstanding the paucity of information on the discontinuous transcription process, it is anticipated that structural genes have their own promoter for transcription, i.e., a site from which the (re?)initiation of transcription is directed. A search for this entity is especially interesting in the case of VSG genes, for it could potentially reveal the mechanism of

VSG gene activation and provide insight into the control of expression sites. Early observations on the transcripts of two ELC genes showed that the cotransposed segment yields discrete, steady-state RNAs, albeit at very low levels and with an unknown relation to the mature VSG mRNA (Pays *et al.,* 1982, 1983b; Van der Ploeg *et al.,* 1982c). Similar transcripts derive from the upstream area of the telomeric gene for VSG 221 (Bernards *et al.,* 1985). Analysis of steady-state RNA appeared an inadequate way for finding the start of transcription, however, because very long primary transcripts could escape detection due to rapid degradation.

Therefore, we turned to the analysis of labeled nascent RNAs synthesized in isolated nuclei. Two genes were studied in this fashion: the ELC of the nontelomeric gene for VSG 117 in the dominant expression site of trypanosome strain 427 and the *in situ* activated telomeric gene for VSG 221 (De Lange *et al.,* 1985; Bernards *et al.,* 1985). From the 117 ELC gene, nascent RNAs were synthesized that hybridize to the complete transposed gene segment. Furthermore, an impressive amount of nascent RNAs are produced that have 70-bp repeat sequences, suggesting that a large part of the 5' barren region is transcribed. We inferred that transcription of this ELC gene starts in the expression site; i.e., the transposition of nontelomeric VSG genes activates these genes by a process akin to promoter addition. In support of this interpretation, we have found that transcription of 70-bp repeats shows the resistance to α-amanitin, characteristic for VSG gene transcription units (Kooter and Borst, 1984). Where transcription starts and how the mini-exon sequence is juxtaposed onto the transcript may become clear when sequences beyond the 5' barren region become available as cloned recombinant DNA.

Likewise, the region as far as 22 kb upstream of the 221 gene, including sequences beyond the short 5' barren region, are transcribed at approximately the same rate as the coding region by an RNA polymerase that is insensitive to α-amanitin *in vitro* (Bernards *et al.,* 1985; Kooter and Borst, 1984; J. M. Kooter, R. Wagter, and P. Borst, unpublished). This indicates that the initiation of transcription occurs at least 22 kb upstream of the coding region. Since the promoter of this VSG gene has not been located yet, it is not clear how this gene and others like it are activated in their own telomere.

VII. Outlook

The study of the genetics of antigenic variation started in 1979 with the isolation of the first cDNAs of VSG mRNAs. Now, some 5 years later, the field has come of age: it has turned out to be a rewarding model system for sophisticated gene switching events. No other genome has yet yielded such an overwhelming number of DNA rearrangements, linked to development and evolution, that are worthwhile to study in detail. Furthermore, the study of VSG genes has allowed a closer look at the structure and behavior of chromosome ends. And finally, the unique, discontinuous transcription system of trypanosome protein-coding genes has given a glimpse of unexpected possibilities in genome usage.

There are still a number of questions one would like to see answered. What is the nature of the activation process of telomeric VSG genes and how is the activity of different expression sites regulated? What is the precise mechanism of gene conversion in the expression site and the frequent extensive conversion of telomeres? How do the ends of trypanosome chromosomes grow and why is there a difference in growth rate and stability of actively transcribed telomeres? What is the mechanism of discontinuous transcription and why has it evolved?

These questions can be answered partly by techniques now available, but many aspects will need the development of *in vitro* or (preferably) *in vivo* systems in which manipulated DNA segments can be tested for their biological activity. It is high time that trypanosome molecular biology lost these last growing pains.

ACKNOWLEDGMENTS

I am grateful to Drs. P. Borst, A. Bernards, K. Osinga, W, Gibson, A, Cornelissen, P. Johnson, L. Van der Ploeg, M. Guyaux, E. Pays, J. Kooter, and P. Laird for communication of unpublished data and to Drs. P. Borst, A. Bernards, K. Osinga, W. Gibson, P. Johnson, and G. A. M. Cross for valuable comments on the manuscript. This work was supported in part by a grant from the Foundation for Fundamental Biological Research (BION) which is subsidized by the Netherlands Organization for the Advancement of Pure Research (ZWO).

REFERENCES

Bach, R., Allet, B., and Crippa, M. (1981). *Nucleic Acids Res.* **9**, 5311.
Barry, J. D., Hadjuk, S. L., and Vickerman, K. (1979). *Trans. R. Soc. Trop. Med. Hyg.* **73**, 205.
Barry, J. D., Crowe, J. S., and Vickerman, K. (1983). *Nature (London)* **306**, 699.
Bernards, A., Van der Ploeg, L. H. T., Frasch, A. C. C., Borst, P., Boothroyd, J. C., Coleman, S., and Cross, G. A. M. (1981). *Cell* **27**, 497.
Bernards, A., Michels, P. A. M., Lincke, C. R., and Borst, P. (1983). *Nature (London)* **303**, 592.

Bernards, A., De Lange, T., Michels, P. A. M., Huisman, M. J., and Borst, P. (1984a). *Cell* **36,** 163.
Bernards, A., Van Harten-Loosbroek, N., and Borst, P. (1984b). *Nucleic Acids Res.* **12,** 4153.
Bernards, A., Kooter, J. M., and Borst, P. (1985). *Mol. Cell. Biol.,* in press.
Blackburn, E. H., and Challoner, P. B. (1984). *Cell* **36,** 447.
Bloom, B. R. (1979). *Nature (London)* **279,** 21.
Boothroyd, J. C., and Cross, G. A. M. (1982). *Gene* **20,** 281.
Boothroyd, J. C., Cross, G. A. M., Hoeijmakers, J. H. J., and Borst, P. (1980). *Nature (London)* **288,** 624.
Boothroyd, J. C., Paynter, C. A., Cross, G. A. M., Bernards, A., and Borst, P. (1981). *Nucleic Acids Res.* **9,** 4735.
Borst, P. (1983). *In* "Mobile Genetic Elements" (J. A. Shapiro, ed.), p. 621. Academic Press, New York.
Borst, P., and Cross, G. A. M. (1982). *Cell* **29,** 291.
Borst, P., and Cross, G. A. M. (1982). *Cell* **29,** 291.
Borst, P., Fase-Fowler, F., Frasch, A. C. C., Hoeijmakers, J. H. J., and Weijers, P. J. (1980). *Mol. Biochem. Parasitol.* **1,** 221.
Borst, P., Frasch, A. C. C., Bernards, A., Van der Ploeg, L. H. T., Hoeijmakers, J. H. J., Arnberg, A. C., and Cross, G. A. M. (1981). *Cold Spring Harbor Symp. Quant. Biol.* **45,** 935.
Borst, P., Van der Ploeg, M., Van Hoek, J. F. M., Tas, J., and James J. (1982). *Mol. Biochem. Parasitol.* **6,** 13.
Borst, P., Bernards, A., Van der Ploeg, L. H. T., Michels, P. A. M., Liu, A. Y. C., De Lange, T., and Kooter, J. M. (1983). *Eur. J. Biochem.* **137,** 383.
Borst, P., Bernards, A., Van der Ploeg, L. H. T., Michels, P. A. M., Liu, A. Y. C., De Lange, T., and Sloof, P. (1984). *In* "Molecular Biology of Host-Parasite Interactions" (N. Agabian and H. Eisen, eds.), p. 205. Liss, New York.
Bridgen, P. J., Cross, G. A. M., and Bridgen, J. (1976). *Nature (London)* **263,** 613.
Brown, R. C., Evans, D. A., and Vickerman, K. (1973). *Int. J. Parasitol.* **3,** 691.
Buck, G. A., Longacre, S., Raibaud, A., Hibner, U., Giroud, C., Baltz, T., Baltz, P., and Eisen, H. (1984). *Nature (London)* **307,** 563.
Campbell, D. A., Van Bree, M., and Boothroyd, J. C. (1984a). *Nucleic Acids Res.* **12,** 2759.
Campbell, D. A., Thornton, D. A., and Boothroyd, J. C. (1984b). *Nature (London)* **311,** 350.
Capbern, A., Giroud, C., Baltz, T., and Mattern, P. (1977). *Exp. Parasitol.* **42,** 6.
Cardoso de Almeida, M. L., and Turner, M. (1983). *Nature (London)* **302,** 349.
Chambon, P. (1975). *Annu. Rev. Biochem.* **44,** 613.
Ciliberto, G., Castagnoli, L., and Cortese, R. (1983). *Curr. Top. Dev. Biol.* **18,** 59.
Cross, G. A. M. (1975). *Parasitology* **71,** 393.
Cross, G. A. M. (1977). *Ann. Soc. Belge Med. Trop.* **57,** 389.
Cross, G. A. M. (1978). *Proc. R. Soc. London Ser. B* **202,** 55.
Cross, G. A. M., and Johnson, J. G. (1976). *In* "Biochemistry of Parasites and Host-Parasite Relationships" (H. van den Bossche, ed.), p. 413. Elsevier, Amsterdam.
Crowe, J. S., Barry, J. D., Luckins, A. G., Ross, C. A., and Vickerman, K. (1983). *Nature (London)* **306,** 389.
De Lange, T., and Borst, P. (1982). *Nature (London)* **299,** 451.
De Lange, T., Kooter, J. M., Michels, P. A. M., and Borst, P. (1983a). *Nucleic Acids Res.* **11,** 8149.
De Lange, T., Liu, A. Y. C., Van der Ploeg, L. H. T., Borst, P., Tromp, M. C., and Van Boom, J. H. (1983b). *Cell* **34,** 891.

De Lange, T., Michels, P. A. M., Veerman, H. J. G., Cornelissen, A. W. C. A., and Borst, P. (1984a). *Nucleic Acids Res.* **12**, 3777.

De Lange, T., Berkvens, T. M., Veerman, H. J. G., Frasch, A. C. C., Barry, J. D., and Borst, P. (1984b). *Nucleic Acids Res.* **12**, 4431.

De Lange, T., Luirink, J., Kooter, J. M., and Borst, P. (1985). *EMBO J.*, in press.

Donelson, J. E., Young, J. R., Dorfman, D. M., Majiwa, P. A. O., and Williams, R. O. (1982). *Nucleic Acids Res.* **10**, 6581.

Dorfman, D. M., and Donelson, J. E. (1984). *Nucleic Acids Res.* **12**, 4907.

Doyle, J. J. (1977). *In* "Immunity of Blood Parasites of Animals and Man" (L. H. Miller, J. A. Pino, and J. McKelvey, Jr., eds.), p. 31. Plenum, New York.

Doyle, J. J., Hirumi, H., Hirumi, K., Lupton, E. N., and Cross, G. A. M. (1980). *Parasitology* **80**, 359.

Duvillier, G., Nouvelot, A., Richet, C., Baltz, T., and Degand, P. (1983). *Biochem. Biophys. Res. Commun.* **114**, 119.

Englund, P. T., Hadjuk, S. L., and Marini, J. C. (1982). *Annu. Rev. Biochem.* **51**, 695.

Ferguson, M. A. J., and Cross, G. A. M. (1984). *J. Mol. Chem.* **259**, 3011.

Ferguson, M. A. J., Haldar, K., and Cross, G. A. M. (1985). *J. Biol. Chem.*, in press.

Financzek, I., Mizumoto, K., and Muramatsu, M. (1982). *Gene* **18**, 115.

Frasch, A. C. C., Borst, P., and Van den Burg, J. (1982). *Gene* **17**, 197.

Gibson, W. C., Marshall, T. F. de C., and Godfrey, D. G. (1980). *Adv. Parasitol.* **18**, 175.

Gibson, W. C., Osinga, K. A., Michels, P. A. M., and Borst, P. (1985). *Mol. Biochem. Parasitol.*, in press.

Hoeijmakers, J. H. J., Borst, P., Van den Burg, J., Weissmann, C., and Cross, G. A. M. (1980). *Gene* **8**, 391.

Holder, A. A. (1983). *Biochem. J.* **209**, 261.

Johnson, J. G., and Cross, G. A. M. (1976). *J. Protozool.* **24**, 587.

Kooter, J. M., and Borst, P. (1984). *Nucleic Acids Res.* **12**, 9457.

Kooter, J. M., De Lange, T., and Borst, P. (1984). *EMBO J.* **3**, 2387.

Kosinski, R. J. (1980). *Parasitology* **80**, 343.

Laurent, M., Pays, E., Magnus, E., Van Meirvenne, N., Matthijssens, G., Williams, R. O., and Steinert, M. (1983). *Nature (London)* **302**, 263.

Laurent, M., Pays, E., Van der Werf, A., Aerts, D., Magnus, E., Van Meirvenne, M., and Steinert, M. (1984). *Nucleic Acids Res.* **12**, 8319.

Liu, A. Y. C., Van der Ploeg, L. H. T., Rijsewijk, F. A. M., and Borst, P. (1983). *J. Mol. Biol.* **167**, 57.

Liu, A. Y. C., Michels, P. A. M., Bernards, A., and Borst, P. (1984). *J. Mol. Biol.* **182**, 383.

Longacre, S., Hibner, U., Raibaud, A., Eisen, H., Baltz, T., Giroud, C., and Baltz, D. (1983). *Mol. Cell. Biol.* **3**, 399.

Lumsden, W. H. R., and Evans, P. A., eds. (1976). "Biol Kinetoplastida," Vol. 1. Academic Press, London.

Lumsden, W. H. R., and Evans, P. A., eds. (1979). "Biol Kinetoplastida," Vol. 2. Academic Press, London.

McConnel, J., Gurnett, A. M., Cordingley, J. S., Walker, J. E., and Turner, M. J. (1981). *Mol. Biochem. Parasitol.* **4**, 226.

McConnel, J., Turner, M. J., and Rovis, L. (1983). *Mol. Biochem. Parasitol.* **8**, 119.

Majiwa, P. A. O., Young, J. R., Englund, P. T., Shapiro, S. Z., and Williams R. O. (1982). *Nature (London)* **297**, 514.

Majumder, H. K., Boothroyd, J. C., and Weber, H. (1981). *Nucleic Acids Res.* **9**, 4745.

Matthyssens, G., Michiels, F., Hamers, R., Pays, E., and Steinert, M. (1981). *Nature (London)* **293**, 230.

Merrit, S. C., Tschudi, C., Konigsberg, W. H., and Richards, F. F. (1983). *Proc. Natl. Acad. Sci. U.S.A.* **80**, 1536.

Michels, P. A. M., Bernards, A., Van der Ploeg, L. H. T., and Borst, P. (1982). *Nucleic Acids Res.* **10**, 2353.

Michels, P. A. M., Liu, A. Y. C., Bernards, A., Sloof, P., Van der Bijl, M. M. W., Schinkel, A. H., Menke, H. H., Borst, P., Veeneman, G. H., Tromp, M. C., and Van Boom, J. H. (1983). *J. Mol. Biol.* **166**, 537.

Michels, P. A. M., Van der Ploeg, L. H. T., Liu, A. Y. C., and Borst, P. (1984). *EMBO J.* **3**, 1345.

Miller, E. N., and Turner, M. J. (1981). *Parasitology* **82**, 63.

Miller, E. N., Allan, L. M., and Turner, M. J. (1984). *Mol. Biochem. Parasitol.*, in press.

Mount, S. M. (1982). *Nucleic Acids Res.* **10**, 459.

Myler, P., Nelson, R. G., Agabian, N., and Weber, H. (1984). *Nature (London)* **309**, 282.

Nelson, R. G., Parsons, M., Barr, P. J., Stuart, K., Selkirk, M., and Agabian, N. (1983). *Cell* **34**, 901.

Nelson, R. G., Parsons, M., Selkirk, M., Newport, G., Barr, P. J., and Agabian, N. (1984). *Nature (London)* **308**, 665.

Opperdoes, F. R. (1984). *Br. Med. Bull.* **41**, 130.

Overath, P., Czichos, J., Stock, U., and Nonnengaesser, C. (1983). *EMBO J.* **2**, 1721.

Parsons, M., Nelson, R. G., Watkins, K. P., and Agabian, N. (1984a). *Cell* **38**, 309.

Parsons, M., Nelson, R. G., Stuart, K., and Agabian, N. (1984b). *Proc. Natl. Acad. Sci. U.S.A.* **81**, 684.

Pays, E., Delronche, M., Lheureux, M., Vervoort, T., Bloch, J., Gannon, F., and Steinert, M. (1980). *Nucleic Acids Res.* **8**, 5965.

Pays, E., Lheureux, M., and Steinert, M. (1981a). *Nature (London)* **292**, 265.

Pays, E., Van Meirvenne, N., LeRay, D., and Steinert, M. (1981b). *Proc. Natl. Acad. Sci. U.S.A.* **78**, 2673.

Pays, E., Lheureux, M., Vervoort, T., and Steinert, M. (1981c). *Mol. Biochem. Parasitol.* **4**, 349.

Pays, E., Lheureux, M., and Steinert, M. (1982). *Nucleic Acids Res.* **10**, 3149.

Pays, E., Laurent, M., Delinte, K., Van Meirvenne, N., and Steinert, M. (1983a). *Nucleic Acids Res.* **11**, 8137.

Pays, E., Van Assel, S., Laurent, M., Dero, B., Michiels, F., Kronenberger, P., Matthyssens, G., Van Meirvenne, N., Le Ray, D., and Steinert, M. (1983b). *Cell* **34**, 359.

Pays, E., Van Assel, I., Laurent, M., Darville, M., Vervoort, T., Van Meirvenne, N., and Steinert, M. (1983c). *Cell* **34**, 371.

Pays, E., Delauw, M. F., Van Assel, S., Laurent, M., Vervoort, T., Van Meirvenne, N., and Steinert, M. (1983d). *Cell* **35**, 721.

Pays, E., Delauw, M. F., Laurent, M., and Steinert, M. (1984). *Nucleic Acids Res.* **12**, 5235.

Penncavage, N. A., Julius, M. A., and Marcu, K. B. (1983). *Nucleic Acids Res.* **11**, 8343.

Raibaud, A., Gaillard, C., Longacre, S., Hibner, U., Buck, G., Bernardi, G., and Eisen, H. (1983). *Proc. Natl. Acad. Sci. U.S.A.* **80**, 4306.

Rice-Ficht, A. C., Chen, K. K., and Donelson, J. E. (1981). *Nature (London)* **294**, 53.

Rice-Ficht, A. C., Chen, K. K., and Donelson, J. E. (1982). *Nature (London)* **298**, 676.

Rifkin, M. R. (1978). *Proc. Natl. Acad. Sci. U.S.A.* **75**, 3450.

Rovis, L., and Dube, D. K. (1981). *Mol. Biochem. Parasitol.* **4**, 77.

Schwartz, D. C., and Cantor, C. R. (1984). *Cell* **37**, 67.

Seebeck, T., Whittaker, P. A., Imboden, M. A., Hardman, N., and Braun, R. (1983). *Proc. Natl. Acad. Sci. U.S.A.* **80,** 4634.

Silverman, M., and Simon, M. (1983). *In* "Mobile Genetic Elements" (J. A. Shapiro, ed.), p. 537. Academic Press, New York.

Tait, A. (1980). *Nature (London)* **287,** 536.

Thomashow, L. S., Milhausen, M., Rutter, W. J., and Agabian, N. (1983). *Cell* **32,** 35.

Van der Ploeg, L. H. T., and Cornelissen, A. W. C. A. (1984). *Philos. Trans. R. Soc. London Ser. B* **307,** 13.

Van der Ploeg, L. H. T., Valerio, D., De Lange, T., Bernards, A., Borst, P., and Grosveld, F. G. (1982a). *Nucleic Acids Res.* **10,** 5905.

Van der Ploeg, L. H. T., Bernards, A., Rijsewijk, F. A. M., and Borst, P. (1982b). *Nucleic Acids Res.* **10,** 593.

Van der Ploeg, L. H. T., Liu, A. Y. C., Michels, P. A. M., De Lange, T., Borst, P. Manjumder, H. K., Weber, H., Veenemen, G. H., and Van Boom, J. (1982c). *Nucleic Acids Res.* **10,** 3591.

Van der Ploeg, L. H. T., Schwartz, D. C., Cantor, C. R., and Borst, P. (1984a). *Cell* **37,** 77.

Van der Ploeg, L. H. T., Liu, A. Y. C., and Borst, P. (1984b). *Cell* **36,** 459.

Van der Ploeg, L. H. T., Cornelissen, A. W. C. A., Barry, J. D., and Borst, P. (1984c). *EMBO J.* **3,** 3109.

Van der Ploeg, L. H. T., Cornelissen, A. W. C. A., Michels, P. A. M., and Borst, P. (1984d). *Cell* **39,** 213.

Van Meirvenne, N., Janssens, P. G., and Magnus, E. (1975a). *Ann. Soc. Belge Med. Trop.* **55,** 1.

Van Meirvenne, N., Janssens, P. G., Magnus, E., Lumsden, W. H. R., and Herbert, W. J. (1975b). *Ann. Soc. Belge Med. Trop.* **55,** 25.

Verbeet, M. P., Klootwijk, J., Van Heerikhuizen, H., Fontijn, R. D., Vreugdenhil, E., and Planta, R. J. (1984). *Nucleic Acids Res.* **12,** 1137.

Vickerman, K. (1969). *J. Cell Sci.* **5,** 163.

Vickerman, K. (1978). *Nature (London)* **273,** 613.

Vickerman, K., and Barry, J. D. (1982). *In* "Immunology of Parasitic Infections" (S. Cohen and K. S. Warren, eds.), p. 204. Blackwell, Oxford.

Williams, R. O., Marcu, K. B., Young, J. R., Rovis, L., and Williams, S. C. (1978). *Nucleic Acids Res.* **5,** 3171.

Williams, R. O., Young, J. R., and Majiwa, P. A. O. (1979). *Nature (London)* **282,** 847.

Williams, R. O., Young, J. R., Majiwa, P. A. O., Doyle, J. J., and Shapiro, S. Z. (1980). *Am. J. Trop. Med. Hyg.* **29,** 1037.

Williams, R. O., Young, J. R., and Majiwa, P. A. O. (1982). *Nature (London)* **299,** 417.

Young, J. R., Donelson, J. E., Majiwa, P. A. O., Shapiro, S. Z., and Williams, R. O. (1982). *Nucleic Acids Res.* **10,** 803.

Young, J. R., Shah, J. S., Matthyssens, G., and Williams, R. O. (1983a). *Cell* **32,** 1149.

Young, J. R., Miller, E. N., Williams, R. O., and Turner, M. J. (1983b). *Nature (London)* **306,** 196.

INTERNATIONAL REVIEW OF CYTOLOGY, VOL. 99

Kinetoplast DNA in Trypanosomid Flagellates

LARRY SIMPSON

Department of Biology and Molecular Biology Institute, University of California, Los Angeles, California

119

I. Introduction

This review will emphasize recent developments in the application of recombinant DNA technology to problems of the structure, replication, and transcription of the unusual mitochondrial DNA in the kinetoplastid protozoa known as kinetoplast DNA. Kinetoplast DNA will be discussed in terms of its two molecular components, minicircles and maxicircles, and species-dependent variations will be emphasized.

II. Minicircle DNA: Structure and Complexity

The minicircle component of the kinetoplast DNA is the most unusual aspect of this mitochondrial genetic system. Circular DNA molecules are not uncommon in nature nor in other mitochondrial genetic systems, but nowhere in nature does one find thousands of small circles catenated together into a single giant network of DNA such as found in the kineto-plastid protozoa. The function of minicircle DNA is still a mystery, but much has been learned recently about the structure and replication of this DNA in several kinetoplastid species: *Trypanosoma brucei, Trypanosoma equiperdum, Leishmania tarentolae, Crithidia fasciculata,* and *Crithidia luciliae.* In general minicircle DNA from all these species has the following general characteristics: (1) circles exist catenated with each other, although there is a small percentage of unattached circles that may have a functional significance; (2) within any one species the circles are of fairly uniform size although the size varies from species to species; (3) there is a species-dependent variable amount of sequence heterogeneity among the minicircles from any one clonal population; (4) the sequence heterogeneity is expressed on the level of the individual minicircle as a variable region and a constant region and sequence changes among the minicircle population in a kinetoplastid species occur rapidly in nature; and (5) there is no apparent sequence homology between the minicircles of a given species and the maxicircle DNA. In addition to these general common properties of minicircle DNA, there are many differences between the various kinetoplastid species, and these differences may prove to have phylogenetic significance. Therefore, I will discuss results from each species in succession.

A. *Trypanosoma brucei*

From the initial optical reassociation studies of Steinert *et al.* (1976) it was realized that the kDNA of *T. brucei* was perhaps the most complex in terms of sequence heterogeneity and the most rapidly changing in terms of evolution. The DNA reassociated with a single copy complexity of 300 times the size of the minicircle, and no hybridization between different *T. brucei* strains was observed. The high complexity of kDNA from *T. brucei* was confirmed by the S_1 C_0t analysis of Donelson *et al.* (1979) in which [³H]kDNA renatured with a single copy $C_0t_{1/2}$ 316 times the $C_0t_{1/2}$ when a cloned 1.1-kb minicircle (pkT51) was the excess driver (Fig. 1). However, there was a minor component (15%) of the kDNA which renatured at a rate suggestive of repetitive sequences. Renaturation of a cloned minicircle, pkT51, in the presence of excess DNA from two other cloned minicircles, pkT3 and pkT204, showed that approximately 25% of the sequence of the pkT51 insert is homologous to sequences in both the pkT3 and pkT204 inserts (Fig. 1). Stuart (1979) confirmed that there are approximately 300 different minicircle sequence classes in *T. brucei* 164 kDNA and showed that the minicircle classes differ in relative abundance (Stuart and Gelvin, 1982). For example, a *Hind*III cut minicircle cloned into the *Hind*III site of pBR322 renatured at a rate implying the existence of 500 copies per network, whereas *Bam*HI cleaved uncloned minicircles renatured at a rate implying the existence of 60 copies per network (Fig. 2).

However, assuming a *T. brucei* network mass of 4.2×10^9 Da that is composed of 15% maxicircle DNA of 22-kb molecules and 85% minicircle DNA of 1.1-kb molecules, one can calculate that there are approximately 45 maxicircles and 5500 minicircles (Stuart and Gelvin, 1980). Furthermore, assuming that there are about 300 different minicircle sequence classes, this would imply about 18 copies of each sequence class if the classes are of equal size. The finding (Stuart and Gelvin, 1980) that one cloned minicircle is present in 500 copies implies that there is a wide divergence in frequency of the different sequence classes in *T. brucei*.

The sequence heterogeneity of *T. brucei* minicircle DNA shown by C_0t analysis is evidenced also by restriction enzyme analysis on gels (Fig. 3). Digestion of *T. brucei* kDNA with most restriction enzymes yields low release of unit-length linearized minicircles with no smaller bands (Borst *et al.,* 1980a, 1981a; Brunel *et al.,* 1980a; Fairlamb *et al.,* 1978; Simpson and Simpson, 1980). Most of the DNA remains catenated in the network that does not enter the gel. This low release of mainly once-cut minicircles by most restriction enzymes is due both to the presence of many minicircle sequence classes and to the high percentage of AT in the minicircle DNA of *T. brucei*. However, digestion with *Taq*I, *Alu*I, or *Mbo*I does release most of the minicircles from the network and produce fragments of less than unit length, implying that sites for these enzymes are present more than once in conserved regions of the minicircles. The apparently single bands of unit-length minicircle DNA released by digestion with enzymes such as *Eco*RI, *Pst*I, *Hpa*II, *Hae*III, *Hind*III, and *Hha*I were

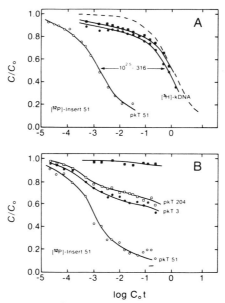

FIG. 1. Renaturation kinetics of the *Trypanosoma brucei* kDNA insert of pkT51 which had been labeled *in vitro* with ^{32}P. (A) The renaturation of ^{32}P-labeled pkT51 insert DNA in the presence of an excess of pkT51 DNA (open circles) or an excess of total network kDNA (solid circles). The network kDNA was labeled with ^{3}H *in vitro* prior to shearing so that its renaturation could also be monitored (solid squares). The broken line is a model curve generated by the standard renaturation equation when it is assumed that a single-copy DNA sequence has a log $C_0t_{1/2} = 0.15$. (B) The renaturation of ^{32}P-labeled pkT51 insert DNA (open circles), pkT3 DNA (solid circles), pkT204 DNA (open squares), or pBR322 DNA (solid squares). The initial concentration, C_0, of the excess unlabeled driver DNAs refers to the concentration of only the inserted kDNA sequence when a pkT recombinant plasmid is the driver DNA. In all of the experiments C_0 was chosen so that the final time point was taken 11 hours after the start of the reaction. About 4000 ^{32}P cpm were sampled at each time point. The pBR322 sequence does not drive the renaturation of the kDNA insert 51 within the time span of the renaturation experiments (solid squares). Reprinted from Donelson *et al.* (1979) with permission.

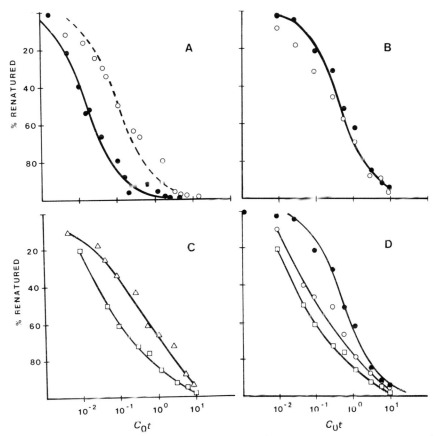

FIG. 2.   Renaturation kinetics of nick-translated kDNA sequences of *T. brucei*. (A) kDNA and homologous driver (open circles) compared to lambda DNA standard (solid circle). Both lines are theoretical single component curves. (B) Maxicircle (open circles) and cloned maxicircle fragment (solid circles) driven by kDNA. (C) Cloned *Hin*dIII minicircle (open squares) and *Bam*H1 minicircle (open triangles) sequences driven by kDNA. (D) Cloned *Hin*dIII minicircle (open squares), kDNA (open circles), and cloned maxicircle segment (solid circles) driven by kDNA. The curve through the cloned maxicircle segment points is a theoretical single component curve. Reprinted from Stuart and Gelvin (1980) with permission.

shown to be composed of 5–10 closely spaced bands by acrylamide gel electrophoresis (Simpson and Simpson, 1980) (Fig. 4), implying either a substantial minor length heterogeneity or the existence of gel mobility anomalies, as shown to exist in *L. tarentolae* minicircle DNA (Challberg and Englund, 1980; Simpson, 1979).

Analysis of cloned minicircle DNA has greatly assisted our understand-

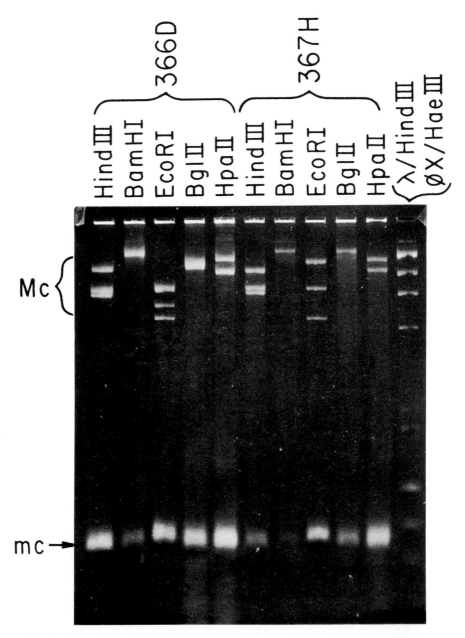

FIG. 3. Comparison of digestions of kDNA from two clonal strains (366D and 367H) of *T. brucei* procyclics in 0.8% agarose. The minicircle (mc) and maxicircle (Mc) regions are indicated. The molecular weight references are λ/*Hind*III and φXRF/*Hae*III fragments. Reprinted from Simpson and Simpson (1980) with permission.

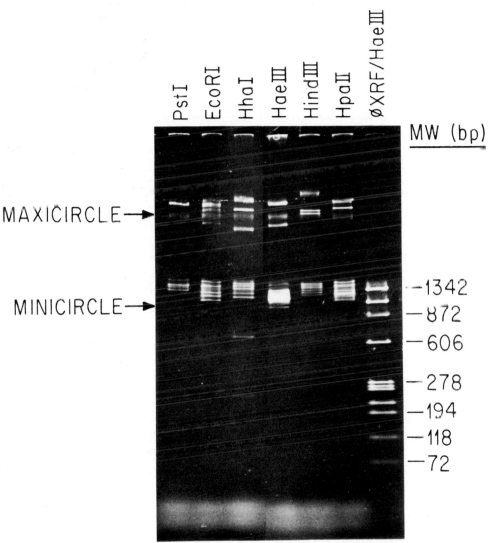

FIG. 4. Demonstration of heterogeneity of the major unit-length minicircle band from 366D procyclic *T. brucei* kDNA in an acrylamide gradient gel. Conditions: 3.5–10% acrylamide with a 3.0% stacking region. The molecular weight references are φXRF/*Hae*III fragments. Reprinted from Simpson and Simpson (1980) with permission.

ing of the observed sequence heterogeneity. Donelson *et al.* (1979) cloned minicircles from *T. brucei* clone 18E2 either by C-tailing minicircles released by digestion with *Hae*III, or with a mixture of *Hpa*II, *Hha*I, and *Alu*I and cloning into the *Pst*I site of pBR322, or by direct ligation of *Hind*III- or *Pst*I-released minicircles into the appropriate sites in pBR322. One minicircle, pkT3, was labeled *in vitro* and used as a probe in a Southern blot hybridization of kDNA digested with several enzymes. Hybridization was observed in all minicircle bands released by digestion with six enzymes. Similar results were obtained by colony hybridization, using three minicircle clones as probes. In general the probes hybridized strongly to themselves and weakly to all the other minicircle clones. Brunel *et al.* (1980a,b) cloned *Hind*III minicircles from *T. brucei* 427–60 in pBR322 and also *Eco*RI minicircles in λgtWES/λB (together with a maxicircle fragment, or with λB) and tested the cloned inserts for digestion with nine enzymes. The only sites found were single sites for *Eco*RI, *Hha*I, *Hae*III, and two to three sites for *Taq*I; all four of the clones studied contained different patterns of restriction sites. However, when the λ clones were spotted onto filters and probed with five of the pBR322 clones, some hybridization was observed with all clones, implying common sequences.

Simpson and Simpson (1980) cloned *Bam*HI-released minicircles from *T. brucei* 366D into pBR322 and tested 25 clones by colony hybridization using one, pTb7, as a probe. Four of the 25 clones showed strong hybridization, but all showed some hybridization, implying the existence of more than one semihomologous minicircle sequence class with a single *Bam*HI site.

Some heteroduplex studies of *T. brucei* minicircles have been reported, but the small size of these molecules precludes obtaining much information from this method (Brunel *et al.,* 1980a; Donelson *et al.,* 1979).

A direct confirmation and extension of the hybridization results was reported by Chen and Donelson (1980) in a sequence analysis of two cloned minicircles from *T. brucei* (Fig. 5). pk51 was derived from a digest of kDNA with a mixture of *Hha*I, *Hpa*II, and *Alu*I, tailed with G and inserted into the *Pst*I site of pBR322. pk201 was released from kDNA by digestion with *Pst*I and ligated directly into the *Pst*I site of pBR322. Both molecules contained approximately 72% AT, as expected from the low buoyant density of the kDNA (ρ = 1.690 g/ml). The most striking aspect was a nearly perfect homology of 122 bp. There were also 12 other common regions containing 50% AT. pk201 contained a decanucleotide sequence repeated 5 times in tandem. The only region of dyad symmetry was an 11-bp sequence in pk51. There was a high frequency of termination codons, TAA, TGA, and TAG, in all reading frames in both mole-

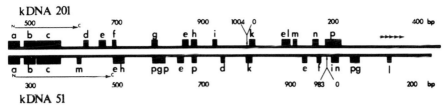

kDNA 201

kDNA 51

FIG. 5.   Diagram showing the relative locations of perfect homologies between *T. brucei* cloned minicircle kDNAs 201 and 51 that are equal to or greater than 10 bp. The two circular kDNA sequences are represented by two lines aligned so that each begins with the first position of the 122-bp near-perfect homology. The other letters indicate the other smaller corresponding regions of homology. The numbers indicate nucleotide positions. The horizontal arrows with N at the left end and C at the right show the largest regions in the two kDNAs that could potentially code for a polypeptide. The five tandem arrows above the lines on the right indicate five repeats of a decanucleotide sequence in kDNA 201. Reprinted from Chen and Donelson (1980) with permission.

cules; the longest open reading frames were 52 amino acids in pk201 and 71 amino acids in pk51. Interestingly, these small open reading frames encompassed the 122-bp common region. The sequence data nicely confirmed the evidence from the C_0t analysis that the different minicircle classes had approximately 25% of their sequences in common, since summation of all the perfect homologies of 10 bp or greater gave a value of 27%.

B. *Trypanosoma gambiense*

Riou and Barrois (1981) found that the minicircles from the kDNA networks of *T. gambiense* were similar to those from *T. brucei* in terms of cleavage by restriction enzymes. The only enzymes that cleaved the majority of the network minicircles and produced fragments of less than unit length were *Taq*I, *Hin*fI, and *Xba*I. The patterns with these enzymes indicated extensive minicircle sequence heterogeneity. They also found that the minicircles hybridized somewhat with minicircle DNA from the Pasteur Institute strain of *T. equiperdum*.

C. *Trypanosoma equiperdum*

Unlike most other kinctoplastid species, the kDNA minicircles from *T. equiperdum* are homogeneous in base sequence. This was first shown by Riou and Barrois (1979) for a strain of *T. equiperdum* from the Pasteur Institute and by Frasch *et al.* (1980) for a strain from the ATCC (30019). However, these two strains differ signficantly in the maxicircle components, indicating a possible molecular pleomorphism among strains of this species.

A unique circular restriction enzyme map for the uncloned minicircle from the Pasteur *T. equiperdum* strain was derived by Riou and Barrois (1979) and the complete nucleotide sequence was reported by Barrois *et al.* (1982). The molecule contained 73% AT, had a 6-fold tandem repeat of a 12-bp sequence, and had three small dyad symmetries of 9–11 bp distributed equidistantly from each other. In terms of possible codogenic function, the longest open reading frames were 18 and 22 amino acids (Fig. 6). A sequence of 130 bp was strikingly homologous to the 122-bp common sequence to *T. brucei* minicircles pkT51 and pkT201; it differed from pkT201 by 12 bp and from pkT51 by 15 bp. The remainder of the *T. equiperdum* molecule had no significant homology with the *T. brucei* sequences. This cross-species partial sequence conservation is intriguing in terms of the functional significance of the minicircle. Both Chen and Donelson (1980) and Barrois *et al.* (1982) speculated that the conserved

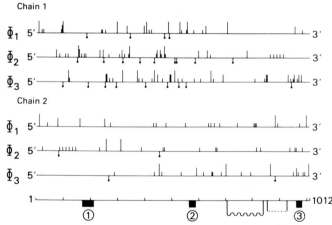

FIG. 6. Diagram showing the distribution of nonsense (vertical bars) and initiation (inverted bars with solid circles) codons in the six reading frames of *T. equiperdum* minicircle DNA. Three reading frames called ϕ_1, ϕ_2, and ϕ_3 were initiated from the 5' end of each DNA strand. On chain 1, ϕ_1 is initiated by its first triplet AAT, ϕ_2 by ATC, and ϕ_3 by TCA. On chain 2, ϕ_1 is initiated by its first triplet ATT, ϕ_2 by TTC, and ϕ_3 by TCT. The length of the kDNA minicircle does not correspond to a multiple of three nucleotides. Therefore, going through *Hin*fI site, phase ϕ_1 is changed into phase ϕ_2. Vertical bars have different lengths, depending on the stop codon (small-sized vertical bar, TAA; medium-sized vertical bar, TAG; large-sized vertical bar, TGA). (Bottom) Schematic representation of the minicircle, with the locations of the three dyad symmetries (circled numbers 1, 2, and 3), the DNA region homologous to *T. brucei* minicircles (wavy line), and the six repeats of 12 bp each (dashed line). Reprinted from Barrois *et al.* (1982) with permission.

region may contain the origin of replication. *T. equiperdum* is apparently a recent derivative of *T. brucei* that has lost the ability to undergo cyclical transmission through the insect host [I⁻ (Opperdoes *et al.*, 1976)], and hence it is not entirely surprising that the *T. brucei* minicircle common region is conserved. It is of great interest, however, that minicircle sequence heterogeneity does not occur in *T. equiperdum*, and this may be related to defects in maxicircle DNA of *T. equiperdum*, that will be discussed below (Frasch *et al.*, 1980).

D. *Trypanosoma evansi*

This is another example of a species that contains kDNA minicircles of a single sequence class. In addition, the networks appear to lack maxicircle DNA (Fairlamb *et al.*, 1978). Frasch *et al.* (1980) reported that the *Mbo*I agarose gel restriction profiles of kDNA from *T. evansi* strains were very similar to that from *T. equiperdum* ATCC 30019 and suggested that there is a basic minicircle sequence characteristic of the *T. brucei* type trypanosomes, that is, obscured in *T. brucei* itself by the extensive sequence heterogeneity.

E. *Trypanosoma mega*

T. mega kDNA minicircles show extensive sequence heterogeneity by both restriction enzyme analysis and reassociation kinetics. Steinert and Van Assel (1980) have calculated that the network may contain approximately 70 different semihomologous sequence classes.

F. *Trypanosoma lewisi*

Two minicircles from *T. lewisi* bloodstream forms were cloned in M13 vectors and completely sequenced (Ponzi *et al.*, 1984). The molecules contain 1018 bp and show a conserved region of 95 bp that is repeated twice in the molecule. The conserved regions are oriented approximately symmetrically within the minicircle. A 15-mer sequence (5'-GGGGTTG-GTGTAATA-3') within the conserved region was found to be also present in the published *T. brucei* and *T. equiperdum* minicircle sequences. The longest polypeptides that could be encoded by the conserved regions of the two *T. lewisi* minicircles are 27 and 26 amino acids, and these amino acid sequences show no relation to the amino acid sequences derived from the *T. brucei* and *T. equiperdum* minicircle sequences (Ponzi *et al.*, 1984).

G. *Trypanosoma cruzi*

The first evidence on kDNA minicircle sequence heterogeneity came from studies on *T. cruzi* (Riou and Yot, 1975, 1977). The minicircle of *T. cruzi* kDNA is intermediate in size between *T. brucei* and *Crithidia* and exhibits a sequence heterogeneity that is extensive but less than that of *T. brucei* or *T. mega* minicircle DNA. The *T. cruzi* minicircle has the unique property of an intramolecular repetitive sequence (conserved region) that is repeated four times with some variation. The initial evidence for this was the fact that several restriction enzymes give major bands with similar mobilities corresponding to multiples of one quarter of the unit minicircle length (Leon *et al.*, 1980; Mattei *et al.*, 1977; Morel *et al.*, 1980; Riou and Gutteridge, 1978; Riou and Yot, 1975, 1977). The existence of a quarterly repetitive conserved sequence has been confirmed by direct sequence analysis of minicircle DNA cloned into the *EcoRI* site of pBR325 and M13mp7 (Van Heuverswyn *et al.*, 1982).

Saucier *et al.* (1981) reported that network kDNA from log phase *T. cruzi* culture forms has a low-density shoulder in CsCl when the kDNA was isolated without use of a proteolytic digestion step (Fig. 7). The kDNA from stationary phase cells lacked this low-density shoulder. They attributed this density shift to a bound protein since the shift was eliminated by pronase digestion of the DNA preparation and they suggested that the bound protein was involved in kDNA replication or in the organization of the kDNA nucleoid body *in situ*. Since 95% of the kDNA is minicircle DNA, it is likely that the bound protein is linked to minicircle DNA. It should be of great interest to extend these findings to other kDNA systems, to localize the position of the bound protein, and to determine the functional significance of this phenomenon.

H. *Crithidia* SPECIES

kDNA from *C. fasciculata*, *C. luciliae*, and *C. acanthocephali* has been studied in detail. *Crithidia* spp. contain the largest kDNA minicircles (2.3 kb) in the kinetoplastid protozoa (Renger and Wolstenholme, 1972; Simpson *et al.*, 1974). The sequence heterogeneity of the minicircle DNA, however, is less than that found in *T. brucei*. Reassociation kinetic analysis has yielded values for the complexity of network DNA from *Crithidia* of 1–2.5 times the unit minicircle size (Fouts *et al.*, 1975; Hoeijmakers *et al.*, 1982b; Kleisen and Borst, 1975b), with no evidence for the presence of two or more components with different complexities (Kleisen and Borst, 1975b). In addition, the decrease in melting temperature of a self-annealed sonicated kDNA preparation from *C. luciliae* (Kleisen *et al.*,

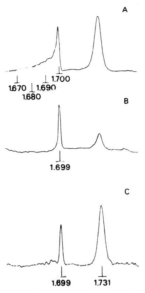

FIG. 7. Equilibrium density centrifugation in CsCl gradients of kDNA from *T. cruzi* and *Micrococcus luteus* DNA (ρ = 1.731 g/cm^3). (A) Untreated sample extracted from trypanosomes grown for 5 days; (B) same sample dialyzed against SSC to remove CsCl and incubated with pronase (1 mg/ml for 2 hours at 37°C); (C) untreated sample extracted from trypanosomes grown for 9 days. Reprinted from Saucier *et al.* (1981) with permission.

1976a) was only 1°C, implying that only 2% of the base pairs were involved in the sequence heterogeneity.

A qualitative estimate of more than 13 semihomologous minicircle sequence classes has been made from analysis of restriction profiles in gels (Kleisen *et al.*, 1976a). An apparent minor length heterogeneity of unit-length linearized minicircles has also been reported (Kleisen *et al.*, 1976a), although it has not been shown that this length heterogeneity is real and is not due to the gel mobility anomalies first observed with *Leishmania* minicircle DNA.

Hoeijmakers *et al.* (1982b) reported the results of an informative electron microscopic heteroduplex analysis of denatured and annealed minicircles of *C. luciliae* that were released from the networks by *Hind*III digestion. In approximately 40% of the reannealed minicircles they found heteroduplex structures characteristic of sequence rearrangements. The majority of these rearrangements were interpreted as inversions and the remainder as translocations, insertions, and deletions. The rearrangements were not at random locations but were localized to four minicircle segments (Fig. 8). In addition, there was a region (65–95% of the molecule

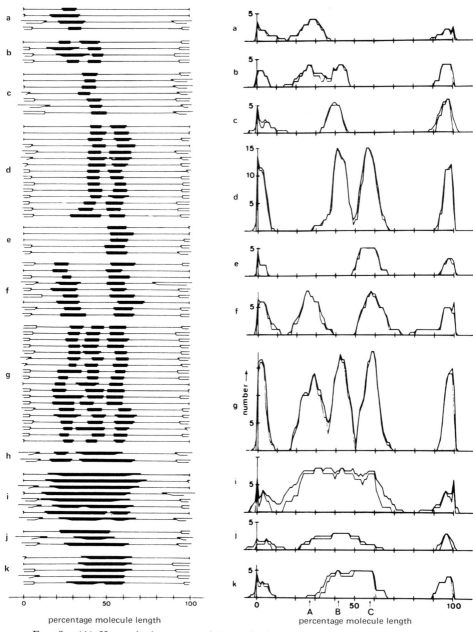

FIG. 8. (A) Heteroduplex maps of 81 *C. luciliae* renatured minicircle molecules of classes I[a,b,c] and III. Ninety molecules of the heteroduplex categories I[a,b,c] and, in part, III (only molecules with apparent deletions replaced by inserts of different size) were traced and divided into groups on the basis of overlap of hetero- and homoduplex segments. Eighty-one molecules (90%) could be classified in the groups shown (a–k). Thin single line,

in the map in Fig. 8) that showed no rearrangements and could correspond to the "constant region" such as found in *T. brucei* minicircles. They concluded that "specific segmental rearrangements form the main basis of the minicircle sequence heterogeneity in *Crithidia*" (Hoeijmakers and Borst, 1982). In order to correlate these results with the previous report of complete renaturation of *Crithidia* kDNA and only a 2% base pair mismatch of the self-annealed DNA (Kleisen *et al.*, 1976a), it must be assumed that the minicircle fragments used in the renaturation analysis were smaller than the rearranged segments found in the electron microscope study. The rearrangement model for the generation of minicircle diversity implies the existence of extensive recombination between minicircles in the network. Some evidence for this was previously obtained by a density shift experiment using *C. acanthocephali* kDNA (Manning and Wolstenholme, 1976). In addition "fused dimer" type minicircle structures have been observed in electron micrographs of minicircle DNA from several kinetoplastid species (Simpson and da Silva, 1971), which are similar to structures involved in genetic recombination.

The nature of the minicircle sequence heterogeneity in *Crithidia* (Hoeijmakers and Borst, 1982) was tested by redigestion of specific bands with a second enzyme. By this method it was shown that the bands in the primary digests are largely homogeneous, implying again that the sequence heterogeneity is caused by rearrangements rather than by accumulation of point mutations.

I. *Leishmania tarentolae*

The minicircles in the *Leishmania* are the smallest reported to date. *L. tarentolae* minicircles are about 870 bp in size, show extensive sequence heterogeneity by restriction enzyme analysis, and renature with a $C_0t_{1/2}$

---

homologous duplex region; parallel thin lines, single strands of a heteroduplex terminal segment; black box, internal heterologous region. The longest strand of a heteroduplex region is always shown above the base line, the smallest below. All molecules are normalized to a uniform length by putting the mean size of the upper and lower strands of each molecule (respectively the longest and shortest combination of strands) to a value of 100%. For absolute measurements the size of a *C. luciliae* minicircle was used (0.76 μm). The correction factor for dsDNA was determined from size measurements of circular ds minicircles (0.76 μm ± 0.03 SD) present in the same spreading. For ss regions the correction factor was based on the average size of ss circles and linears present in the spread preparation (0.72 μm ± 0.07 SD). (B) Histogram of heteroduplex regions of the groups of (A). Number of molecules versus percentage minicircle size. The designation of the histograms corresponds with that of the groups of (A). The two molecules constituting group h were not included here. The arrows labeled A, B, and C represent the three sites where the small heteroduplex areas are preferentially localized. Reprinted from Hoeijmakers *et al.* (1982b) with permission.

corresponding to a complexity between one and two times the minicircle length (Wesley and Simpson, 1973b). Reannealed open minicircles showed no decrease in the melting temperature and contained no regions of strand separation in the electron microscope (Wesley and Simpson, 1973a). Open circles did exhibit a multiphasic melting curve, however, that implied intramolecular sequence heterogeneity (Wesley and Simpson, 1973a). Intermolecular sequence heterogeneity is evidenced by restriction enzyme analysis (Challberg and Englund, 1980; Simpson, 1979) and also by cloning and sequencing of several sequence classes. The sequence heterogeneity of minicircle kDNA from *L. tarentolae* can best be visualized by high-resolution acrylamide gradient gel electrophoresis (Simpson, 1979) as shown in Fig. 9. The rich complexity of the banding pattern is almost obliterated by electrophoresis in agarose, but the major bands are visible (Figs. 10 and 11). Challberg and Englund (1980) concluded from analysis of minicircle fragments in agarose gels that there were three major sequence classes (Table I). Class I minicircles were linearized by *Hpa*II and represented 70% of the total DNA; class II minicircles were linearized by *Hind*II and represented 15% of the DNA; and class III minicircles were linearized by *Hpa*II, could be separated from linearized class I by agarose gel electrophoresis, and represented 7% of the DNA. A remaining 4% of the DNA consisted of minor classes and

TABLE I

COMPONENTS OF *L. tarentolae* KINETOPLAST DNA NETWORKS[a]

Component	Approximate percentage of total kDNA	*Hpa*II fragments	Enzyme used to obtain full-length fragment
Class I minicircles	70	α	*Hpa*II
Class II minicircles	15		*Hind*II
Class III minicircles	7	β	*Hpa*II
Other minicircles	4	—	—
Maxicircles	4	—	—

[a] To determine the percentage of kDNA in each component, a *Hpa*II digest of [³H]kDNA was electrophoresed on a 1–4% agarose tube gel. The gel was stained with ethidium bromide to locate the bands and sliced, and the slices were counted. The percentages in each component were calculated from the total radioactivity recovered (about 75% of the original sample). Approximately 4% was found in three high-molecular-weight fragments presumed to be derived from maxicircles. Another 4% was spread throughout the gel and is assumed to correspond either to low-molecular-weight maxicircle fragments or to fragments from minor classes of minicircles. No corrections were made for possible variations in base composition of the different components. Reprinted from Challberg and Englund (1980) with permission.

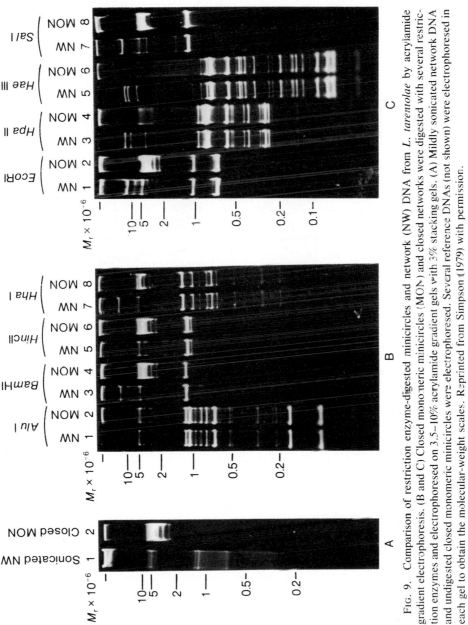

FIG. 9. Comparison of restriction enzyme-digested minicircles and network (NW) DNA from *L. tarentolae* by acrylamide gradient electrophoresis. (B and C) Closed monomeric minicircles (MON) and closed networks were digested with several restriction enzymes and electrophoresed on 3.5–10% acrylamide gradient gels with 3% stacking gels. (A) Mildly sonicated network DNA and undigested closed monomeric minicircles were electrophoresed. Several reference DNAs (not shown) were electrophoresed in each gel to obtain the molecular-weight scales. Reprinted from Simpson (1979) with permission.

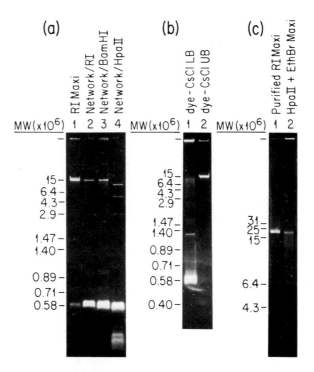

FIG. 10. Agarose gel electrophoresis of digested kDNA. (a) A 0.8% agarose gel of purified *L. tarentolae* EcoRI-digested maxicircle DNA and total network kDNA digested with *EcoRI*, *BamHI*, and *HpaII*. This *EcoRI* maxicircle preparation was purified by a single cycle of Hoechst 33258/CsCl separation and has some contaminating minicircle linear molecules migrating as if their molecular weight is approximately 0.6×10^6. The reference DNAs (not shown) were λ/*HindIII* fragments and ϕX174 RF/*HaeIII* fragments. (b) A 0.8% agarose gel of the lower band (LB) and upper band (UB) from the Hoechst 33258/CsCl gradient of *EcoRI*-digested kDNA shown in Fig. 18; same reference fragments as in (a). (c) A 0.5% agarose gel of purified *EcoRI* maxicircle DNA and once-cleaved, permuted maxicircle linear molecules produced by digestion of kDNA with *HpaII* in the presence of ethidium bromide (EthBr). The reference DNAs (not shown) were λ DNA, T7 DNA, and λ/*HindIII* fragments. Reprinted from Simpson (1979) with permission.

another 4% of maxicircle DNA. Evidence for the homogeneity of the sequence classes came from the construction of unique restriction enzyme maps for class I and class II uncloned gel-isolated linearized fragments for the enzymes *HaeIII*, *MboI*, and *HpaII* (Fig. 12). In addition, both class I and class II fragments reassociated with a $C_0t_{1/2}$ equivalent to a complexity of about 850 bp, and the reassociated molecules melted off hydroxyapetite at the same temperature as the native molecules. Furthermore, when class I linears were used as the driver and class II linears as the tracer, there was an acceleration of the rate of annealing of class II molecules, indicating sequence homology between class I and class II molecules. The heterologous hybrid showed a decrease in the melting temperature, indicating a 10–15% base pair mismatch. Recent Maxam–Gilbert sequence analysis of class I and class II linears isolated from a gel

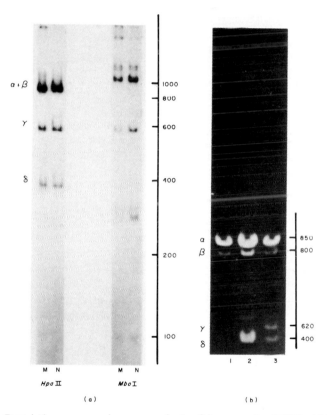

FIG. 11.   Restriction enzyme cleavage products of *L. tarentolae* kDNA. (a) Fragments produced from digestion of purified minicircles (M) or intact networks (N) were 5′ terminally labeled with ^{32}P and electrophoresed on an 8% polyacrylamide gel which was then autoradiographed. (b) Fragments were electrophoresed on a 1.4% agarose slab gel which was then stained and photographed. Lane 1 is an *Hpa*II digest of the *Hin*dII-resistant component of networks. Lane 2 is an *Hpa*II *Hin*dII double digest of intact networks. Appropriate controls showed that all digests are complete. The base-pair scales were established using restriction fragments of SV40 DNA, but many kDNA frgments do not electrophorese as expected from their size. Control experiments showed that *Hpa*IIα and *Hpa*IIβ migrate together on the 8% polyacrylamide gel. Numbers of base pairs are indicated on the right-hand sides of the gels. Reprinted from Challberg and Englund (1980) with permission.

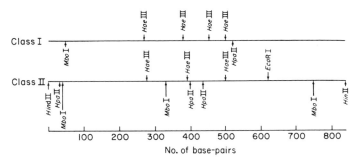

FIG. 12.   Restriction enzyme cleavage site maps for *L. tarentolae* class I and class II minicircles. The maps are aligned arbitrarily to maximize similarities. Class I minicircles do not contain *Hind*II or *Eco*RI sites. Reprinted from Challberg and Englund (1980) with permission.

has confirmed the homogeneity of class II but has shown class I to be heterogeneous (P. T. Englund, personal communication). The reason that this heterogeneity was not evidenced in the earlier analysis of Challberg and Englund (1980) is probably due to the limited number of enzymes used to construct the restriction map. Heterogeneity is clearly apparent in high-resolution acrylamide gels of digested kDNA from *L. tarentolae*, but it is nevertheless obvious that the sequence diversity in *Leishmania* is less than that in *T. brucei* and probably less than that in *Crithidia*.

The minicircle sequence diversity in *L. tarentolae* was also demonstrated by cloning different sequence classes in *Escherichia coli* and direct sequence analysis (Simpson *et al.*, 1979, 1980; Kidane *et al.*, 1984). Total kDNA was digested with *Bam*HI and cloned into pBR322. One unit-length minicircle clone, pLt12, was selected and used as a probe in colony hybridization to detect nonhomologous minicircle clones. Two *Hind*III minicircle clones in pBR322 were selected by this method, pLt26 and pLt154. An *Eco*RI minicircle, pKSR1, and a *Pst*I minicircle, pLtP1, were also cloned into pBR322. All minicircle inserts were subcloned into M13mp7, M13mp8, or M13mp9, in both orientations. The cloned minicircles were shown to be semihomologous by colony and blot hybridization. Restriction maps were constructed and sequence analysis performed. pKSR1 and pLt19 were shown to be the identical minicircle cloned at the *Eco*RI or the *Bam*HI site. It was shown by sequence analysis (Kidane *et al.*, 1984) that pLt19-KSR1, pLt26, and pLt154 contained a conserved sequence of 160–270 bp (Fig. 13). Barker *et al.* (1982) have also cloned and sequenced a minicircle from *L. tarentolae*, which has almost exactly the same sequence as the pLt19 clone derived in our laboratory.

A 14-mer sequence (5′-AGGGGTTGGTGTAA-3′) within the conserved region of the pKSR1, pLt26, and pLt154 minicircles was noted by

Kidane *et al.* (1984) to be conserved in the minicircle sequences reported from *T. brucei, T. equiperdum,* and *T. cruzi.* As noted above, this sequence is present also in the two cloned minicircles from *T. lewisi* and may represent a universal conserved minicircle sequence in all trypanosomatids. Evidence from Ntambi and Englund indicates that this "universal" minicircle sequence may represent an origin of replication. They found that newly replicated *T. equiperdum* minicircles which have recatenated to the network contain a single gap about 10 nt in size which overlaps the "universal" sequence.

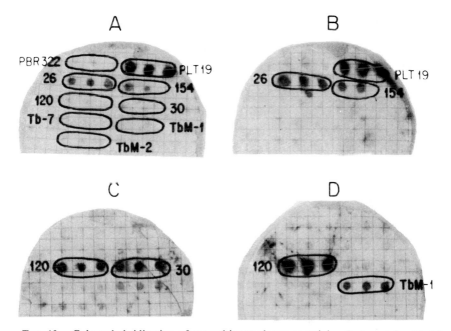

FIG. 13. Colony hybridization of recombinant clones containing *L. tarentolae* kDNA fragments and 366D procyclic *T. brucei* kDNA fragments. The clones were spotted in triplicate on each of four filters as indicated. The ^{32}P-labeled probes were (A) total kDNA from *L. tarentolae;* (B) total closed monomeric minicircle DNA from *L. tarentolae;* (C) purified *Bam*HI-linearized maxicircle DNA from *L. tarentolae;* (D) purified insert DNA from pLt120 plasmid, which contains a 6.6-kb fragment of *L. tarentolae* maxicircle DNA containing the 9 and 12 S RNA genes. pLt19, 26, and 154 are *L. tarentolae* kDNA unit-length minicircle clones; pLt120 and 30 are *L. tarentolae Eco*RI/*Bam*HI and *Hin*dIII maxicircle clones. TbM-1 is the *Hin*dIII C fragment maxicircle clone from *T. brucei;* TbM-2 is the *Hin*dIII B fragment maxicircle clone from *T. brucei.* Tb-7 is a unit-length minicircle clone from *T. brucei.* The original filters were inserted under the autoradiographs for photography. The relative positions of the colonies are as indicated in A by circling the three colonies of each type. In B, C, and D only those colonies showing strong positives are circled. Reprinted from Simpson and Simpson (1980) with permission.

Kidane *et al*. (1984) showed by a combination of quantitative gel electrophoresis and examination of T ladders of minicircles from *L. tarentolae* kinetoplast DNA randomly cloned into M13 at several sites that approximately 50% of the total minicircle DNA could be assigned into five sequence classes. The pKSR1-Lt19 sequence class was shown to represent 26% of the total minicircle DNA and a portion of the sequence was identical to the revised sequence (Marini *et al*., 1983) of the cloned 414-bp *Mbo*I fragment of Marini *et al*. (1982), implying that the pKSR1 minicircle probably was identical to the class II minicircle of Challberg and Englund (1980).

Arnot and Barker (1981) have analyzed kDNA from *Leishmania tropica major, Leishmania aethiopica,* and *Leishmania SP48* by agarose gel electrophoresis and by blot hybridization and have concluded that the minicircles from *L. tropica major* (700 bp) are smaller than those from *L. tarentolae* and that rapid minicircle sequence evolution is occurring in these Old World leishmanias.

An interesting anomaly in electrophoretic behavior of minicircle DNA from *L. tarentolae* was noted in the initial studies (Challberg and Englund, 1980; Simpson, 1979). The observation was that the mobility of full-length uncloned linearized minicircle DNA and certain fragments was different in acrylamide than in agarose gels, leading to large discrepancies in apparent molecular weights. Similar behavior was observed with cloned minicircles and minicircle fragments. Simpson *et al*. (1980) and Kidane *et al*. (1984) showed that the larger *Sma*I fragment of the cloned minicircle, pLt19, had an apparent size in 5% acrylamide 1.5 times the apparent size in 1.5% agarose. However, cleavage of this fragment at the *Eco*RI site produced two fragments which ran properly in acrylamide. This phenomenon was studied in detail by Marini *et al*. (1982), Kidane *et al*. (1984), Wu and Crothers (1984), and Hagerman (1984). The *Mbo*I fragment A from class II minicircles of *L. tarentolae* was cloned in pBR322 as pPE103 and found by sequence analysis to contain 414 bp (Marini *et al*., 1982, 1983). This fragment had an apparent size in 1% agarose of 450 bp and an apparent size in 12% acrylamide of 1380 bp. It eluted from Sephacryl 500 with an apparent size of 375 bp and behaved like a fragment of less than 309 bp in electric dichroism experiments. Marini *et al*. (1983) concluded that the molecule had an unusually compact configuration and proposed that it contains a region of smoothly bent B-DNA which may be caused by sequence periodicities. The sequence of this fragment was shown to contain a significant 10-nt periodicity of pdApdA, which could cause a bending of the molecule. They suggested that the curvature of the minicircle DNA is functionally significant in terms of binding a protein to the bent region, or of facilitating condensation of the network into the kDNA nucleoid body.

This argument was strengthened by the experiments of Wu and Crothers (1984), who constructed circularly permuted fragments of the same *Mbo*I minicircle fragment by cloning a double tandem repeat into a pBR322 plasmid and digesting with different restriction enzymes which only cut once. The relative gel mobilities of the permuted linearized plasmids were compared in acrylamide and agarose. By this method the "bend" was mapped to a site 148 nt from the *Mbo*I site, which corresponds to approximately 20 nt from the *Eco*RI site in pKSR1. This localization agrees well with that obtained by Kidane *et al.* (1984) by restriction enzyme digestion of circular pKSR1 minicircle DNA and comparative agarose–acrylamide gel electrophoresis. Wu and Crothers (1984) point out a $CA_{5-6}T$ tract repeated at a 10-nt interval four times at that locus in the sequence, and they attribute the bending of the molecule to this sequence periodicity. Hagerman (1984), however, has proposed an alternative physical explanation for the observed sequence dependent bend in the pKSR1 minicircle, which is based on the "purine clash" model of Calladine (1982). Further work is needed to distinguish between these theories.

Similar anomalies in electrophoretic behavior were noted for *T. brucei* minicircle DNA by Chen and Donelson (1980) and for *T. equiperdum*, *Herpetomonas muscarum*, and *C. fasciculata* by Ntambi *et al.* (1984) and may be a general property of kDNA minicircles.

J. *Phytomonas davidi*

Phytomonas davidi kDNA minicircles, which are 1.1 kb in size, renature with a $C_0t_{1/2}$ corresponding to a complexity of approximately 1.3 times the size of the minicircle and the reannealed circles melt 0.7–1.8°C below the native melting temperature, indicating no more than 2–4% base pair mismatch (Cheng and Simpson, 1978). Physically purified closed monomeric minicircles exhibit at least 13 bands of superhelical molecules in acrylamide–agarose gels in a distinctly non-Gaussian banding pattern which could be explained by the existence of a minor minicircle length heterogeneity or mobility abnormalities, as seen in *L. tarentolae*.

III. Evolution of Minicircle Sequence Heterogeneity

Minicircle sequence evolution has been studied both by examination of different strains or isolates from nature and by comparison of the same strain kept under nonselective laboratory conditions. Steinert *et al.* (1976) showed that two isolates of *T. brucei* did not cross hybridize. Borst *et al.* (1980a, 1981a) showed that the *Taq*I agarose restriction profiles of kDNA

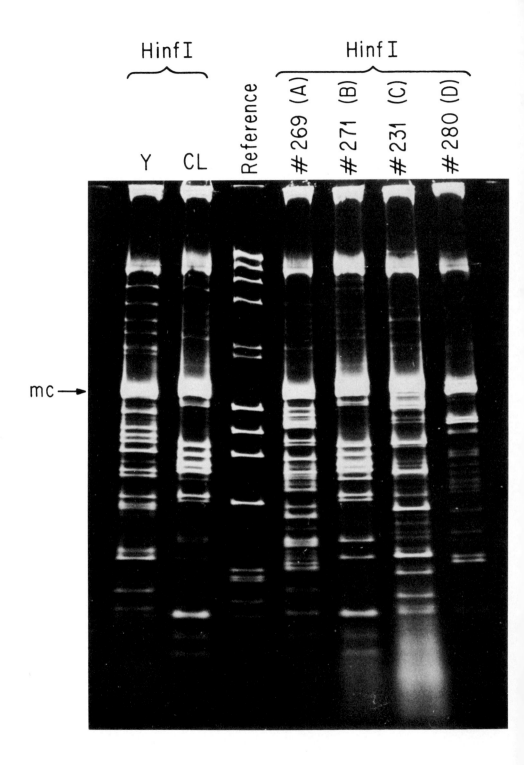

digests from nine strains of *T. brucei* were clearly different, but that four clones of *T. brucei* strain 427 yielded identical restriction profiles.

A comparison of the minicircle restriction profiles of kDNA from *C. fasciculata* and *C. luciliae* in high-resolution acrylamide gradient gels showed hardly any bands in common (Hoeijmakers and Borst, 1982). Cross hybridization of blots of *C. luciliae* kDNA digests with a labeled specific gel isolated fragment of *C. fasciculata* minicircle DNA showed weak hybridization of a specific subset of bands. Use of total *C. fasciculata* kDNA as a probe gave weak hybridization to most *C. luciliae* bands. Since the maxicircle DNA from these species were shown to be greater than 90% homologous, Hoeijmakers and Borst (1982) suggested that *C. luciliae* and *C. fasciculata* represent different strains of the same species which have undergone extensive minicircle sequence evolution.

The actual rate of change of minicircle sequences in both *C. fasciculata* and *C. luciliae* was monitored qualitatively by examination of kDNA from cells grown in the laboratory over a 2-year period (Hoeijmakers and Borst, 1982) at intervals of 9–12 months. kDNA from the uncloned *C. fasciculata* culture showed many small changes in restriction bands and the cloned *C. luciliae* showed a few changes. The changes were not due to methylation differences or to the presence of gaps or nicks in the molecules and were not caused by different maxicircle DNA concentrations in the networks.

Camargo *et al.* (1981) analyzed kDNA restriction profiles and other properties of 13 species of monoxenic trypanosomatids. They found interspecific but not intraspecific differences and concluded that this method can be used for species identification of insect trypanosomatids. However, the absence of restriction profile differences between three laboratory strains of *C. fasciculata* does not necessarily imply that this species shows no natural minicircle sequence evolution, since the actual history of these three laboratory strains is uncertain. Examination of *Crithidia* strains freshly isolated from nature is necessary to answer this important question.

Different strains or stocks of *T. cruzi* isolated from human patients and animals in Brazil were shown to exhibit distinctly different restriction profiles in high-resolution acrylamide gradient gels (Morel *et al.*, 1980) (Fig. 14). Control experiments showed that strain CL cultures showed no

FIG. 14.   Acrylamide gradient gel profiles of *Hin*fI digests of kDNA from the Y and CL strains of *T. cruzi* and from four stocks of *T. cruzi* (from patients) that represent zymodeme groups (A–D). The reference DNA is a mixture of λ/*Hin*dIII and φXRF/*Hae*III. The unit-length minicircle band is indicated by mc. Reprinted from Morel *et al.* (1980) with permission.

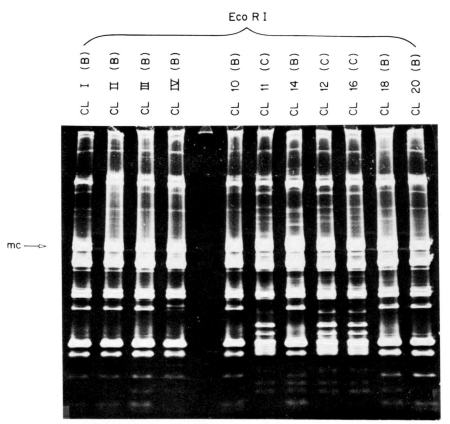

FIG. 15.   Acrylamide gradient gel profiles of *Eco*RI digests of *T. cruzi* kDNA from several CL control cultures and clones. I, CL cells kept in serial culture for 2 years (1978–1980) and harvested in stationary phase; II, CL cells harvested in logarithmic phase; III, CL cells from a culture kept at −70°C for 2 years (1978–1980); IV, CL strain kept in mice for 2 years (1978–1980) and grown for 10 passages in culture before analysis; CL 10, 11, 14, 12, 16, 18, and 20 clones from parental CL strain. The zymodeme groups are given in parentheses. Reprinted from Morel *et al.* (1980) with permission.

large-scale changes in restriction profiles with several enzymes after 2 years of continuous laboratory passage either in culture or in mice (Fig. 15). There were also no differences in restriction profiles of kDNA from cells in early log phase or in stationary phase, or from cells which had a large proportion of trypomastigotes or epimastigotes. In addition, six randomly isolated human-derived stocks of *T. cruzi* from the same area that belonged to the zymodeme group C showed identical restriction profiles. Morel *et al.* (1980) concluded that *T. cruzi* minicircle sequences change rapidly enough in nature to produce differences between strains, but not so rapidly as to preclude establishment of a stable hemoculture. Morel *et al.* (1980) proposed to designate subpopulations of *T. cruzi* having similar minicircle DNA restriction profiles by the term "schizodemes."

Frasch *et al.* (1981) have used the kDNA restriction profile method to compare stocks of *T. cruzi* and the presumed closely related *Trypanosoma rangeli* in agarose gels. They found that *Bsp*RI and *Msp*I gave patterns that allowed the differentiation of these species and concluded that these enzymes cut within the variable region of the minicircles. On the other hand, *Taq*I cut within the repetitive constant region, yielding profiles that were identical for the two species. Sanchez *et al.* (1984) extended this analysis to seven additional *T. cruzi* isolates and introduced the use of cloned minicircle and maxicircle fragments as hybridization probes. No correlation between different schizodeme patterns and infectivity for mice was apparent.

The rate of change of minicircle sequences in *L. tarentolae* was studied by examination of kDNA restriction profiles from cells frozen at intervals of 4 and 2 years (Simpson *et al.*, 1980). Profiles for the enzymes *Hha*I, *Hae*III, and *Msp*I were examined in acrylamide gradient gels (Fig. 16). The only changes observed were approximately two *Hha*I bands out of 19 during the period 1973–1977.

Arnot and Barker (1981) have attempted, with some success, to distinguish different species of Old World *Leishmania* and an unknown human isolate by blot hybridization of digested kDNA with labeled *L. aethiopica* kDNA. Lopes *et al.* (1984) have likewise attempted, with some success, to distinguish different species and strains of New World *Leishmania* by comparison of restricted profiles in acrylamide gels.

Wirth and Pratt (1982) found no detectable hybridization between the kDNA of two subspecies of *Leishmania brasiliensis* and three subspecies of *Leishmania mexicana*, nor between *L. mexicana* and *L. tropica*. They also showed that whole organisms could be spotted directly onto nitrocellulose filters and hybridized with nick-translated kDNA probes. This led to the development of a technique termed "touch blots," in which fragments of lesions from infected animals are touched to the filters and then the filters are hybridized with the labeled kDNA probe. Sufficient amastigotes are transferred to the filter by this method to provide a strong hybridization signal (Wirth and Pratt, 1982). This technique may prove to have clinical importance in the diagnosis of dermal leishmaniasis.

Handman *et al.* (1983) employed schizodeme analysis in acrylamide gels to characterize four infective and noninfective clones of *L. tropica* derived from a stock that is infective for mice. They found that the clones formed two schizodemes that were closely related in terms of cross hybridization of minicircle sequences, implying that the noninfective clones were derived from the original infective presumptive parental clone.

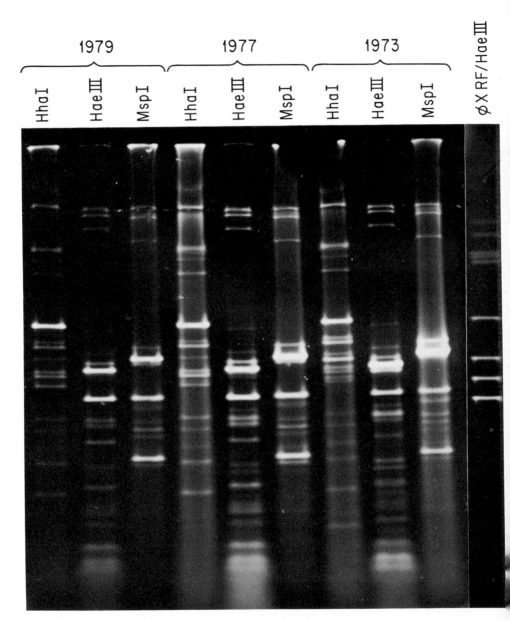

FIG. 16. Comparison of restricted kDNA from *L. tarentolae* cells kept in serial culture from 1973–1979 versus kDNA from stabilates frozen in 1973 and 1977. The gels are 3.5–10% acrylamide gradient gels with a 3% stack. The gel was stained with ethidium bromide. Reprinted from Simpson *et al.* (1982b) with permission.

As stated previously, the kDNA minicircles in two strains of *T. equiperdum* show an absence of sequence heterogeneity and at least one of these minicircles has a 130-bp region almost identical to the common region in *T. brucei* minicircle clones pkT51 and 201 (Barrois *et al.*, 1982). The suggestion has been made that the absence of minicircle sequence heterogeneity is somehow related to the observed deletion in the maxicircle DNA seen in strain ATCC 30019 (Frasch *et al.*, 1980). However, the Pasteur strain of *T. equiperdum* seems to contain an intact maxicircle at least at the level of overall size and restriction map (Riou and Saucier, 1979).

T. evansi represents another I⁻ trypanosome that is unable to synthesize mitochondrial enzymes and differentiate into the procyclic form, and the kDNA of this species was shown by restriction analysis to contain catenated minicircles of an apparent single-sequence class and no maxicircle DNA at all (Fairlamb *et al.*, 1978). *T. equinum* is also unable to differentiate into the procyclic form in nature and contains an altered form of kDNA. As visualized by 4',6-diamidino-2-phenylindole (DAPI) staining and electron microscopy, the mitochondrial DNA appears as scattered as clumps of material instead of the well-organized kinetoplast nucleoid body (Cuthbertson, 1981). In addition, a cellular DNA is present at the same buoyant density as kDNA, but it consists of a heterogeneous collection of circular molecules from 0.1 to 9.7 μm in size. Similar molecular forms have been observed in several strains of dyskinetoplastic *T. equipardum* (Frasch *et al.*, 1980; Hajduk and Cosgrove, 1979; Riou and Pautrizel, 1977), and it was shown that there was no sequence homology of this DNA with kDNA from *T. brucei* or normal *T. equiperdum*. However, in another dyskinetoplastic strain of *T. equiperdum*, a complete absence of kDNA was reported (Riou *et al.*, 1980). Furthermore, there is some evidence that the DNA in dyskinetoplastic *T. equiperdum* that bands in CsCl at the low kDNA buoyant density is a nuclear satellite DNA present in all brucei-type trypanosomes that is normally obscured by the kDNA band. Borst *et al.* (1980b) found that there is a small fraction of *T. brucei* nuclear DNA that is resistent to cleavage by most enzymes, but is digested by *Alu*I or *Hind*II into 180-bp repeats and which bands at the same density in CsCl as kDNA. They also showed that the low-density DNA in one strain of dyskinetoplastic *T. equiperdum* that banded at the density of kDNA was similarly digested by *Alu*I into 180-bp fragments and multimers. It was suggested (Borst *et al.*, 1980b) that the presence of this nuclear satellite DNA could explain the observations of apparently abnormal kDNA in dyskinetoplastic trypanosomes. However, the presence of DAPI-staining clumps in the mitochondria of certain dyskinetoplastic trypanosomes remains to be explained. In general, there is a clear correlation of the presence of a single minicircle sequence class with defects or absence of functional maxicircle DNA, but the meaning of this correlation is not clear. Frasch *et al.* (1980) suggested that minicircle sequence heterogeneity is related to recombination between maxicircle DNA and minicircle DNA, but there is as yet no evidence for this type of interaction. It is clear that in normal strains extensive minicircle sequence changes occur under natural conditions to produce strain differences. Some changes can be seen in periods of several years under laboratory conditions, but cultures are stable enough for purposes of strain differentiation by comparison of minicircle restriction profiles. The actual mechanism of minicircle sequence change is uncertain, although there is some evidence for segmental rearrangements being important, at least in *Crithidia*. Studies are required on the relative stabilities of the constant region and the variable region, on the exact nature of sequence changes between strains and species, and on the possible role of maxicircle–minicircle interactions in minicircle sequence changes.

IV. Transcription of Minicircle DNA

There is one report of a 240-nt transcript of the minicircle DNA of *C. acanthocephali* (Fouts and Wolstenholme, 1979), but this has not yet been confirmed by Northern blot or S_1 nuclease analysis. Attempts to demonstrate a minicircle transcript in *T. brucei* and *C. luciliae* by hybridization of *in vitro* labeled total cell RNA to kDNA blots have been negative (Hoeijmakers and Borst, 1978; Hoeijmakers *et al.*, 1981). In all cases, only maxicircle fragments showed hybridization. In addition, hybridization of kinetoplast RNA blots with *in vitro* labeled minicircle clones has also proved negative both for *L. tarentolae* (L. Simpson, G. Kidane, and A. Simpson, unpublished results) and *T. brucei* (Stuart and Gelvin, 1980). In the Northern blots there often is a band with the mobility of minicircle DNA that hybridizes with the minicircle probes, but it was shown that this band represents contaminating minicircle DNA in the RNA preparations.

The evidence against minicircle DNA having a codogenic function includes the apparent lack of stable transcripts, the rapid rate of sequence changes in nature, and the presence of many termination codons in the *T. brucei* and *T. equiperdum* sequences reported to date. However, the small open reading frames that cover the conserved regions of the two *T. brucei* clones (Chen and Donelson, 1980) and the three *L. tarentolae* clones (Kidane *et al.*, 1984) leave open the possibility that minicircle DNA is transcribed but that transcription is limited to certain points in the life cycle, or that the transcripts are unstable. Further work is needed to settle this important question.

There is, however, a recent report by Shlomai and Zadok (1984) that suggests that minicircles of *C. fasciculata* contain long open reading frames which are expressed in the cell. kDNA minicircle fragments obtained by *FnuDII* and *AluI* restriction enzyme digestion were inserted into the *SmaI* site of the pORF expression vectors. These vectors contain the 5' end of the *E. coli ompF* gene, which provides a strong promotor, translation initiation site, and a signal sequence for export. In addition, they contain the out-of-frame *E. coli lacZ* gene lacking the 5' end. Several kDNA minicircle fragments were able to realign the *ompF* and *lacZ* genes in frame and induce the expression of β-galactosidase activity in the *E. coli* cells. The size of the kDNA inserts was from 550 to 720 nt and all showed cross hybridization. Expression of a tribrid fusion protein in the bacterial cells was established by observation of a new protein species of 135,000 Da in SDS-polyacrylamide gels. The new fusion protein was extracted from gels, and antibodies were raised in rabbits. The sera were absorbed with bacterial extracts carring pORF vectors lacking the kDNA inserts but expressing β-galactosidase. The presence of a reactive protein within the *Crithidia* cells was determined by indirect immunofluorescence. The site of reactivity was the flagellar pocket. Western blots of *Crithidia* total cell proteins revealed three polypeptides which reacted with the antisera. Two of these polypeptides contained sugar residues, as seen by labeling with glucosamine. These results imply that certain *Crithidia* kDNA fragments are transcribed and translated and that the protein product may be associated with the kDNA itself, which is situated next to the flagellar pocket. However, this must be confirmed by a sequence analysis of the kDNA fragments cloned, and it must be established that these are derived from minicircles and not from maxicircles which are known to be transcribed and translated.

V. Replication of Minicircle Kinetoplast DNA

The hypothesis of Englund (1978) that closed minicircles are removed from the network at random by a specific topoisomerase, replicated as free circular molecules, and then recatenated as nicked circles at the periphery of the network, perhaps at the two sites described in *Crithidia* by autoradiography experiments (Simpson and Simpson, 1976; Simpson *et al.*, 1974), has been adequately reviewed (Englund, 1981; Englund *et al.*, 1982). The hypothesis predicts the existence of specific topoisomerases which can decatenate closed circles and recatenate nicked circles. Marini *et al.* (1980) and Kayser *et al.* (1982) showed that both T_4 phage topoisomerase and DNA-gyrase can decatenate protein network DNA from *C. fasciculata* and *T. cruzi* into monomeric minicircles and maxicircles. Riou *et al.* (1982) have partially purified two types of topoisomerase activities from a nuclear fraction of *T. cruzi*, the first of which relaxes supercoiled circles and the second of which induced catenation of circles in the presence of ATP.

Shlomai and Zadok (1983) reported a DNA topoisomerase activity in *C. fasciculata* which catalyzes a reversible interlocking of duplex circles into huge catenane forms resembling naturally occurring networks. There was a specific requirement for covalently closed circles as substrates for decatenation, but no preference for closed circles for catenane formation. These properties suggested to Shlomai and Zadok (1983) a potential role for this topoisomerase in the replication of kinetoplast DNA minicircles. However, the intracellular localization of this or any other topoisomerase from hemoflagellates has not yet been established.

Recent work from Englund's laboratory (Kitchin *et al.*, 1984, 1985) has shown more details of the minicircle replication pattern in the case of *C. fasciculata*. After θ-type replication of released minicircles, the two progeny molecules have different structures. One contains a nascent strand with a single nick or gap, while the other contains a nascent strand composed of small fragments which are separated by gaps. The former are nicked several times and the latter are partially repaired prior to recatenation to the network. The two progeny are attached to the network at different rates. Once attached, convalent closure ensues. In a related study done with *T. equiperdum* minicircles, Ntambi and Englund (1985) found that newly replicated reattached minicircles possess a gap of about 10 nt which overlaps the "universal" 14-mer sequence and that appears to be bounded by a ribonucleotide, implying that this site represents an origin of replication of the minicircle.

VI. Maxicircle DNA: Isolation, Cloning, and Restriction Mapping

The maxicircle component of the kinetoplast DNA appears to represent the homolog of the informational mitochondrial DNA found in other eukaryotic cells. The maxicircle DNA is present as 20–50 apparently identical circular molecules catenated to the network. The molecular size varies from 20 kb in *T. brucei* to almost 39 kb in *T. cruzi*. This DNA has been the object of investigation in several laboratories and the results will be discussed on a species basis.

A. *Trypanosoma brucei*

The kDNA network in *T. brucei* is comparatively small (4.5 × 3 μm) when spread by the Kleinschmidt method for electron microscopy, and maxicircle molecules can often be visualized as long edge loops extending out from the network. Fairlamb *et al.* (1978) showed that digestion of the network with *Pst*I or nuclease S₁ would selectively remove the edge loop molecules and not affect the network structure, implying that minicircle catenation represents the major structural stabilizing force of the network.

The size of the maxicircle was calculated both by summation of fragments released from network DNA by digestion with several enzymes and by measurement in the electron microscopy of the *Pst*I linearized maxicircle to be 20–22 kb. Stuart (1979) showed that *Bam*HI digestion would release linearized maxicircle molecules and a small percentage of linearized minicircles from the network kDNA of *T. brucei* 164 and he used this method to enrich for maxicircle DNA by isopycnic banding in CsCl–EthBr. Stuart (1979), Simpson and Simpson (1980), and Borst and Fase-Fowler (1979) showed that linearized maxicircle DNA from several *T. brucei* strains banded in CsCl at a lower density than the kDNA network or the minicircle DNA (1.682 versus 1.690 g/ml), implying that the maxicircle DNA is lower in %GC than the minicircle DNA. This difference in %GC has been used to preparatively isolate linearized maxicircle DNA by addition of the AT-intercalating dye, Hoechst 33258, to the CsCl density gradient (Simpson and Simpson, 1980). The release of a low-density band in CsCl gradients has also been used to calculate the percentage of maxicircle DNA in total kDNA from *T. brucei*. Values reported range from 10 (Borst and Fase-Fowler, 1979) to 20 (Stuart, 1979) to 39% (Simpson and Simpson, 1980). Despite the range in these values, the relative percentage of maxicircle DNA is clearly greater in *T. brucei* kDNA than in kDNA from other species, in agreement with the greater ease of edge loop visualization in network preparation.

Unique circular restriction maps have been constructed for maxicircle DNA from several *T. brucei* strains: strain 427-60 (Borst and Fase-Fowler, 1979), strain 164 (Stuart, 1979), and strain 366D (Simpson and Simpson, 1980) (Fig. 17). The most striking aspect of these maps was the relative absence of restriction sites from a 5- to 6-kb region of the 20- to 22-kb molecule, termed the "silent region" or "variable region" by Borst

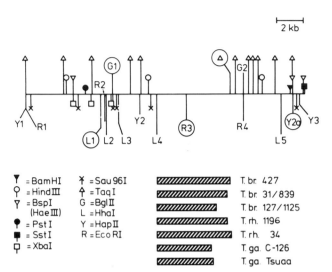

FIG. 17. The linearized physical map of the maxicircle from *T. brucei* kDNA. The length is 20.5 kb. Polymorphic sites are circled; circled open triangle is site Q6. The hatched bars represent the size of the variable region. The *Taq*I map is incomplete and there are several additional fragments in the span R4–R1. T. br., *T. brucei brucei;* T. rh., *T. brucei rhode-siense;* T. ga., *T. brucei gambiense.* Reprinted from Borst *et al.* (1980a) with permission.

and Fase-Fowler (1979) or the "divergent region" by Muhich *et al.* (1983). Direct gel comparisons of kDNA digests from nine stocks have revealed that there are two basic types of sequence changes (Borst *et al.*, 1980a, 1981a):

1. Restriction site polymorphisms, apparently randomly localized.
2. Deletions or insertions of up to 1.5 kb within the divergent region.

Site polymorphisms were also observed between strains 366D and 367H (Fig. 3) versus strain 164 (Simpson, 1979; Stuart, 1979). Four human infective stocks of *T. brucei* (*rodesiense* and *gambiense*) were also examined by Borst *et al.* (1981a). They found that the human infective stocks could not be distinguished from the *T. brucei brucei* stocks on the basis of maxicircle DNA site changes or deletions and they concluded that the human infective forms are host-range variants of the same species by this criterion. In terms of the rates of nucleotide substitution in maxicircle DNA from all *T. brucei* stocks examined, the variation (0.01–0.03 substitutions per base pair) was similar to that observed with mitochondrial DNA within a single species of mice or gophers. However, the nature of the deletions and insertions within the divergent region is not yet understood and this may affect these calculations.

Reassociation complexity analysis of maxicircle DNA from *T. brucei* has confirmed the sequence homogeneity of these molecules as compared to the minicircle DNA and the kinetic complexity compares well to the calculated size of the molecule (Stuart and Gelvin, 1980).

Much of the maxicircle DNA of *T. brucei* has been cloned in bacterial plasmids and phages. Brunel *et al.* (1980a) cloned the two smaller *Eco*RI fragments (*Eco*2 and *Eco*3) from strain 427-60 in λgt WES/λB, but were unable to clone the largest *Eco*RI (*Eco*1) fragment, although this is the optimal size for this cloning vector. Simpson and Simpson (1980) cloned the two smaller *Hind*III fragments from strain 366D in pBR322. These fragments cover essentially the same region of the maxicircle as the cloned *Eco*RI fragments of Brunel *et al.* (1980a). Stuart and Gelvin (1982) cloned 80% of the maxicircle from strain 164 as small fragments in pBR322 and pBR325; the only fragment that Stuart and Gelvin (1982) were unable to clone was the R4-R1 (Fig. 17) fragment from the divergent region.

Riou and Barrois (1981) examined the kDNA from a strain of *T. gambiense* and showed that maxicircle DNA of density 1.684 g/ml was liberated from network DNA by digestion with *Pst*I and that this represented 17% of the network DNA. A restriction map was constructed for seven

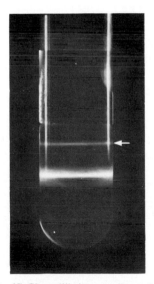

Fig. 18.   Hoechst 33258 dye/CsCl equilibrium gradient of *Eco*RI-digested *L. tarentolae* network kDNA. Ti60 rotor, 48 hour, 40,000 rpm, 20°C; long-wave-length ultraviolet illumination. Arrow indicates low-density band. Reprinted from Simpson (1979) with permission.

enzymes and reassociation kinetic analysis gave a complexity that agreed well with the size of the molecule.

B. *Trypanosoma evansi*

Isolated kDNA networks from this species, which is normally I⁻ in the terminology of Opperdoes *et al.* (1976) and does not undergo development in an insect vector, appear to completely lack maxicircle DNA edge loops in the electron microscope (Fairlamb *et al.*, 1978).

C. *Trypanosoma equiperdum*

This is another I⁻ species which does not undergo development in an insect vector and which has unusual kDNA molecular species. As discussed previously, the minicircle component of the kDNA is homogeneous in terms of sequence. The maxicircle component of the kDNA has been studied from two strains of *T. equiperdum*. The Pasteur strain was shown by Riou and Saucier (1979) to contain maxicircle DNA that is 23 kb in size and to be homogeneous by reassociation kinetic analysis, with a density in CsCl of 1.685 g/ml. A restriction enzyme map of this molecule was similar but not identical to those constructed for *T. brucei* maxicircle DNA. A comparison by C_0t analysis of the extent of sequence homology between the maxicircle DNA from a *T. gambiense* strain and this *T. equiperdum* strain (Riou and Barrois, 1981) indicated approximately 10% base sequence homology. On the other hand, maxicircle DNA isolated from the ATCC strain 30019 of *T. equiperdum* (Frasch *et al.*, 1980) was found to have a size of only 12.6 kb and to have significant sequence homology only with the large *T. brucei Eco*RI fragment (*Eco*1) and a portion of the smallest *Eco*RI fragment (*Eco*3). The cross-species homology was determined by hybridization of Southern blots of *T. equiperdum* maxicircle DNA with the two cloned *T. brucei Eco*RI maxicircle fragments, *Eco*2 and *Eco*3, and with total poly(A)⁺ RNA from *T. brucei*. The conservation of the *Eco*1 *T. brucei* fragment was deduced only from the conservation of the *Eco*RI, *Hpa*II, and *Hin*dIII sites and the overall size; however, new restriction sites are present in the *T. equiperdum* sequence. The differences between the findings of Riou and Saucier (1979) and Frasch *et al.* (1980) possibly indicate an inherent pleomorphism of *T. equiperdum* with regard to the maxicircle DNA. However, it is clear that in general there is a strong correlation between the I⁻ state and defects in the maxicircle DNA.

There is one report by Hadjuk and Vickerman (1981) of the absence of any detectable alteration in the maxicircle DNA of a *T. brucei* strain that was rendered I^- by repeated passage in mice. They were not able to detect any gross changes in restriction sites or fragment sizes, as compared to the I^+ parental strain. This observation does not invalidate the hypothesis since the supposed maxicircle damage may be in the form of point mutations or the damage may be to a nuclear gene.

D. *Trypanosoma mega*

Maxicircle DNA has been identified in *T. mega* (Borst *et al.*, 1977) by the release of restriction fragments from digested kDNA and by the rare appearance of free 24-kb circles in kDNA preparations visualized in the electron microscope. Linearized 26-kb maxicircle molecules could also be released from the network DNA by nuclease S_1 digestion. Large circles of intermediate length between minicircular and maxicircular size were also seen rarely, the identity of which is unknown.

E. *Trypanosoma cruzi*

Maxicircle DNA in *T. cruzi* networks (Leon *et al.*, 1980) was shown to have a size of approximately 39 kb by summation of fragments released by digestion with several enzymes. Direct visualization of intact circular maxicircle DNA in the electron microscope was reported by Kayser *et al.* (1982), who used topoisomerase II from *Micrococcus luteus* to decatenate *T. cruzi* kDNA networks into monomeric minicircles and maxicircles. The size of the released maxicircle molecules was 38 kb, which agreed well with the size determined from gel analysis. *T. cruzi* kDNA networks appear to be relatively fragile to handling in the laboratory, unlike those from *T. brucei* or *Crithidia*, but similar to those from *Phytomonas* and *Leishmania*.

F. *Leishmania tarentolae*

The first indication of the existence of maxicircle DNA in the kDNA of *L. tarentolae* was from the reassociation analysis of Wesley and Simpson (1973b), who showed by C_0t fractionation that total kDNA contained a component that had a complexity more than 10 times that of the minicircle DNA and represented approximately 5% of the total kDNA. Gel analysis of kDNA digests showed the maxicircle size to be approximately 30 kb and to represent 5.4% of the total kDNA network. Maxicircle DNA has been isolated from *L. tarentolae* network DNA by digestion with a single cleaving restriction enzyme, such as *Eco*RI or *Bam*HI, and separated from network DNA and released minicircles by centrifugation in CsCl in

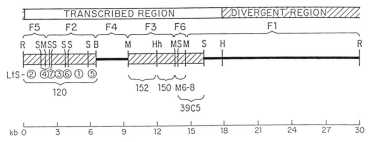

FIG. 19. Partial restriction map of the maxicircle DNA of *L. tarentolae*. The cloned and sequenced fragments within the transcribed region are indicated by cross-hatched boxes and the plasmid numbers 120, 152, 150, M6-8, and 39C5. Restriction site symbols: B, *Bam*HI; H, *Hae*III; Hh, *Hha*I; M, *Msp*I; R, *Eco*RI; S, *Sau*3A. The *Msp*I/*Bam*HI subfragments of the *Eco*RI-linearized maxicircle are designated as F1–F7. Reprinted from de la Cruz *et al.* (1984) with permission.

the presence of Hoechst 33258 (Fig. 18). This method has been proposed as a general method for maxicircle DNA isolation due to the universal density difference found between maxicircle and minicircle DNA (Simpson, 1979). The density difference in *L. tarentolae* is larger than that in *T. brucei* since the minicircle DNA in *L. tarentolae* has a higher %GC than the minicircle DNA in *T. brucei*.

A unique circular restriction enzyme map was constructed for maxicircle DNA liberated from the network by either *Eco*RI or *Bam*HI digestion (Masuda *et al.*, 1979) (Fig. 19). Intramolecular heterogeneity of AT composition was demonstrated first by CsCl buoyant analysis of several *Hpa*II maxicircle fragments (Simpson, 1979) and more precisely by partial denaturation mapping using the electron microscope (Simpson *et al.*, 1982a). The most striking features of the partial denaturation map are the large AT-rich regions located 10–20 kb from the single *Eco*RI site and the six to seven short AT-rich regions located within the divergent region (Fig. 20).

The *in situ* closed circular nature of the maxicircle DNA of *L. tarentolae* was demonstrated by the ability of ethidium bromide to restrict the digestion by *Hpa*II of the intact molecule to only one of the several *Hpa*II sites (Simpson, 1979).

Several maxicircle fragments have been cloned in bacterial plasmids, but a large region has proved so far relatively refractory to cloning. A 6.6-kb *Eco*RI/*Bam*HI fragment (pLt120) was cloned in pBR322, and two adjacent 2.6-kb and 1.9-kb *Hha*I subfragments of the *Msp*I/*Bam*HI fragment 3 were cloned in pBR322 using *Eco*RI linkers (pLt150 and pLt152) (de la Cruz *et al.*, 1984). In addition, the 1.8-kb *Hin*dIII fragment 4 was also cloned in pBR322 (pLt30) (Masuda *et al.*, 1979) and the 0.9-kb *Hin*dIII fragment 5 was cloned in M13mp8 (pDAH54) (D. Hughes and L. Simpson, unpublished results). The pLt120, pLt150 pLt152, pLt30, and PDAH54 inserts were subcloned in M13 and sequenced (de la Cruz *et al.*, 1984; Muhich *et al.*, 1985). The *Msp/Bam* fragments F4, F6, and a portion of F1 have also been cloned and sequenced (L. Simpson, A. M. Simpson, V. de la Cruz, and N. Neckelmann, unpublished results).

G. *Crithidia* SPECIES

Maxicircle DNA was first visualized as rare circular molecules 33 kb in size in *C. luciliae* kDNA preparations spread for electron microscopy (Steinert and Van Assell, 1975). Unequivocal demonstration of maxicircle

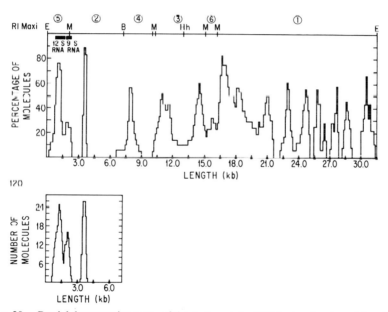

FIG. 20.   Partial denaturation maps of the *L. tarentolae* RI Maxi molecule and the pLt120 insert. The histogram of the RI Maxi is aligned with that of the 120. Both maps were determined by electron microscopy. The 120 map is presented in terms of the numbers of full-length 120 insert molecules showing denaturation at each length increment. However, in the case of the RI Maxi, very few instances of full-length partially denatured molecules were found. Usually molecules were fragmented, frequently in high AT regions. Twenty-eight molecules ranging in size from 17 to 34 kb were measured, length-normalized by the internal standards, and aligned so as to maximize correlation of the pattern of denaturation loops. The percentage of molecules denatured at increments along the length is plotted, since the absolute number of molecules at each length varied due to this fragmentation. The localizations of the *Msp*I and *Bam*HI sites are indicated as are the localizations of the 9 and 12 S RNA genes. Reprinted from Simpson *et al.* (1980) with permission.

DNA as a minor component of kDNA networks was first accomplished by Kleisen *et al.* (1976b) using gel restriction analysis of kDNA from *C. luciliae.* Liberation of once-cleaved maxicircle DNA by nuclease S_1 digestion confirmed the closed circular nature of the molecule *in situ.* The size of 32 kb obtained by restriction analysis agreed well with that obtained by electron microscopy.

From the proportion of minor high-molecular-weight bands in a digest of total kDNA, Kleisen *et al.* (1976b) calculated that 3–5% of the kDNA was maxicircle DNA. An accurate measurement of the relative number of maxicircles and minicircles in a network from *C. fasciculata* was performed by Marini *et al.* (1980) who counted the numbers of such molecules after complete decatenation by T_4 topoisomerase. They found 25 maxicircles and 5785 minicircles in a field of decatenated form I network DNA. Since it can be calculated from the value for DNA per network and the size of the minicircle that there are approximately 5000 *Crithidia* minicircles per network, this follows that there are about 22 maxicircles per network (about 6% of the kDNA). Marini *et al.* (1980) did not observe any intermediate-size circles in their preparation. Weislogel *et al.* (1977) showed that selective removal of maxicircle DNA from the network of *C. luciliae* by digestion with *Pst*I, which cuts the maxicircle DNA once and cleaves only 6% of the minicircle DNA, had little effect on the structural appearance of the network in the electron microscope, showing, as in the case of *T. brucei,* that the network is held together mainly by minicircle catenation.

A restriction map of the maxicircle DNA from *C. luciliae* (Hoeijmakers *et al.,* 1982a) (Fig. 21) has been reported. An interesting observation in this work was that a 6-kb restriction fragment had different electrophoretic mobility in the presence and absence of ethidium bromide (Fig. 21).

A comparison of restriction sites in the maxicircle DNA from *C. luciliae* and *C. fasciculata* showed a sequence homology of more than 96% (Hoeijmakers *et al.,* 1982a), which is similar to the diversity found between maxicircle DNA from different strains of *T. brucei* (Borst *et al.,* 1981a). From this evidence Hoeijmakers *et al.* (1982a) proposed that *C. luciliae* and *C. fasciculata* represent different varieties of the same species rather than different species. This is an interesting approach to kinetoplastid taxonomy, but the conclusion ignores the extensive biological, morphological, and enzymatic evidence that these are separate species.

Maslov *et al.* (1982) have constructed a restriction map of the 24.5-kb maxicircle DNA of *Crithidia oncopelti* and have cloned a 12-kb *Bam*HI fragment in a bacterial plasmid (pCo52). Maslov *et al.* (1984) have constructed detailed restriction maps of maxicircle DNA from *C. luciliae,*

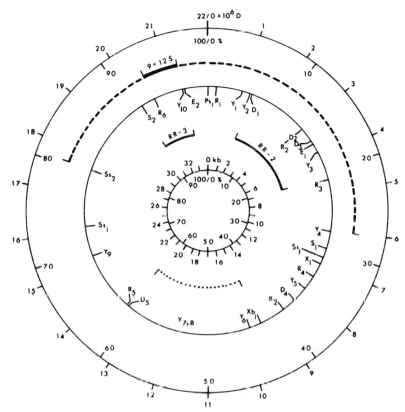

FIG. 21.   The physical map of the maxicircle of *C. luciliae*. The exact order of the *Hap*II fragments Y_6Y_7, Y_7Y_8, and Y_8Y_9 is not determined. Regions that show hybridization with total cellular RNA in Southern blots are indicated on the outside of the circular map. The solid line represents strong hybridization due to the 9 and 12 S RNAs; the dashed line indicates regions showing weak hybridization. The regions hybridizing with the cloned *T. brucei* maxicircle fragments RR-2 and RR-3 (λTbR-2 and TbR-3) are indicated on the inside. The maxicircle segment that is associated with the difference in fragment electrophoretic mobility in the presence or absence of ethidium bromide is indicated by a dotted line on the inside. The map of *C. fasciculata* maxicircle is identical to that of *C. luciliae* except for an additional *Sst*I site (*Sst*$_f$) and some of the *Hap*II sites. A complete *Hap*II map for the *C. fasciculata* maxicircle has not been constructed, but most if not all of the differences are located between map units 30 and 70. D, *Hind*III; E, *Hae*III; Ps, *Pst*I; R, *Eco*RI; S, *Sal*I; Ss, *Sst*I; X, *Xho*I; Xb, *Xba*I; Y, *Hap*II (*Msp*I). Reprinted from Hoeijmakers *et al.* (1982a) with permission.

Leptomanas pessoai, and *Leishmania gymnodactyli* and have compared these molecules by cross hybridization (see Section X,C).

H. *Herpetomonas samuelpessoa*

S. Gomes, H. Van Heuversvyn, R. Mueller, L. Simpson, A. Simpson, and C. Morel (unpublished results) have cloned approximately 50% of the *Pst*I-linearized maxicircle of *H. samuelpessoa* as *Eco*RI fragments in pBR322. One fragment was shown to contain the 9 and 12 S RNA genes and these were subcloned into M13. Gomes and Morel (1982) reported that the cloned *Eco*RI maxicircle fragments of *H. samuelpessoa* exhibit similar anomalous mobility properties in gel electrophoresis as minicircle DNA. All of these fragments gave larger apparent sizes in acrylamide than in agarose, ranging from 25 to 400% increases.

VII. Replication of Maxicircle DNA

A. REPLICATION BY ROLLING-CIRCLE MODEL

Riou and Barrois (1981) reported linear DNA molecules connected to maxicircle edge loops in 6% of the kDNA networks isolated from a strain of *T. gambiense*. They also observed three double-branched edge loop molecules. These structures could represent either broken Cairn's forms or rolling-circle intermediates in the replication of the maxicircle. Similar results were reported by Hajduk *et al.* (1984) with *C. fasciculata* kDNA. They observed rolling-circle-type edge loops in replicating networks and also showed that a population of free maxicircle linears are preferentially labeled by [³H]thymidine in a 20-minute pulse. They suggested that maxicircles replicate by a rolling-circle intermediate, which produces free linear molecules that then recircularize and become catenated with network molecules. The putative origin of replication of the leading strand was localized by the pulse-labeling results to a site within the divergent or nontranscribed region.

B. AUTONOMOUS REPLICATION SEQUENCES IN MAXICIRCLE DNA

Fragments exhibiting autonomous replicating sequence (ars) activity in yeast have been isolated from the maxicircle DNA of *T. brucei* (Davison and Thi, 1982), *C. fasciculata* (Koduri and Ray, 1984), and *L. tarentolae* (Hughes *et al.,* 1984). The *C. fasciculata* fragment has been subcloned to a 189-nt fragment (Cf 189ars) that still retains strong ars activity (Kim and

Ray, 1984). The Cf 189ars fragment was found to be strikingly similar in nucleotide sequence (78% similarity) to a portion of one of the four fragments from the *L. tarentolae* maxicircle that showed strong ars activity (Hughes *et al.*, 1984). The relative localization of this sequence in the *L. tarentolae* maxicircle with respect to the 9 and 12 S ribosomal RNA genes was similar to that in the *C. fasciculata* maxicircle, implying an overall conservation of gene order in these species. Hughes *et al.* (1984) suggested that this sequence represented the origin of replication of the lagging strand of the maxicircle.

VIII. Transcription of Maxicircle DNA

A. *Trypanosoma brucei*

Transcription of the maxicircle DNA of *T. brucei* has been studied by two methods: hybridization of *in vitro* labeled total cell RNA or labeled cDNA to DNA blots and hybridization of labeled cloned maxicircle DNA fragments to RNA blots. In addition, three cDNA clones were shown to hybridize to DNA blots and in two cases the transcripts were identified. Hoeijmakers *et al.* (1981) showed that *in vitro* labeled total cellular RNA (strain 427) hybridized mainly to a 2.0-kb maxicircle fragment from sites D3 to E2. Stuart and Gelvin (1982) showed that labeled cDNA synthesized from total cell RNA templates hybridized mainly to the H1-H2 segment, but in addition to all segments of the maxicircle (strain 164) except the 1.2-kb III-R1 segment. A cDNA probe made from RNA from bloodstream *T. brucei* gave identical results except for a weaker signal.

Northern blot experiments were performed by both Hoeijmakers *et al.* (1981) and Stuart and Gelvin (1982). Hoeijmakers *et al.* (1981) hybridized the two cloned *Eco*RI fragments, *Eco*2 and *Eco*3, which cover approximately 50% of the molecule, to RNA blots and were able to visualize approximately eight RNA species. The *Eco*3 clone hybridized to two major and four minor RNA species, whereas the *Eco*2 clone hybridized to two minor RNAs. The two major RNAs in the *Eco*3 fragment correspond to the 9 and 12 S RNAs first isolated from a purified kinetoplast fraction of *L. tarentolae* (Simpson and Simpson, 1978) and represent the mitochondrial ribosomal RNAs. The minor transcripts were enriched by poly(A) selection and the transcript sizes varied from 1.1 to 0.36 kb. Stuart and Gelvin (1982) confirmed and extended these results, and, by using as probes six smaller cloned maxicircle fragments representing approximately 80% of the molecule and one gel-isolated maxicircle fragment representing the remaining 20%, were able to visualize approximately 18

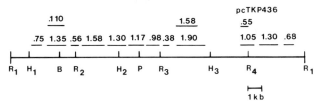

FIG. 22.   Diagram summarizing *T. brucei* 164 maxicircle transcript sizes and the approximate location of their coding sequences on the maxicircle. pcTKP 436 is a cDNA clone containing a 550-bp insert. It hybridized to a 1.05-kb transcript and the H3-R4 maxicircle segment but not to the R4-R1 segment. Reprinted from Stuart and Gelvin (1980) with permission.

total transcripts, some of which are probably identical (Fig. 22). The transcript sizes ranged from 1.9 to 0.45 kb, implying that the RNA preparation examined by Hoeijmakers *et al.* (1981) was somewhat degraded. In no case was any evidence obtained for transfer RNAs, although it is possible that hybridization conditions employed were too stringent.

Three total cell RNA cDNA clones were found by Hoeijmakers *et al.* (1981) to hybridize to maxicircle fragments. Clones Tck-1 and Tck-2 proved to be clones of the same transcript, the 570-bp R-2b transcript from the *Eco*2 region. Another clone, Tck-3, hybridized to a maxicircle region 5′ to the 12 S gene, but no transcript could be seen on Northern blots. Stuart and Gelvin (1982) identified one cDNA clone, pcTKP 436, that represented part of the 1.05-kb transcript which overlaps the R4 site. None of the remaining seven transcripts of Hoeijmakers *et al.* (1981) were localized within the *Eco*2 and *Eco*3 segments, but Stuart and Gelvin (1982) were able to localize the transcripts more precisely since they used smaller DNA probes and several of the transcripts appeared to overlap two adjacent DNA probes (Fig. 22).

No major differences between maxicircle transcripts from bloodstream forms and procyclic forms were observed by Hoeijmakers *et al.* (1981), but conclusions to this effect are precluded by the low resolution of the gels and the apparent degradation of the RNA preparations used. It is clear, however, that the poly(A)$^+$ maxicircle transcripts are significantly lower in abundance in bloodstream forms than in procyclic forms. Stuart and Gelvin (1982) were able to detect a 0.98-kb transcript that seemed to be procyclic-specific. Simpson and Simpson (1980) showed that 9 and 12 S RNAs could be isolated from kinetoplast fractions of both stages of the life cycle of *T. brucei* strain 366D, and they mapped the location of these genes to the identical segment shown by Hoeijmakers *et al.* (1981) and Stuart and Gelvin (1982) to give the major hybridization with total cell RNA.

The relative orientation of the 9 and 12 S RNA genes in *T. brucei* maxicircle DNA was determined by Hoeijmakers *et al.* (1981). This orientation implied a transcription from the 12 S gene to the 9 S gene, a situation unlike all other adjacent ribosomal transcription units known, in which the direction of transcription is from small rRNA to large rRNA.

Expression of the cloned 6-kb *Eco*2 maxicircle fragment of *T. brucei* in *E. coli* minicells was demonstrated by Brunel *et al.* (1980b). Two polypeptides, 10.3 and 13.5 Da, were synthesized under control of a promotor localized within the maxicircle fragment. However, it is possible that these products are prematurely terminated due to the different pattern of codon usage in the mitochondrial genetic system versus that in *E. coli* (Benne *et al.*, 1983; de la Cruz *et al.*, 1984).

B. *Leishmania tarentolae*

Transcription of maxicircle DNA was first demonstrated by hybridization of 9 and 12 S RNAs isolated from a kinetoplast fraction to a discrete region of the maxicircle DNA of *L. tarentolae* (Simpson and Simpson, 1980). These RNAs represented the major steady-state transcripts found in purified kinetoplast fractions and were also the major species labeled *in vivo* in a 1-hour pulse of [³H]uridine. The 9 and 12 S RNAs lacked poly(A) tails, and *in vivo* transcription of these species was inhibited by acriflavin and rifampin. RNA species with identical mobilities in denaturing gels were observed in kinetoplast fractions of *P. davidi* (Cheng and Simpson, 1978) and *T. brucei* (Simpson and Simpson, 1980). Despite the small size of these transcripts, it is fairly certain that these represent the mitochondrial ribosomal RNAs. The evidence for this statement is that they represent the major maxicircle transcripts, that these genes represent the most conserved region of the maxicircle molecule in all species so far examined, and that secondary structures can be constructed which conform well to the *E. coli* rRNA models (de la Cruz *et al.*, 1985a).

At least 10 polyadenylated *L. tarentolae* maxicircle transcripts have been identifed by Northern blots (Simpson *et al.*, 1982a) (Figs. 23 and 24). Transcript sizes ranged from 1.8 to 0.25 kb. The DNA probes were M13 *Sau*3A subclones of the pLt120 region, and gel isolated *Msp*I/*Bam*HI restriction fragments of the purified RI Maxi DNA. Most of the maxicircle was transcriptionally active, although there was an absence of transcripts directly upstream of the 12 S gene and fewer transcripts from the 14.5-kb *Msp/Bam* fragment 1. Muhich *et al.* (1983) have localized the transcription within the *Msp*I/*Bam*HI fragment 1 to the 2.4-kb *Msp*I/*Hae*III subfragment adjacent to fragment 6, with the remaining 12.1 kb of fragment 1 being inactive. Transcription of the pLt120 region and most of the *Msp/*

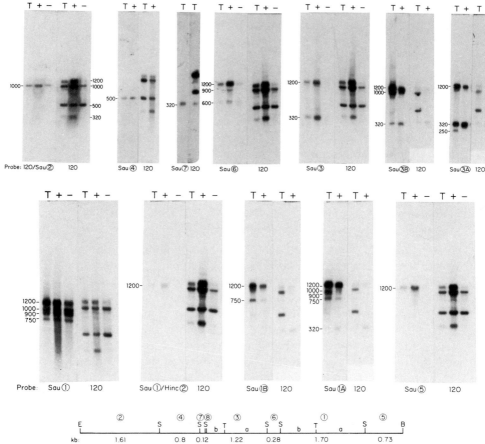

FIG. 23.   Detailed Northern blot analysis of the 120 region of the maxicircle of *L. tarento-lae*. Total kRNA (T), poly(A)⁺-kRNA (+), and, in some cases, poly(A)⁻-kRNA (−) were run in 1.5% agarose formaldehyde gels and blotted onto S∂S BA83 filters which were probed with the indicated DNA fragments. All the probes but Sau (7) were gel-isolated fragments of the pLt120 cloned insert DNA. The Sau (7) probe was the RF from an M13mp7 clone of 120/*Sau*3A fragment 7. Identical results were obtained using as probes RFs of all *Sau*3A M13 clones of the 120 region. The filters were boiled to remove the probes and rehybridized with 120 insert probe and reexposed. Note that several RNA preparations were used for different blots and the extent of poly(A)⁺ selection varied as can be seen by the 120 control hybridizations. The blots are presented in the same sequence as the fragments are localized in the molecule. Reprinted from Simpson *et al.* (1982a) with permission.

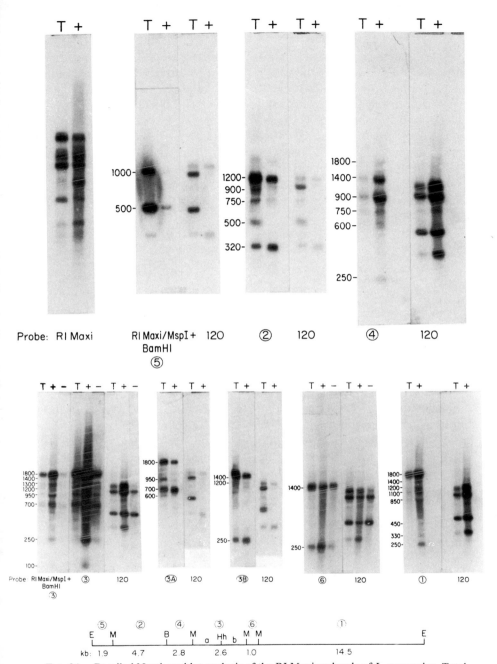

FIG. 24. Detailed Northern blot analysis of the RI Maxi molecule of *L. tarentolae*. Total kRNA (T), poly(A)⁺-kRNA (+), and, in two cases, poly(A)⁻-kRNA (−) were run in 1.5% agarose–formaldehyde gels and blotted onto S∂S BA83 filters which were probed with the indicated *MspI/BamHI* gel-isolated fragments of the RI Maxi. The blots are presented in the same sequence as the fragments are localized in the molecule. The blot of probe F3 was reexposed so as to visualize the low-molecular-weight RNA species. Rehybridizations of the blots were performed with the 120 probe as in Fig. 23. Reprinted from Simpson *et al.* (1982a) with permission.

Bam fragment 3 region was shown to be from one strand and in the direction 12 to 9 S, as in the case of *T. brucei*. However, transcription of the 1.8-kb RNA (the mRNA for the *COI* gene) in the *Msp/Bam* fragment 3 was from the opposite strand (de la Cruz *et al., 1984).*

Identification of specific stable RNAs as the transcripts of identified maxicircle structural genes was accomplished by Northern blot hybridizations using synthetic oligonucleotides, and the 5' ends of the RNAs were determined by primer runoff experiments (Simpson *et al.,* 1985). The 5' ends of the RNAs are located 20–64 nt from the putative translation initiation codons. The extent of RNA processing, if any, is not known.

IX. Kinetoplast Ribosomes

A. ISOLATION OF RIBOSOMES

Isolation of mitochondrial ribosomes from kinetoplastid protozoa has proved to be a difficult and as yet unsolved problem, in spite of earlier reports to the contrary. Hanas *et al.* (1975) reported that mitochondrial ribosomes in *T. brucei* sedimented at 72 S, as compared to the cytoplasmic 84 S ribosomes, and that mitochondrial protein synthesis was blocked selectively by chloramphenicol. Laub-Kuperszteijn and Thirion (1974) reported the mitoribosome of *C. luciliae* sedimented at 60 S, and that 50% of total cellular protein synthesis was chloramphenicol-sensitive. Kleisen and Borst (1975a) were unable to confirm this report, and Spithill *et al.* (1981) showed that chloramphenicol had nonspecific effects on cellular respiration and produced secondary inhibition of both cytoplasmic and putative mitochondrial protein synthesis. The existence of 68 S kinetoplast ribosomes and 23 and 16 S kinetoplast ribosomal RNAs from *C. oncopelti* and *C. fasciculata* was reported in a series of papers from the laboratory of G. N. Zaitseva (Zaitseva *et al.,* 1979a,b). However, there is no confirmatory evidence that these RNAs are transcribed from maxicircle DNA, and, on the other hand, there is evidence for the ribosomal nature of the 9 and 12 S RNAs. Several laboratories have attempted to isolate mitoribosomes from *Crithidia, T. brucei,* and *L. tarentolae* without success. Further work must be performed to resolve this question.

B. DNA SEQUENCES OF 9 AND 12 S RNA GENES

The DNA sequences of the 9 and 12 S RNA genes from *T. brucei* and *L. tarentolae* have been reported (Eperon *et al.,* 1983; de la Cruz *et al.,* 1985a,b). Little sequence homology to known ribosomal genes was observed, although tentative secondary structures could be formed for both species, which agreed well with the *E. coli* 16 and 23 S rRNA models (see below). Comparison of the *T. brucei* and the *L. tarentolae* sequences shows a nucleotide similarity of 84% for the 9 S RNA and 81% for the 12 S RNA. With regard to the unmatched bases, the ratio of transversions to transitions for the 9 S RNAs is 1.66 and for the 12 S RNAs is 2.81; assuming no large base ratio effect, this possibly implies that these two species are separated in evolution by a greater distance than previously presumed, since Cann *et al.* (1984) showed that in the case of primate mitochondrial genes, transitions predominate in recently diverged species, whereas transversions predominate in distantly diverged species.

C. Secondary Structures of 9 and 12 S RNAs

Eperon *et al.* (1983) proposed a tentative partial secondary structure for domain VI of the 12 S RNA from *T. brucei*, in terms of the *E. coli* model of Noller *et al.* (1981). De la Cruz *et al.* (1985a,b) have established the precise 5' and 3' ends of the 9 and 12 S RNAs of both *L. tarentolae* and *T. brucei* and proposed a complete secondary structure for the 9 S (Fig. 25) and a partial structure for the 12 S RNAs which agreed well with the *E. coli* models. The striking aspect of these structures is that many domains are severely reduced or even completely missing, but the overall secondary structure is well conserved in spite of little actual absolute sequence similarity. It is presumed that an analysis of the specific conserved regions in these minimal ribosomal RNAs may lead to a better understanding of the detailed function of ribosomal RNA in general.

X. Genomic Organization of Maxicircle DNA

A. Maxicircle Structural Genes

The first evidence for the existence of nonribosomal structural genes in the maxicircle was obtained by low stringency heterologous hybridization of *L. tarentolae* maxicircle fragments with labeled yeast petite mitochondrial DNA probes known to contain part of all of the genes for cytochrome oxidase subunits I (COI), II (COII), III (COIII), F_0–F_1 ATPases 6 (ATP6) and 9 (ATP9), and cytochrome *b* (CY*b*) (Simpson *et al.*, 1982b). Regions homologous to COI, COII, and CY*b* were localized fairly accurately. The ATP6 probe hybridized to a broad region of the maxicircle and no hybridization was observed with the ATP9 probe. The COIII hybrid melted at a low temperature, implying extensive mismatch. Candidates

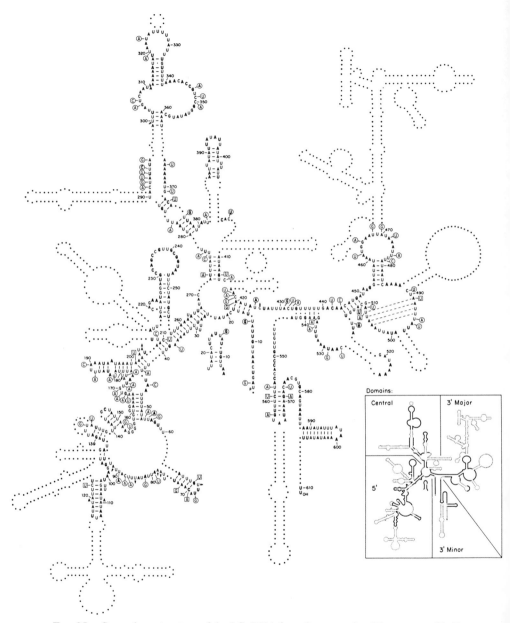

FIG. 25.   Secondary structure of the 9 S rRNA from *L. tarentolae*. The pattern of helices was initially determined by a comparison of the *T. brucei* (Eperon *et al.*, 1983) and *L. tarentolae* sequences using the reversed homology matrix of Pustell and Kafatos (1982) and was confirmed by a comparison with the generalized secondary structure pattern for the *E. coli* 16 S small subunit rRNA (Noller *et al.*, 1982; Maly and Brimacombe, 1983; Woese *et al.*,

for transcripts of these presumptive structural genes were suggested (Simpson *et al.*, 1982a). Johnson *et al.* (1982) used maize and yeast petite COII probes to localize the presumptive structural gene for COII on the maxicircle of *T. brucei*. This localization, however, did not agree with the cross-species hybridization of specific *L. tarentolae* maxicircle fragments performed by Muhich *et al.* (1983), nor with a subsequent sequence analysis by Benne *et al.* (1983). The cross-species hybridization indicated the presence of the *CYb* gene rather than the *COII* gene and this was substantiated by the sequence results of Benne *et al.* (1983) and Johnson *et al.* (1984). The reason for this discrepency is unknown.

The first nonribosomal maxicircle structural gene sequence to be obtained was that for the *CYb* gene of *T. brucei* (Benne *et al.*, 1983; Johnson *et al.*, 1984). Benne *et al.* (1983) and Hensgens *et al.* (1984) have reported the sequence of the entire transcribed region of the *T. brucei* maxicircle. The *COI*, *COII*, and *CYb* genes were identified, in addition to 10 unidentified reading frames and three genes homologous to unidentified reading frames 1, 4, and 5 from the human mitochondrial genome (HURF1, HURF4, and HURF5). The genomic organization, shown in Fig. 26, is similar to that of the *L. tarentolae* maxicircle, as expected from the cross-species hybridization results of Muhich *et al.* (1983).

In the case of *L. tarentolae*, the DNA sequence of approximately 80% of the transcribed region of the *L. tarentolae* maxicircle is known (de la Cruz *et al.*, 1984; L. Simpson, A. M. Simpson, V. de la Cruz, and N. Neckelmann, unpublished results). Seven structural genes were identified in this sequence; the genomic localizations of the *COI*, *COII*, *COIII*, and *CYb* genes agreed with those predicted by the heterologous hybridizations. In addition to the genes for COI, COII, COIII, and CYb, three open reading frames were identified as homologous to unidentified reading frames 1, 4, and 5 from the human mitochondrial genome (HURF1, HURF4, and HURF5). Several unidentified open reading frames were also observed. The *L. tarentolae* genes were localized by comparison of the translated amino acid sequences with those of known mitochondrial genes. The *ATP6*, *ATP8*, and *ATP9* genes have not been identified within

---

1983). Nucleotides conserved between the 9 S sequences and *E. coli* are shown in bold face. Nucleotide positions that are not represented by the 9 S RNA on the *E. coli* model are indicated by small circles. The numbers refer to the *L. tarentolae* sequence, the 5' nt being position one. Mismatches in the *T. brucei* sequence are indicated adjacent to the *L. tarento-lae* sequence. Nucleotides within circles are noncompensatory changes with regard to base pairing, and nucleotides within boxes are compensatory changes. Deletions in the *T. brucei* sequence are indicated by solid triangles and insertions are indicated by lines pointing between the appropriate *L. tarentolae* nucleotides. The locations of the four domains are indicated in the inset. Reprinted from de la Cruz *et al.* (1985a) with permission.

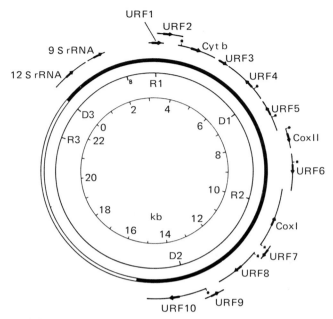

FIG. 26. Partial gene map of *T. brucei* maxicircle DNA. The arrangement of genes is derived from various studies. Arrows indicate direction of transcription of each gene. The black bar indicates the sequenced area (B, *Bam*HI). Reprinted from Hensgens *et al.* (1984) with permission.

the maxicircle sequences reported to date. The maxicircle genes showed different levels of sequence conservation as compared to mammalian and fungal sequences, with *COI* being the most conserved and *COIII* the least conserved. The genomic organization of the published, identified, and sequenced *L. tarentolae* maxicircle genes is shown in Fig. 27. This organization is unique and unlike that in either mammalian, insect, or fungal systems.

From a comparison of the translated amino acid sequences of the maxicircle genes with the sequences of known proteins from other organisms, Benne *et al.* (1983), Hensgens *et al.* (1984), and de la Cruz *et al.* (1984) concluded that UGA = tryptophan is the only apparent deviation from the universal genetic code.

A striking feature of the identified maxicircle structural genes is a pronounced strand asymmetry; the T:A ratio of the coding strand is approximately 2.0 for the six identified genes and also for several open reading frames (de la Cruz *et al.*, 1984).

Transcripts of the structural genes in both *L. tarentolae* and *T. brucei* have been identified and the 5' ends mapped (Benne *et al.*, 1983; Simpson *et al.*, 1982a, 1985). Unlike the human mitochondrial situation where the 5′ end of the transcript is flush with the initiation codon, in *L. tarentolae* there are substantial 5′-untranslated sequences in all the maxicircle transcripts. There are no indications of intervening sequences in any of the identified structural genes.

There is, however, a −1 frameshift in the C-terminal portion of the COII reading frame, which is present at the same location in the maxicircle DNA sequences from both *T. brucei* and *L. tarentolae* (Hensgens *et al.*, 1984; de la Cruz *et al.*, 1984). This conserved frameshift is not understood. Possible explanations include the presence of a pseudogene, or a specific insertion of U's at the RNA level (R. Benne, personal communication).

FIG. 27. Geonomic organization of the maxicircle DNA of *L. tarentolae*. The nontranscribed divergent region is indicated by cross-hatching and the identified genes are indicated by dark shading. Unidentified open reading frames (ORFs) are blank. The portions of the transcribed region that have not yet been sequenced are indicated by stipling. All identified genes are transcribed left to right except for COI, which is transcribed in the reverse direction (arrow). The identified genes are the large and small ribosomal RNAs (12 and 9 S RNAs), subunits I, II, and III of the cytochrome oxidase complex (COI, COII, and COIII), cytochrome *b* (CY*b*), and sequences homologous to human mitochondrial unidentified reading frames 4 and 5 (HURF4 and HURF5). Open reading frames which are putative genes are indicated as ORF1–10. Reprinted from de la Cruz *et al.* (1984) with permission.

B. Evolution of Maxicircle DNA: Structural and Ribosomal Genes

Conservation of the 9 and 12 S ribosomal genes has been demonstrated for several species including *L. tarentolae, T. brucei, C. fasciculata,* and *C. luciliae.* Cross-species hybridization of maxicircle sequences outside the 9 and 12 S regions has also been demonstrated, but the homologies are less striking. Hoeijmakers *et al.* (1982a) used the *T. brucei Eco2* and *Eco3* maxicircle probes and found adjacent regions in the *C. luciliae* maxicircle showing hybridization with the *T. brucei* probes. Muhich *et al.* (1983) examined in detail sequence homologies between maxicircle DNA of *T. brucei* and *L. tarentolae,* which differ by almost 10 kb in size. They found that the genomic organizations of the transcribed regions were similar, but that the nontranscribed regions were completely nonhomologous. Comparison of the *CYb* gene sequences from the *L. tarentolae* and *T. brucei* maxicircles shows a nucleotide similarity of 83% and an amino acid identity of 84%. Of the nonidentical codons, 26% are third base substitutions, 96% of which are silent. In regard to nucleotide changes, the transversion/transition ratio is 1.0, again possibly implying a distant divergence time for these species.

One striking difference between the *L. tarentolae* and the *T. brucei* maxicircle genomes is the absence of the *COIII* gene in *T. brucei.* De la Cruz *et al.* (1984) were unable to detect this gene in *T. brucei* maxicircle DNA by hybridization, using a *L. tarentolae* maxicircle gene probe, and the gene has not yet been identified in the *T. brucei* maxicircle sequences (Benne *et al.,* 1983; Hensgens *et al.,* 1984). It may be relevant that the *L. tarentolae COIII* gene is the most diverged of the identified maxicircle genes, as compared to the human and yeast homologous genes.

C. Evolution of Maxicircle DNA: Divergent Region

The first indication of the existence of a region in the maxicircle that served a noncoding function was the finding that maxicircle DNA from different strains of *T. brucei* differed by as much as 1 kb, and that this size variation was localized in one region of the maxicircle that was termed the "variable" region (Borst *et al.,* 1980a, 1981a). This region appeared to be "nonclonable" in bacterial plasmids or phage vectors and was found to show little if any transcription (Brunel *et al.,* 1980; Stuart and Gelvin, 1980; Simpson and Simpson, 1980).

Muhich *et al.* (1983) compared the genomic organization of the maxicircle DNA of *T. brucei* and *L. tarentolae* and found that there was one large region of nonhomology. This region was within the *Msp/Bam* fragment 1, which corresponds in relative location to the *T. brucei* variable region. They concluded that evolutionary changes in maxicircle DNA occur mainly in one region, which they termed the "divergent region," and probably involve insertions and deletions. A similar conclusion was reached by Maslov *et al.* (1984) who cross hybridized maxicircle fragments from *C. oncopelti, C. luciliae, Lp. pessoai,* and *L. gymnodactyli,* using labeled maxicircle fragments from *C. oncopelti* as probes. Maslov *et al.* (1984) constructed a general model of the structural organization of maxicircles in which the molecule is composed of a 17-kb conservative region common to all species and a divergent region represented by the remainder of the molecule. They conclude that the major size differences between the maxicircle molecules from different species are due to length variability of the divergent regions.

Two contiguous fragments totaling 2.76 kb from the divergent region of *L. tarentolae* have been cloned and sequenced (M. Muhich, N. Neck-elmann, and L. Simpson, unpublished results). Extensive tandem repeats were seen, both short (AAATT, AATAATAT) and long (ATAT-TAAAACAAGTTATTCCC), some of which are present throughout the divergent region as shown by blot hybridization. The presence of these repetitive sequences is consistent with the lack of transcription of this region and the observed rapid rate of sequence change in nature. Similar conclusions were reached by an electron microscopic analysis of hetero-duplexes formed with maxicircle DNA of *T. brucei* (Borst *et al.*, 1982). The molecular mechanism for the rapid rate of sequence change of the divergent region is unknown, but any mechanism must take into account the existence of multiple tandem repeats.

XI. Unusual Kinetoplast DNAs

The minicircle/maxicircle organization pattern of kinetoplast DNA appears to be correlated to the existence of the highly organized kinetoplast DNA nucleoid bodies observed in thin sections of intact cells by electron microscopy. Several kinetoplastid protozoa, however, lack this characteristic nucleoid body structure in the mitochondrion and instead possess a structure reminiscent of the bacterial nucleus consisting of one or more masses of "isotropically arranged bundles of isotropic fibrils" (Vickerman, 1977). This condition has been termed "pankinetoplastic" by Vickerman (1977) and has been observed in the free-living kinetoplastid, *Bodo,* as well as in the parasitic kinetoplastids, *Cryptobia vaginalis* and *Herpetomonas ingenoplastum.* The mitochondrial DNA of *H. ingenoplastum* has been isolated and characterized (Borst *et al.*, 1981b). It consists of two components, a minor class of 36-kb circles, which show sequence homology with the 9 and 12 S genes of *T. brucei,* and a major class of 17- and 23-kb circles, which exhibit sequence heterogeneity and have no sequences in common with *Herpetomonas muscarum* maxicircle or minicircle DNA. It was hypothesized that the 36-kb circles represent defective maxicircles and the 17- and 23-kb circles represent an evolutionary precursor of the small minicircles found in other kinetoplastids.

XII. Conclusions

The minicircle represents the unique aspect of the mitochondrial DNA of the kinetoplastid protozoa, but the significance of this DNA is not yet understood, although much is now known about its structure and replication. In spite of the lack of understanding of the function of the minicircle DNA, the extensive sequence divergence observed to occur in nature has been employed to distinguish different strains and species of these parasites. The maxicircle DNA, on the other hand, represents the informational molecule in the mitochondrion of the kinetoplastid protozoa and has been shown to contain sequences homologous to several known mitochondrial structural genes. The basic genomic organization of the maxicircle DNA has been shown to be conserved in representative species from several kinetoplastid genera, with evolutionary changes probably occurring mainly by insertions/deletions and rearrangements within one region of the molecule, which we have termed the "divergent region."

The genomic organization of the maxicircle DNA has some similarities with the organization of the human mitochondrial genome and the yeast mitochondrial genome, but it is unique in several respects, implying that the evolution of mitochondrial genomes has progressed along several pathways in different organisms but that the basic informational content has remained fairly constant. The high divergence of the identified maxicircle genes from those of mammalian, fungal, and insect systems implies a relatively ancient branching of the kinetoplastid protozoa from other eukaryotic lines, an interpretation consistent with the phylogenetic analysis from cytochrome *c* data by Schwartz and Dayhoff (1978).

There has been a large amount of progress in our understanding of the kinetoplast genome in the last few years, but much remains to be investigated. I would make the following suggestions as possible fruitful lines for future research:

1. Evolution of minicircle sequence diversity. Comparison of rates of change of the constant and variable regions. Mechanisms for the generation of sequence changes (e.g., recombination, mutation).

2. Examination of the possibility of recombination between minicircles and maxicircles, especially within the divergent region of the maxicircle.

3. Investigation of the role, if any, of the minicircle sequence-dependent "bending" in terms of packaging of DNA in the nucleoid body or interaction with specific proteins.

4. Investigation of the protein apparently bound to minicircle DNA in terms of sequence localization and function.

5. Isolation and characterization of the mitochondrial topoisomerases responsible for decatenation of closed minicircle and catenation of nicked minicircles.

6. Investigation of the *Crithidia* minicircle open reading frames apparently responsible for the production of three polypeptides localized to the vicinity of the flagellar pocket.

7. Investigation of the replication of maxicircle DNA.

8. Comparative study of the detailed maxicircle genomic organization in representative species of several genera.

9. Examination of the evolution of specific mitochondrial genes, such as the ribosomal genes, by comparison of sequences from representative species of several genera.

10. Examination of the possibility of regulation of maxicircle transcription in different stages of the life cycle, particularly in the African trypanosomes.

11. Examination of the possibility of processing of maxicircle transcripts.

12. Examination of the possibility of interactions between the kinetoplast and nuclear genomes during the mitochondrial biogenesis that occurs in the life cycle of *T. brucei*.

13. Molecular events occurring during the development of dyskinetoplastic cells after treatment with dyes.

14. Characterization of the mitochondrial DNA in "pankinetoplastic" cells, including free-living species.

Acknowledgments

This work was supported by grants from the NIH and the NSF-INT. I would like to thank K. Stuart, J. Davidson, P. Englund, and C. Morel for sending me preprints of research papers.

References

Arnot, D., and Barker, D. (1981). *Mol. Biochem. Parasitol.* **3,** 47–56.

Avise, J. C., Giblin-Davidson, C., Learm, J., Patton, J. C., and Lansman, R. A. (1979a). *Proc. Natl. Acad. Sci. U.S.A.* **76,** 6694–6698.

Avise, J. C., Lansman, R. A., and Shade, R. O. (1979b). *Genetics* **92,** 279–295.

Barker, D., Arnot, D., and Butcher, J. (1982). *In* "Biochemical Characterization of *Leishmania*" (M. Chance and B. Walton, eds.), pp. 139–180. UNDP/World Bank/WHO, Geneva.

Barrois, M., Riou, G., and Galibert, F. (1982). *Proc. Natl. Acad. Sci. U.S.A.* **78,** 3323–3327.

Benne, R., De Vries, B., Van den Berg, J., and Klaver, B. (1983). *Nucleic Acids Res.* **11,** 6925–6941.

Borst, P., and Fase-Fowler, F. (1979). *Biochim. Biophys. Acta* **565,** 1–12.

Borst, P., and Hoeijmakers, J. (1979). *Plasmid* **2,** 20–40.

Borst, P., Fase-Fowler, F., Steinert, M., and Van Assel, S. (1977). *Cell Res.* **110,** 167–173.

Borst, P., Fase-Fowler, F., Hoeijmakers, J., and Frasch, A. (1980a). *Biochim. Biophys. Acta* **610,** 197–210.

Borst, P., Fase-Fowler, F., Frasch, A., Hoeijmakers, J., and Weijers, P. J. (1980b). *Mol. Biochem. Parasitol.* **1,** 221–246.

Borst, P., Hoeijmakers, J., Frasch, A., Snijders, A., Janssen, J., and Fase-Fowler, F. (1980c). *In* "The Organization and Expression of the Mitochondrial Genome" (A. Kroon and C. Saccone, eds.), pp. 7–19. Elsevier, Amsterdam.

Borst, P., Fase-Fowler, F., and Gibson, W. (1981a). *Mol. Biochem. Parasitol.* **3,** 117–131.

Borst, P., Hoeijmakers, J., and Hajduk, S. (1981b). *Parasitology* **82,** 81–93.

Borst, P., Weijers, P. J., and Brakenhoff, G. J. (1982). *Biochim. Biophys. Acta* **699,** 272–280.

Brunel, F., Davison, J., Merchez, M., Borst, P., and Weijers, P. (1980a). *In* "DNA-Recombination, Intermediates and Repair (FEBS)" (S. Zadrazil and J. Sponar, eds.), pp. 45–54. Pergamon, Oxford.

Brunel, F., Davison, J., and Merchez, M. (1980b). *Gene* **12,** 223–234.

Calladine, C. R. (1982). *J. Mol. Biol.* **161,** 343–352.

Camargo, E., Mattei, D., Barbieri, C., and Morel, C. (1981). *J. Protozool.* **29,** 251–258.

Cann, R. L., Brown, W. M., and Wilson, A. C. (1984). *Genetics* **106,** 479–499.

Challberg, S., and Englund, P. (1980). *J. Mol. Biol.* **138,** 447–472.

Chen, K., and Donelson, J. (1980). *Proc. Natl. Acad. Sci. U.S.A.* **77,** 2445–2449.

Cheng, D., and Simpson, L. (1978). *Plasmid* **1,** 297–315.

Chugonov, V., Shirshov, A., and Zaitseva, G. (1971). *Biokhimiya* **36,** 630–635.

Cuthbertson, R. S. (1981). *J. Protozool.* **28,** 182–188.

Davison, J., and Thi, V. (1982). *Curr. Genet.* **6,** 19–20.

de la Cruz, V., Neckelmann, N., and Simpson, L. (1984). *J. Biol. Chem.* **251,** 15136–15147.

de la Cruz, V., Lake, J., Simpson, A. M., and Simpson, L. (1985a). *Proc. Natl. Acad. Sci. U.S.A.* **82,** 1401–1405.

de la Cruz, V., Simpson, A., Lake, J., and Simpson, L. (1985b). *Nucleic Acids Res.* **13,** 2337–2356.

Donelson, J., Majiwa, F., and Williams, R. (1979). *Plasmid* **2,** 572–588.

Dressler, D. (1970). *Proc. Natl. Acad. Sci. U.S.A.* **67,** 1934–1942.

Englund, P. (1978). *Cell* **14,** 157–168.

Englund, P. (1981). *In* "Biochemistry and Physiology of Protozoa" (M. Levandowsky and S. H. Hutner, eds.), 2nd Ed., Vol. 4, pp. 333–381. Academic Press, New York.

Englund, P., Hajduk, S., and Marini, J. (1982). *Annu. Rev. Biochem.* **51,** 695–726.

Eperon, I., Janssen, J., Hoeijmakers, J., and Borst, P. (1983). *Nucleic Acids Res.* **11**, 105–125.

Fairlamb, A., Weislogel, P., Hoeijmakers, J., and Borst, P. (1978). *J. Cell Biol.* **76**, 293–309.

Fouts, D., Manning, J., and Wolstenholme, D. (1975). *J. Cell Biol.* **67**, 378–399.

Fouts, D., and Wolstenholme, D. (1979). *Nucleic Acids Res.* **6**, 3785–3804.

Frasch, A., Goijman, S., Cazzulo, J., and Stoppani, A. (1981). *Mol. Biochem. Parasitol.* **4**, 163–170.

Frasch, A., Hajduk, S., Hoeijmakers, J., Borst, P., Brunel, F., and Davison, J. (1980). *Biochim. Biophys. Acta* **607**, 397–410.

Gomes, S., and Morel, C. (1982). *Annu. Meet. Basic Res. Chagas Dis., 9th, Caxambu, Brazil,* p. 92.

Hagerman, P. (1984). *Proc. Natl. Acad. Sci. U.S.A.* **81**, 4632–4636.

Hajduk, S., and Cosgrove, W. (1979). *Biochim. Biophys. Acta* **561**, 1–9.

Hajduk, S., and Vickerman, K. (1981). *Mol. Biochem. Parasitol.* **4**, 17–28.

Hajduk, S., Klein, V., and Englund, P. (1984). *Cell* **36**, 463–492.

Hanas, J., Linden, G., and Stuart, K. (1975). *J. Cell Biol.* **65**, 103–111.

Handman, E., Hocking, R., Mitchell, G., and Spithill, T. (1983). *Mol. Biochem. Parasitol.* **7**, 111–126.

Hensgens, L. A. M., Brakenhoff, J., De Vries, F., Sloff, P., Tromp, M. C., Van Boom, J. H., and Benne, R. (1984). *Nucleic Acids Res.* **12**, 7327–7344.

Hoeijmakers, J., and Borst, P. (1978). *Biochim. Biophys. Acta* **521**, 407–411.

Hoeijmakers, J., and Borst, P. (1982). *Plasmid* **7**, 210–220

Hoeijmakers, J., Snijders, A., Janssen, J., and Borst, P. (1981). *Plasmid* **5**, 329–350.

Hoeijmakers, J., Schoutsen, B., and Borst, P. (1982a). *Plasmid* **7**, 199–209.

Hoeijmakers, J., Weijers, P., Brakenhoff, C., and Borst, P. (1982b). *Plasmid* **7**, 221–229.

Hughes, D., Simpson, L., Kayne, P., and Neckelmann, N. (1984). *Mol. Biochem. Parasitol.* **13**, 263–275.

Johnson, B., Hill, G., Fox, T., and Stuart, K. (1982). *Mol. Biochem. Parasitol.* **5**, 381–390.

Johnson, B., Hill, G., and Donelson, J. (1984). *Mol. Biochem. Parasitol.* **13**, 135–146.

Kayser, A., Douc-Rasy, S., and Riou, G. (1982). *Biochimie* **64**, 285–288.

Kidane, G., Hughes, D., and Simpson, L. (1984). *Gene* **27**, 265–277.

Kitchin, P. A., Klein, V. A., Fein, B. I., and Englund, P. T. (1984). *J. Biol. Chem.* **259**, 15532–15539.

Kitchin, P. A., Klein, V. A., and Englund, P. T. (1985). *J. Biol. Chem.* **260**, 3844–3851.

Kleisen, C., and Borst, P. (1975a). *Biochim. Biophys. Acta* **390**, 78–81.

Kleisen, C., and Borst, P. (1975b). *Biochim. Biophys. Acta* **407**, 473–478.

Kleisen, C., Borst, P., and Weijers, P. (1976a). *Eur. J. Biochem.* **64**, 141–151.

Kleisen, C., Weislogeln, P., Fonck, K., and Borst, P. (1976b). *Eur. J. Biochem.* **64**, 153–160.

Koduri, R., and Ray, D. (1984). *Mol. Biochem. Parasitol* **10**, 151–169.

Laub-Kupersztejn, R., and Thirion, J. (1974). *Biochim. Biophys. Acta* **340**, 314–322.

Leon, W., Frank, A., Hoeijmakers, J., Fase-Fowler, F., Borst, P., Brunel, F., and Davison, J. (1980). *Biochim. Biophys. Acta* **607**, 221–231.

Lopes, U., Momen, H., Grimaldi, G., Marzochi, M., Pachedo, R., and Morel, C. (1984). *J. Parasitol.* **70**, 89–98.

Maly, P., and Brimacombe, R. (1983). *Nucleic Acids Res.* **11**, 7263–7286.

Manning, J., and Wolstenholme, A. (1976). *J. Cell Biol.* **70**, 406–418.

Marini, J., Miller, K., and Englund, P. (1980). *J. Biol. Chem.* **255**, 4976–4977.

Marini, J., Levene, S., Crothers, J., and Englund, P. (1982). *Proc. Natl. Acad. Sci. U.S.A.* **79**, 7664–7668.

Marini, J., Levene, S., Crothers, D., and Englund, P. (1983). *Proc. Natl. Acad. Sci. U.S.A.* **24,** 7678.

Maslov, D., Entelis, N., Kolesnikov, A., and Zaitseva, G. (1982). *Bioorg. Chem.* **8,** 676–685.

Maslov, D., Kolesnikov, A., and Zaitseva, G. (1984). *Mol. Biochem. Parasitol.* **12,** 351–364.

Masuda, H., Simpson, L., Rosenblatt, H., and Simpson, A. M. (1979). *Gene* **6,** 51–73.

Mattei, D. M., Goldenberg, S., Morel, C., Azevedo, H. P., and Roitman, I. (1977). *FEBS Lett.* **74,** 264–268.

Morel, C., Chiari, E., Camargo, E., Mattei, D., Romanha, A., and Simpson, L. (1980). *Proc. Natl. Acad. Sci. U.S.A.* **77,** 6810–6814.

Muhich, M., Simpson, L., and Simpson, A. M. (1983). *Proc. Natl. Acad. Sci. U.S.A.* **80,** 4060–4064.

Muhich, M., Necklemann, N., and Simpson, L. (1985). *Nucleic Acids Res.* **13,** 3241–3260.

Noller, H. F., Kop, J., Wheaton, V., Brosius, J., Gutell, R. R., Kipylov, A. M., Dohme, F., Her, W., Stahl, D. A., Gupta, R., and Woese, C. R. (1981). *Nucleic Acids Res.* **9,** 6167–6189.

Ntambi, J., and Englund, P. (1985). *J. Biol. Chem.* **260,** 5574–5579.

Ntambi, J. M., Marini, J. C., Bangs, J. D., Hajduk, S. L., Jimenez, H. E., Kitchin, P. A., Klein, V. A., Ryan, K. A., and Englund, P. T. (1984). *Mol. Biochem. Parasitol.* **12,** 273–286.

Opperdoes, F., Borst, P., and de Rijke, D. (1976). *Comp. Biochem. Physiol.* **55,** 25–30.

Ponzi, M., Birago, C., and Battaglia, P. A. (1984). *Mol. Biochem. Parasitol.* **13,** 111–119.

Pustell, J., and Kafatos, F. T. (1982). *Nucleic Acids Res.* **10,** 4765–4782.

Renger, H., and Wolstenholme, D. (1972). *J. Cell Biol.* **54,** 346–364.

Riou, G., and Barrois, M. (1979). *Biochem. Biophys. Res. Commun.* **90,** 405–409.

Riou, G., and Barrois, M. (1981). *Biochimie* **63,** 755–765.

Riou, G., and Gutteridge, W. (1978). *Biochimie* **60,** 365–379.

Riou, G., and Pautrizel, R. (1977). *Biochem. Biophys. Res. Commun.* **79,** 1084–1091.

Riou, G., and Saucier, J. (1979). *J. Cell Biol.* **82,** 248–263.

Riou, G., and Yot, P. (1975). *C. R. Acad. Sci. Paris* **280,** 2701.

Riou, G. P., and Yot, P. (1977). *Biochemistry* **16,** 2390–2396.

Riou, G., Baltz, Th., Gabillet, M., and Pautrizel, R. (1980). *Mol. Biochem. Parasitol.* **1,** 97–106.

Riou, G., Gabillot, M., Douc-Rasy, S., and Kayser, A. (1982). *C. R. Acad. Sci. Ser. 3* **294,** 439–442.

Sanchez, D., Frasch, A., Carrasco, A., Gonzalez-Cappa, S., Isola, E., and Stopanni, A. (1984). *Mol. Biochem. Parasitol.* **11,** 169–178.

Saucier, J., Bevard, J., da Silva, J., and Riou, G. (1981). *Biochem. Biophys. Res. Commun.* **101,** 988–994.

Schwartz, R. M., and Dayhoff, M. O. (1978). *Science* **199,** 395–403.

Sholmai, J., and Zadok, A. (1983). *Nucleic Acids Res.* **11,** 4019–4034.

Sholmai, J., and Zadok, A. (1984). *Nucleic Acids Res.* **12,** 8017–8028.

Simpson, A., and Simpson, L. (1980). *Mol. Biochem. Parasitol.* **2,** 93–108.

Simpson, A. M., and Simpson, L. (1974). *J. Protozool.* **21,** 774–781.

Simpson, A. M., and Simpson, L. (1976). *J. Protozool.* **23,** 583–587.

Simpson, A., Neckelmann, N., de la Cruz, V., Muhich, M., and Simpson, L. (1985). *Nucleic Acids Res.* **13,** 5977–5993.

Simpson, L. (1979). *Proc. Natl. Acad. Sci. U.S.A.* **76,** 1585–1588.

Simpson, L., and da Silva, A. (1971). *J. Mol. Biol.* **56,** 443–473.

Simpson, L., and Simpson, A. (1978). *Cell* **14,** 169–178.

Simpson, L., Simpson, A. M., and Wesley, R. D. (1974). *Biochim. Biophys. Acta* **349,** 161–172.

Simpson, L., Simpson, A. M., Masuda, H., Rosenblatt, H., Michael, N., and Kidane, G. (1979). *ICN–UCLA Symp.* **15,** 533–548.

Simpson, L., Simpson, A., Kidane, G., Livingston, L., and Spithill, T. (1980). *Am. J. Trop. Med. Hyg.* **29** (Suppl.), 1053–1063.

Simpson, L., Simpson, A., and Livingston, L. (1982a). *Mol. Biochem. Parasitol.* **6,** 237–257.

Simpson, L., Spithill, T., and Simpson, A. M. (1982b). *Mol. Biochem. Parasitol.* **6,** 253–263.

Sloof, P., Menke, H., Caspers, M., and Borst, P. (1983). *Nucleic Acids Res.* **11,** 3889–3901.

Spithill, T., Shimer, S., and Hill, G. (1981). *Mol. Biochem. Parasitol.* **2,** 235–256.

Steinert, M., and Van Assell, S. (1975). *Exp. Cell Res.* **96,** 406–409.

Steinert, M., and Van Assel, S. (1980). *Plasmid* **3,** 7–17.

Steinert, M., Van Assell, S., Borst, P., and Newton, B. (1976). *In* "The Genetic Function of Mitochondrial DNA" (C. Saccone and A. Kroon, eds.), pp. 71–81. Elsevier, Amsterdam.

Stuart, K. (1979). *Plasmid* **2,** 520–528.

Stuart, K., and Gelvin, S. (1980). *Am. J. Trop. Med. Hyg.* **29** (Suppl.), 1075–1081.

Stuart, K., and Gelvin, S. (1982). *Mol. Cell. Biol.* **2,** 845–852.

Van Heuversvyn, H., Mueller, R., Calcagnotto, A. M., Cardoso, M. A. B., Pereira, J. L. A., and Morel, C. M. (1982). *Abstr. Congr. Arg. Microbiol. 3rd, Buenos Aires, Aug.* No. 152.

Vickerman, K. (1977). *J. Protozool.* **24,** 221–233.

Weislogel, P., Hoeijmakers, J., Fairlamb, A., Kleisen, C., and Borst, P. (1977). *Biochim. Biophys. Acta* **478,** 167–179.

Wesley, R. D., and Simpson, L. (1973a). *Biochim. Biophys. Acta* **319,** 254–266.

Wesley, R. D., and Simpson, L. (1973b). *Biochim. Biophys. Acta* **319,** 267–276.

Wirth, D., and Pratt, D. (1982). *Proc. Natl. Acad. Sci. U.S.A.* **79,** 6999–7003.

Woese, C., Gutell R., Gupta, R., and Noller, H. (1983). *Microbiol. Rev.* **47,** 621–669.

Wu, H., and Crothers, D. (1984). *Nature (London)* **308,** 509–513.

Zaitseva, G., Mett, I., and Mett, A. (1979a). *Cytology* **21,** 310–317.

Zaitseva, G., Mett, I., Maslov, D., Lunina, L., and Kolensnikov, A. (1979b). *Biokhimiya* **44,** 2073–2082.

INTERNATIONAL REVIEW OF CYTOLOGY, VOL. 99

Chlamydomonas reinhardtii: A Model System for the Genetic Analysis of Flagellar Structure and Motility

BESSIE PEI-HSI HUANG

Department of Cell Biology, Baylor College of Medicine, Houston, Texas

Introduction

Eukaryotic cilia and flagella are complex microtubule-based motile organelles whose movements are characterized by the passage of waves of bending along their length. Among the numerous microtubule-based systems of movement that have been studied, the structural organization and mechanochemistry of cilia and flagella have been the most extensively characterized.

Cilia and flagella from different organisms and cell types exhibit a diverse array of propagated bending patterns (for review, see Sleigh, 1974). However, regardless of their origin, they share a common structural organization. The core structure, known as the flagellar axoneme, consists of a cylinder of nine doublet microtubules surrounding a pair of singlet mi-

181

crotubules. This is the classical "9+2" flagellar microtubular organization. Attached to the microtubules are functionally important accessory structures which occur with precise longitudinal periodicities.

A large body of experimental evidence has established that the bending movements of cilia and flagella are the result of localized sliding displacements between the doublet microtubules of the axoneme (Satir, 1968; Summers and Gibbons, 1971, 1973). The process appears to be analogous to the sliding filament model of muscle contraction, and the motive force is believed to be generated by intermittent cross bridging of adjacent doublets by dynein-ATPase-containing arms (Gibbons and Gibbons, 1973, 1974).

Although much progress which has been made over the past three decades in elucidating the basic structure and mechanochemistry of cilia and flagella, there are still many unanswered questions. For example, little is known about the temporal and spatial regulation of ATP-driven sliding within the flagellar axoneme, the mechanics and chemistry of how sliding displacements are converted into propagated flagellar bending, and the regulation of the assembly of flagellar proteins into the precise organization of the axoneme.

In prokaryotic cells, and, more recently in several eukaryotic systems, genetics has been an important approach for understanding the molecular processes involved in the assembly and function of complex multimolecular organelles. Primarily due to the availability of a large number and variety of mutants, some of the highly interdependent processes responsible for bacterial and viral function and development have been elucidated. The feasibility of a similar combined genetic and biochemical approach to understanding flagellar motility and development has been demonstrated with the unicellular biflagellate alga *Chlamydomonas reinhardtii*. The usefulness of *Chlamydomonas* for attacking questions of flagellar structure and function was first indicated more than 25 years ago by Lewin (1954) when he isolated a series of *Chlamydomonas* mutants with specific defects for flagellar motility. Since then, a variety and large number of additional flagellar mutants in *C. reinhardtii* have been isolated. At the same time, a substantial body of information on the structure, biochemistry, assembly, and functional properties of wild-type *Chlamydomonas* flagella has been obtained.

In this article, the characteristics of *C. reinhardtii* that make it uniquely useful for genetic analysis of flagellar function and structure will be summarized. In addition, a review will be presented on how recent analyses of various flagellar mutants have contributed to our knowledge of the properties and interactions of individual flagellar proteins in the structure and function of the flagellum. Other related reviews on this subject have been published recently (Huang, 1984; Luck, 1984).

II. Genetic Analysis and Properties of the Flagella in *Chlamydomonas reinhardtii*

A. GENETIC ANALYSIS IN *Chlamydomonas*

The unicellular green alga *C. reinhardtii* possesses a combination of characteristics which makes it a particularly favorable organism for genetic and biochemical studies. It is a haploid cell which can be grown vegetatively to high densities on defined media, either in liquid cultures or as lawns on agar plates (Sager and Granick, 1953; Gorman and Levine, 1965). Its mitotic division can be synchronized by use of alternating light and dark periods (Bernstein, 1960; Kates and Jones, 1964), and it has a wide temperature growth range of 15–34°C. *C. reinhardtii* has a defined nuclear genetic system, for which 18 linkage groups have been identified (Harris, 1984). The cells undergo a classical meiosis, and methods for the production and isolation of meiotic products in unordered tetrads are well established (Levine and Ebersold, 1960). It is feasible to undertake studies of gene recombination with sufficiently large numbers of progeny to permit high genetic resolution. Tests for dominance and complementation of mutations can be performed by the construction of stable diploid strains (Ebersold, 1967) and, in some cases, in temporary dikaryons formed during the normal mating process (Lewin, 1954; Starling and Randall, 1971).

B. WILD-TYPE FLAGELLAR STRUCTURE AND MOTILITY PROPERTIES

In *Chlamydomonas*, a single pair of motile flagella approximately 10 μm in length is found at the anterior pole of the cell (Fig. 1). The paired flagella exhibit coordinate behavior in waveform (Schmidt and Eckert, 1976; Hyams and Borisy, 1975, 1978) and in flagellar shortening and assembly (Rosenbaum *et al.,* 1969; Lefebvre *et al.,* 1978). In addition to their function in cell motility, the flagella of *Chlamydomonas* also play a critical role in cell–cell recognition and adhesion during gamete mating (reviewed by Goodenough and Thorner, 1983).

The flagella in *Chlamydomonas* have a typical eukaryotic structure (Fig. 2). The 9+2 framework of microtubules runs the length of the flagellum, enclosed in a specialized extension of the cell membrane. In liquid cultures, *Chlamydomonas* cells bear their flagella and are motile throughout interphase of the cell cycle. As in all eukaryotic cells, the flagella in *Chlamydomonas* are assembled onto basal bodies (Fig. 3). Detailed accounts of the ultrastructure of the basal-body complex and flagella have been published (Ringo, 1967; Johnson and Porter, 1968; Cavalier-Smith, 1974; Weiss, 1984). In preparation for cell division, the flagellar pair is

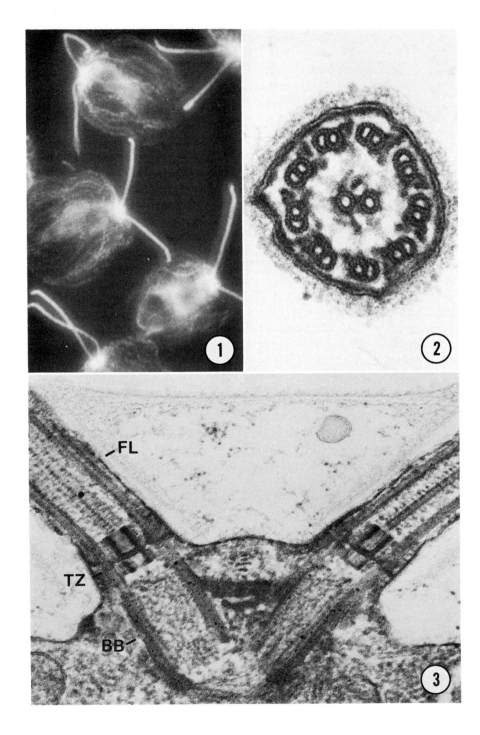

resorbed, as are the transition zones of the basal bodies. Upon completion of cytokinesis, new transition zones are assembled onto the basal bodies and flagellar outgrowth occurs rapidly through a specialized tunnel in the cell wall.

Detailed analyses of the *in vivo* flagellar waveforms in *Chlamydomonas* have been performed utilizing high-speed cinematography and stroboscopic analyses of free-swimming and tethered cells (Schmidt and Eckert, 1976; Hyams and Borisy, 1975, 1978) and computer-assisted image analysis of stroboscopic images of freely rotating uniflagellate cells (Brokaw *et al.*, 1982; Brokaw and Luck, 1983). These studies have confirmed the initial observations of Ringo (1967) that flagellar movements in *Chlamydomonas* constitute a paradigm for studies on flagellar function (for a review of ciliary and flagellar waveforms, see Sleigh, 1974). During normal forward swimming, the flagellar pair beat with a breast stroke-like biphasic "ciliary" pattern of beating. The effective stroke of the two flagella are in approximately the same plane, but in opposite directions (directed outward). In response to an appropriate mechanical or light stimulus, *Chlamydomonas* cells will reverse their direction of swimming. During these normally brief periods of backward swimming, the flagella propagate symmetrical undulatory bends, typical of "flagellar" movements. Conversion between the two forms of motility has been shown to be regulated by intracellular Ca^{2+} levels (Schmidt and Eckert, 1976; Hyams and Borisy, 1975, 1978) and to be an intrinsic property of the flagellar axonemes (Bessen *et al.*, 1980).

In addition to this transient reversal of swimming direction, *Chlamydomonas* cells also display other behavioral responses which are mediated by changes in flagellar activity. *Chlamydomonas* cells exhibit both positive and negative phototaxis (Feinleib and Curry, 1971; Stavis and Hirschberg, 1973), as well as negative geotaxis (Bean, 1977). There is evidence that Ca^{2+} also plays a critical role in phototaxis (Stavis and Hirschberg, 1973; Kamiya and Witman, 1984; for a review of algal phototaxis, see Foster and Smyth, 1980).

FIG. 1. Indirect immunofluorescence photomicrograph of wild-type *Chlamydomonas* cells stained with an anti-α-tubulin mouse monoclonal antibody. The brightly stained paired flagella extend for 10–12 μm from the apical end of the cells. In addition to the flagella, the region of the basal bodies and a cortical array of microtubules are also stained. ×2,000.

FIG. 2. High magnification electron micrograph of a cross section of a wild-type *Chlamydomonas* flagellum fixed *in situ* and stained with tannic acid. The flagellar membrane, a specialized extension of the cell membrane, is seen to enclose a classical eukaryotic axonemal 9+2 organization of microtubules and associated structures. ×200,000.

FIG. 3. Electron micrograph of a midsagittal section through the paired basal bodies (BB) of a wild-type *Chlamydomonas* cell. The flagella (FL) are seen to extend from the basal bodies (BB) beyond a specialized region known as the transition zone (TZ). ×85,000.

As first shown in other flagellar systems (reviewed by Gibbons, 1981), the generation of flagellar movements in *Chlamydomonas* is intrinsic to the flagellar axoneme. Isolated flagellar axomemes from *Chlamydomonas* can be reactivated *in vitro* to produce movements resembling those of intact flagella (Allen and Borisy, 1974a,b; Witman *et al.*, 1976, 1978). *In vitro* systems have also been developed for the analysis of the motility properties of the paired flagella as isolated complexes (Hyams and Borisy, 1975, 1978) or in detergent-extracted, demembranated cell models (Goodenough, 1983; Kamiya and Witman, 1984).

In addition to using its flagella to swim through liquid media, *Chlamydomonas* cells have been shown to glide along the surface of a solid substrate through the activity of the flagellar surface (Lewin, 1952; Bloodgood, 1981). The mechanochemical basis for this gliding form of cell locomotion may also be responsible for flagellar surface-mediated translocation of adherent particles, first described by Bloodgood (1977). This flagellar surface-related motility can occur independently of flagellar beating (Bloodgood, 1977).

C. BIOCHEMISTRY OF *Chlamydomonas* FLAGELLA

A major feature of *Chlamydomonas* that makes it suitable for combined biochemical and genetic analysis is the ease with which flagella can be detached from the cells and recovered in highly purified fractions. Currently, the two methods most commonly used to deflagellate *Chlamydomonas* cells for the mass isolation of flagella are pH shock and exposure of cells to the local anesthetic, dibucaine (for methods see Huang *et al.*, 1979; Witman *et al.*, 1978). It is of interest that these, as well as other deflagellation techniques, invariably detach the flagella just above the transition zone of the basal body. Homogeneous preparations of intact flagella can be easily purified in milligram quantities. Fractionation of isolated flagella into axonemal, membrane, mastigoneme (hair-like, membrane-appended structures), and soluble matrix compartments can be routinely performed (for example, see Witman *et al.*, 1972a). As seen in Figs. 4 and 5, isolated axonemes from *Chlamydomonas* retain their structural integrity. The axonemes can be further fractionated by treatment with high-salt or low-ionic-strength buffers, high temperature, and detergents (for example, see Witman *et al.*, 1972a,b; Piperno and Luck, 1979b; Piperno *et al.*, 1981).

Analysis of the polypeptide composition of isolated *Chlamydomonas* flagella has revealed a complexity which matches their ultrastructure; an estimated 250–300 components have been detected on two-dimensional gels of isolated flagella (Piperno *et al.*, 1977; Piperno and Luck, 1979b). Of these, over 150 have been identified as axonemal components. Figure 6 shows a typical autoradiogram of ^{35}S-labeled wild-type axonemal proteins resolved by a modification of the two-dimensional electrophoretic methods developed by O'Farrell and colleagues (1977). As expected from axonemal structure, ~70% of the axonemal mass is accounted for by the α- and β-tubulin subunits (the heavily exposed components in the 55,000-MW region seen in Fig. 6). The tubulin subunits were overloaded in this gel to reveal the more than 150 other polypeptides constituting the remaining 30% of the axonemal mass. Under appropriate electrophoretic conditions the tubulin subunits present in *Chlamydomonas* axonemes have been resolved into two α-components and one β-component (Lefebvre *et al.*, 1980; Adams *et al.*, 1981). As will be described in detail later, the structural localization of many of the axonemal polypeptides (those indicated in Fig. 6 by numbers and different symbols) has been largely determined by the analysis of mutants defective for the assembly of various axonemal substructures.

A number of enzymatic activities and proteins have been identified and purified from wild-type *Chlamydomonas* flagella. Four to five different ATPases showing the characteristics of flagellar dyneins (Gibbons and Rowe, 1965) have been isolated from the axoneme (Watanabe and Flavin, 1976; Piperno and Luck, 1979b, 1981; Pfister *et al.*, 1982; Pfister and Witman, 1984). A membrane or soluble matrix-associated Ca^{2+}-activated ATPase has been identified (Watanabe and Flavin, 1973, 1976; Bessen *et al.*, 1980), as have other nucleotide-metabolizing enzyme activities, such as adenylate kinase and nucleoside-diphosphate kinase (Watanabe and Flavin, 1973, 1976). In addition, Piperno and Luck (1979a) have purified, using DNase I affinity chromatography, a component from *Chlamydomonas* flagella axonemes that appears to be identical with β-actin. Calmodulin has also been isolated from *Chlamydomonas* and has been localized to the flagellum (Van Eldik *et al.*, 1980; Gitelman and Witman, 1980; Schleicher *et al.*, 1984). Recently Otter and Witman (1984) have obtained evidence that the flagella contain a soluble form of calmodulin present in detergent-soluble membrane and matrix fractions and a form tightly bound to the axoneme. At the present time the functional significance of the presence of β-actin and calmodulin in *Chlamydomonas* flagella is unknown.

In addition to its polypeptide complexity, recent evidence suggests that many of the flagellar components show different forms of posttranslational modifications. For example, *in vivo* pulse labeling with $^{32}PO_4$ has revealed that a large number of axonemal polypeptides are phosphorylated (Piperno *et al.*, 1981; Huang *et al.*, 1981; Adams *et al.*, 1981; Piperno and Luck, 1981; May and Rosenbaum, 1983; Segal *et al.*, 1984). The

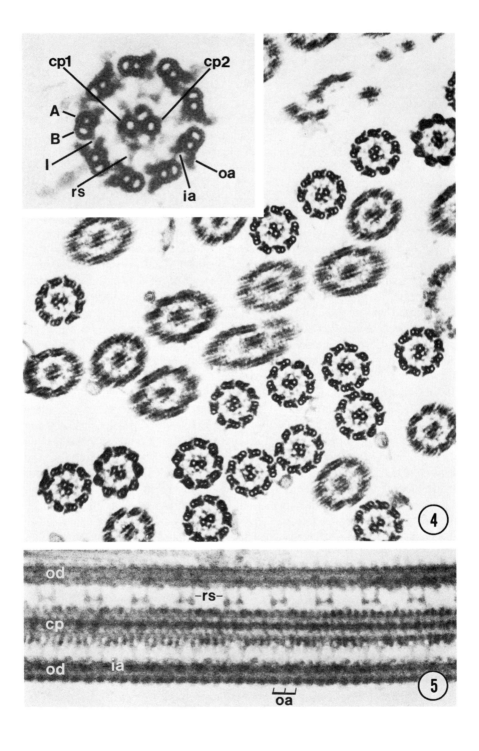

cp1 cp2
A
B
l
rs ia oa

od
−rs−
cp
od ia
oa

4

5

major flagellar membrane protein is a high-molecular-weight glycoprotein (Witman *et al.*, 1972a) which appears to be synthesized as a larger precursor (Lefebvre *et al.*, 1978; Jarvik and Rosenbaum, 1980). The major α-tubulin subunit found assembled into the axoneme has been shown to be acetylated and this posttranslational modification appears to be coupled to the assembly and disassembly of the flagellum (L'Hernault and Rosenbaum, 1983, 1985).

D. FLAGELLAR REGENERATION

In *Chlamydomonas* cell populations will synchronously regenerate new flagella when the flagella are experimentally removed. Flagella regeneration provides an opportunity to study the biosynthetic and assembly processes involved in flagellar morphogenesis at the biochemical and molecular levels.

In *Chlamydomonas* the kinetics of flagellar regeneration have been well characterized (Randall *et al.*, 1967; Rosenbaum *et al.*, 1969). Initiation of flagellar outgrowth occurs almost immediately, and with deceleratory kinetics the new flagella approach their original lengths of 10–12 μm within 90 minutes. Cells will regenerate flagella with similar kinetics even after several consecutive deflagellations (Randall, 1969).

As first shown by utilizing inhibitors of protein synthesis, new protein synthesis is required for complete regeneration of flagella in *Chlamydomonas* (Rosenbaum *et al.*, 1969). In the presence of cycloheximide, only half-length flagella are formed. More recently, taking advantage of the observation that gametes generated by nitrogen starvation have a low basal level of protein synthesis, Lefebvre and colleagues (1978) and Remillard and Witman (1982) were able to label and detect in deflagellated cells the stimulated synthesis of the tubulin subunits and several other presumptive flagellar proteins in the cell body. From *in vitro* translation

---

FIG. 4. Electron micrograph of a preparation of isolated wild-type *Chalmydomonas* axonemes. ×60,000. Inset: A high magnification image of a cross section of an isolated wild-type axoneme. Each of the nine outer doublet microtubules are composed on two subfibers; designated subfiber A (A) and subfiber B (B). Extending from subfiber A of each doublet toward subfiber B of the adjacent doublet are the outer arms (oa) and inner arms (ia). Projecting from the inner surface of subfiber A of each of the doublets toward the central pair microtubules are the radial spokes (rs). Peripheral links (l) connect adjacent outer doublets. The central pair microtubules are designated cp1 and cp2. ×166,800.

FIG. 5. A high magnification electron micrograph of a longitudinal section of an isolated wild-type axoneme. The section passes through the outer doublet microtubules (od) and central pair microtubules (cp). The periodic occurrence of the radial spokes (rs), outer arms (oa), inner arms (ia) along the length of the outer doublets, and projections on the central pair microtubules is seen. ×150,000.

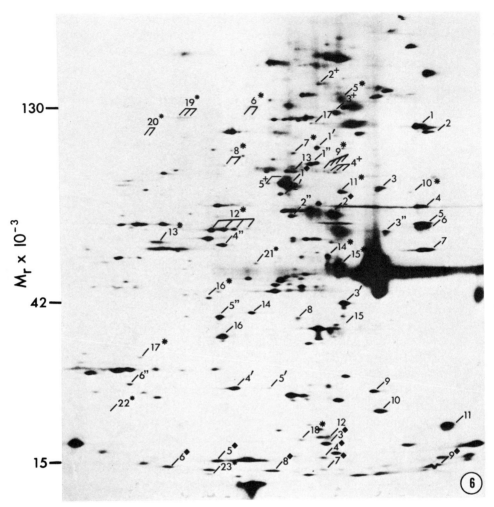

FIG. 6. An autoradiogram of ^{35}S-labeled wild-type axonemal polypeptides resolved by two-dimensional gel electrophoresis. The gel is oriented with basic polypeptides on the left. The designated components refer to polypeptides missing from axonemes of mutants lacking radial spokes (1–17), the central pair microtubules and associated structures (5*–23*), the outer arms (1♦–9♦), the inner arms (1′–5′), and the subfiber B beak projections on outer doublets 5 and 6 (2⁺–5⁺). Components 1″–6″ comprise a radial spoke-specific suppressor system.

studies, Lefebvre *et al.* (1980) identified 30 proteins whose *in vivo* synthesis was stimulated using poly(A) RNA isolated after deflagellation as compared to poly(A) RNA from nondeflagellated cells. Ten of these polypeptides, in addition to the α- and β-tubulins, were identified as flagellar polypeptides based on their comigration on two-dimensional gels with flagellar components. Evidence has been obtained in studies using cDNA probes complementary to tubulin mRNAs that the accumulation of tubulin mRNA is stimulated by deflagellation (Silflow and Rosenbaum, 1981; Minami *et al.*, 1981; Brunke *et al.*, 1982). While tubulin RNA synthesis is partially responsible for this accumulation (Keller *et al.*, 1984), changes in its stability also appears to play an important role (Baker *et al.*, 1984). In addition to the isolation of cDNAs for the tubulin mRNAs, Schloss and collaborators (1984) have isolated a group of cDNA clones for RNAs that change in abundance during the process of flagellar regeneration in *Chlamydomonas*. Some of these "regulated" mRNAs may encode for specific flagellar proteins.

III. Phenotypes and Genetics of *Chlamydomonas* Flagellar Motility Mutants

Flagellar mutants of *Chlamydomonas* were first isolated by Lewin (1954) in the species *Chlamydomonas moewusii*. Lewin also isolated 20 flagellar mutants in *C. reinhardtii* which were designated *pf* (paralyzed flagella) mutants 1–20. Many of these mutants were subsequently mapped to independent loci (Ebersold *et al.*, 1962). Since these initial studies, large numbers of additional flagellar mutants in *C. reinhardtii* have been isolated with various mutagenic agents and enrichment procedures (for methods see, Lewin, 1960; Warr *et al.*, 1966; Huang *et al.*, 1977, 1982a; Jarvik and Rosenbaum, 1980; Adams *et al.*, 1982). Mutants isolated on the basis of impaired cell motility fall into two phenotypic classes; those in which the mutations appear to specifically affect the motility properties of the flagella and those in which the lesions have consequence on the normal assembly of the flagellar pair. The flagellar-function defective mutants that have been isolated include paralyzed flagellar mutants showing little or no movement of the flagella (Lewin, 1954; Warr *et al.*, 1966; McVittie, 1972a,b; Randall and Starling, 1972; Huang *et al.*, 1979, 1981; Adams *et al.*, 1981; Dutcher *et al.*, 1984), mutants in which the flagella beat with abnormal beat frequencies (Huang *et al.*, 1982a; Kamiya, 1984; Mitchell and Rosenbaum, 1985), mutants showing defective bending wave patterns (Warr *et al.*, 1966; Huang *et al.*, 1982a), and mutants in which the flagella beat only in the backward swimming "flagellar" mode (Nakamura, 1979,

1981; Segal *et al.*, 1984). Phototaxis mutants have also been described (Smyth and Ebersold, 1970; Hudock and Hudock, 1973; Hirschberg and Stavis, 1977).

Among the flagellar assembly mutants, nonconditional and thermosensitive strains defective for the formation and stability of the flagellar pair have been reported (Goodenough and St. Clair, 1975; Huang *et al.*, 1977; Jarvik and Rosenbaum, 1980; Adams *et al.*, 1982; Lefebvre *et al.*, 1984). In addition, mutants which assemble only a single flagellum (Huang *et al.*, 1982a) or variable numbers of flagella (Warr, 1968; Kuchka and Jarvik, 1982; Wright *et al.*, 1983; Hoops *et al.*, 1984; Adams *et al.*, 1985) and mutants defective for flagellar length (McVittie, 1972a,b; Starling and Randall, 1971; Kuchka *et al.*, 1983) have been isolated.

To date, genetic analysis of many of these mutants has led to the identification of the map locations of 46 genes which influence flagellar motility and/or assembly. As seen in Table I, these flagellar-related genes have been found to be distributed throughout 13 of the 18 linkage groups described for *C. reinhardtii* (for a recent complete genetic map of *C. reinhardtii*, see Harris, 1984). In several cases, evidence of potentially relevant linkage of flagellar-related genes has been obtained (Huang *et al.*, 1981, 1982a).

IV. Mutations Affecting Specific Axonemal Substructures

As we have previously noted, the generation of flagellar movements has been shown to be intrinsic to the core structure, the flagellar axoneme. Periodic substructures attached to the outer doublet and central pair microtubules are a constant feature of motile cilia and flagella. The most prominent of these appended structures are the radial spokes, outer and inner dynein arms, peripheral links which are attached to the A- and B-subfibers of the outer doublet microtubules, and sets of projections extending from the central pair microtubules (see Figs. 4 and 5).

To date, mutations affecting the assembly and function of the radial spokes, central pair microtubule complex, and outer and inner arms have been identified. In addition, mutants missing specialized beak-like projections found on the B-subfiber of the outer doublets 5 and 6 have been characterized. The following section reviews how recent analysis of these mutants at the ultrastructural, biochemical, and genetic levels has provided new and important information on the polypeptide composition, assembly hierarchy, localization of enzyme activities, and functional roles of specific axonemal substructures in *Chlamydomonas* flagellar motility. A summary of these results appears in Table II.

TABLE I

LINKAGE MAP OF *Chlamydomonas reinhardtii* MAP POSITIONS AND
DESIGNATION OF FLAGELLAR-RELATED GENES[a]

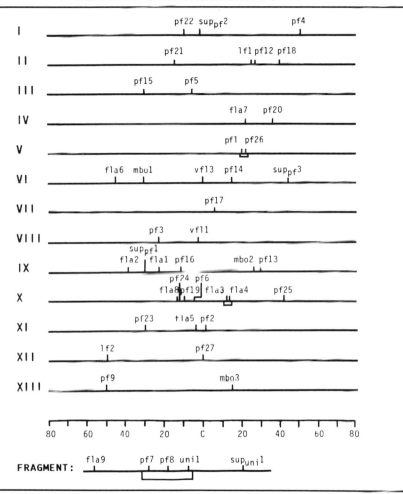

[a] pf, Paralyzed flagella; sup, suppressor mutations; lf, long flagella; fla, flagellarless (temperature-sensitive mutations); mbo, moves backward only; vfl, variable number of flagella; uni, uniflagellate cells.

TABLE II

MUTATIONS AFFECTING SPECIFIC AXONEMAL STRUCTURES

Axonemal substructure (polypeptide components)	Gene (number of mutant alleles)	Linkage group	Mutant phenotype	Axoneme polypeptide deficiencies	Gene product (molecular weight)	References
Radial spokes (RS 1–17)[a]	pf14 (2)	VI	pf (flaccid), missing radial spokes	RS 1–17	RS 3 (86,000)	Luck et al. (1977); Piperno et al. (1977, 1981); Witman et al. (1978); Remillard and Witman (1982)
	pf1 (3)	V	pf (flaccid), missing radial spokeheads, 1 ts allele	RS 1, 4, 6, 9, 10	RS 4 (76,000)	Luck et al. (1977); Piperno et al. (1977, 1981); Jarvik and Rosenbaum (1980); Huang et al. (1981); Remillard and Witman (1982)
	pf17 (1)	VII	pf (flaccid), missing radial spokeheads	RS 1, 4, 6, 9, 10	RS 9 (26,000)	Huang et al. (1981)
	pf24 (1)	X	pf (flaccid), reduced radial spokes	RS 1, 2, 4, 6, 9, 10, 16	RS 2 (118,000)	Huang et al. (1981)
	pf25 (7)	X	abnormal motility	RS 8, 11	RS 11 (22,000)	Huang et al. (1981)
	pf26 (1)	V	pf (flaccid), ts	RS 6	RS 6 (67,000)	Huang et al. (1981)
	pf27 (1)	XII	pf (flaccid), reduced radial spokes	RS 2, 3, 13	Unknown	Huang et al. (1981)
Central pair microtubule (MT) complex (CP 1–23)[b]	pf15 (3)	III	pf (rigid), missing CP MTs	CP 1–23	Unknown	Warr et al. (1966); Witman et al. (1978); Adams et al. (1981)
	pf18 (15)	II	pf (rigid), missing CP MTs	CP 1–23	Unknown	Warr et al. (1966); Adams et al. (1981); Remillard and Witman (1982)

		Phenotype	Missing/altered components	Structural defect	References
*pf*19 (8)	X	pf (rigid), missing CP MTs	CP 1–23	Unknown	Warr et al. (1966); Witman et al. (1978); Adams et al. (1981); Remillard and Witman (1982)
*pf*20 (16)	IV	pf (rigid), missing CP MTs	CP 1–23	Unknown	Warr et al. (1966); Adams et al. (1981)
*pf*6 (6)	X	flagella twitch, missing projections on CP MT 1	CP 9, 12, 18	Unknown	Lutcher et al. (1984)
*pf*16 (3)	IX	flagella twitch, unstable CP MT 1	CP 1, 2, 4, 9, 12, 14, 17–20	CP 14 (57,000)	Lutcher et al. (1984)
Outer arms (OA 1–9, HMW I, II, V, X)[c]					
*pf*13 (2)	IX	pf (flaccid), reduced flagellar length, reduced outer arms	OA 1–9, HMW I, II, V, X	Unknown	Huang et al. (1979); Piperno and Luck (1981)
*pf*22 (2)	I	pf (flaccid), reduced flagellar length, reduced outer arms	OA 1–9, HMW I, II, V, VIII, X	Unknown	Huang et al. (1979)
*pf*28 (1)	ND[d]	1/2 beat frequency, defective backward swimming, missing outer arms	3 HMW bands	Unknown	Mitchell and Rosenbaum (1985)
*oda*38 (1)	ND[d]	1/3 beat frequency, no backward swimming, missing outer arms	4 HMW bands	Unknown	Kamiya (1984)
*sup*_{pf}1 (1)	IX	1/2 beat frequency, suppresses RS and CP mutants	HMW II	HMW II (325,000)	Huang et al. (1982b); Brokaw et al. (1982)

195

(continued)

TABLE II (*Continued*)

Axonemal substructure (polypeptide components)	Gene (number of mutant alleles)	Linkage group	Mutant phenotype	Axoneme polypeptide deficiencies	Gene product (molecular weight)	References
Inner arms (IA 1–5, HMW III, IV, VI–VIII)[c]	*pf*23 (1)	XI	pf (flaccid), reduced flagella length, reduced inner arms	IA 1–5, HMW III, IV, VI–VIII	Unknown	Huang *et al.* (1979); Piperno and Luck (1981)
Beak projections on outer doublets 5 and 6 (B 1, 3, 4, 6–7, IA 2)[e]	*mbo*1 (3)	VI	move backwards only, missing 5 and 6 beaks	B 1, 3, 4, 6–7, IA 2	Unknown	Segal *et al.* (1984)
	*mbo*2 (3)	IX	move backwards only, missing 5 and 6 beaks	B 1–7, IA 2	Unknown	Segal *et al.* (1984)
	*mbo*3 (1)	XIII	move backwards only, missing 5 and <6 beaks	B 1, 3, 4, 6–7, IA 2	Unknown	Segal *et al.* (1984)

[a] See Piperno *et al.* (1981).
[b] See Dutcher *et al.* (1984).
[c] See Huang *et al.* (1979).
[d] ND, Not determined.
[e] See Segal *et al.* (1984).

A. RADIAL SPOKES

The radial spokes in wild-type *Chlamydomonas* axonemes extend from the wall of the A-tubule of each of the outer doublets toward the central pair microtubules (see Figs. 4 and 5). The radial spokes occur in pairs along the length of the doublets, the pairs repeating every 96 nm, and the members of each pair separated by 32 nm (Goodenough and Heuser, 1985). Each spoke is differentiated into a proximal thin stalk and a distal enlarged head region (Hopkins, 1970; Chasey, 1974; Witman *et al.*, 1978). The distal spokehead has been shown at the ultrastructural level to interact with projections on the central pair microtubules (Warner and Satir, 1974). Recent analysis of radial spoke structure in *Chlamydomonas* by the quick-freeze deep-etch technique (Goodenough and Heuser, 1985) has revealed that the spoke heads are made up of four major subunits, two which bind to the spoke stalk and two which interact to a pair of central pair microtubule projections. These connections between the radial spokehead and projections from the central pair microtubules have been interpreted as sites of mechanical constraint and have led to the proposal that the function of the radial spokes may be to restrain outer doublet sliding and to convert it to bending (Warner and Satir, 1974).

Two paralyzed flagellar mutant alleles for the *pf*14 locus were shown to be specifically missing the radial spokes (Fig. 7) (Witman *et al.*, 1976, 1978; Piperno *et al.*, 1977). Witman and colleagues (1976, 1978) studied the motility properties of isolated *pf*14 mutant axonemes in *in vitro* functional assays of reactivation (propagation of bending movements) and disintegration (unrestrained doublet sliding). Although the isolated mutant axonemes could not be reactivated in the presence of ATP, they were observed to actively disintegrate in trypsin-digested preparations. These studies led to the proposal that the radial spokes might be required for bend formation, but not for doublet-to-doublet sliding.

In 1977, Luck and colleagues (Piperno *et al.*, 1977) published the first in a series of papers on the genetic and biochemical analysis of radial spoke mutants. Utilizing the two-dimensional polypeptide-mapping procedure illustrated in Fig. 6, Luck's laboratory identified 17 different polypeptides normally present in wild-type axonemes to be missing in radial spokeless *pf*14 mutants (Piperno *et al.*, 1981). These components are indicated in Fig. 6 as polypeptides 1–17. Utilizing this molecular signature for the radial spoke, mutants representing six additional genetic loci were identified (Table II, Huang *et al.*, 1981).

Two of these mutants, *pf*1 and *pf*17, were found to lack a subset of 5 of the 17 putative radial spoke components. When their axonemal morphology was studied, both mutants showed the absence of radial spokeheads, while the spokestalks were detected, occurring with normal periodicity

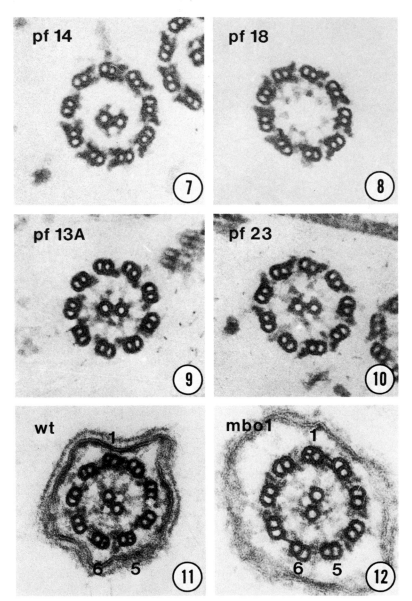

FIGS. 7–10. Electron micrographs illustrating the morphology of isolated axonemes from mutants missing different axonemal substructures. Fig. 7, the radial spokeless mutant *pf*14; Fig. 8, the central pairless mutant *pf*18; Fig. 9, the outer armless mutant *pf*13A; Fig. 10, the inner armless mutant *pf*23. Figs. 7–10, ×120,000.

FIGS. 11–12. Electron micrographs comparing the morphology of the proximal flagellar regions from the wild type (wt) (Fig. 11) and the mutant *mbo*1 (Fig. 12). The *mbo*1 flagella is missing subfiber B peak projections normally found on outer doublets 5 and 6. Fig. 11, ×120,000; Fig. 12, ×135,000.

(Piperno *et al.,* 1977; Huang *et al.,* 1981). This finding suggested that these 5 polypeptides (radial spoke (RS) components 1, 4, 6, 9, and 10) are structural subunits of the radial spokehead and that the remaining 12 polypeptides missing in the *pf*14 mutants are likely to be localized to the proximal spokestalk. As seen in Table II, the other identified radial spoke mutants (*pf* 24, 25, 26, and 27) showed different deficiencies or alterations for radial spoke polypeptides.

In many of the radial spoke mutants, a single gene mutation was found to affect the assembly of multiple polypeptides into the axoneme. In these cases, there was the possibility that one of the affected polypeptides was the mutant gene product and that the deficiency for other components was a consequence of an interdependent assembly of polypeptides into structure. To identify mutants in which the gene product might be one of the structural components of the radial spokes, Luck and colleagues applied two experimental approaches. The methods utilized included a molecular analysis of *in vivo* complementation in temporary dikaryons termed "dikaryon rescue" and the isolation and characterization of induced intragenic revertants (Luck *et al.,* 1977; Huang *et al.,* 1981).

The first approach was based on an assumption that in *pf* mutants only the gene product would be missing from the cellular precursor pool of axonemal proteins known to exist (Rosenbaum *et al.,* 1969). To identify pool components Luck and colleagues took advantage of a feature of the regular mating cycle of *Chlamydomonas.* In the mating reaction, mating type plus and minus biflagellate gametes fuse efficiently to give a population of quadriflagellate temporary dikaryons. Lewin (1954) observed that in some dikaryons derived from the mating of *pf* and wild-type gametes, the paralyzed flagella recovered function after fusion. He interpreted the rescue of function as *in situ* complementation of the defective mutant pool by a wild-type component or components. To adapt this method to gene product diagnosis Luck and colleagues labeled the mutant cells by growth on $^{35}SO_4$ and carried out fusion with nonradioactive wild-type cells in the presence of anisomycin to inhibit protein synthesis (Luck *et al.,* 1977; Huang *et al.,* 1981). The expectation was that in restoration of function to the *pf* flagella, mutant (^{35}S-labeled) and wild-type polypeptides would have an equal chance of being incorporated into the newly assembled radial spoke structure, except in the case of the mutant gene product where only polypeptides from the wild-type pool (unlabeled) could be incorporated. The expectation for axonemes derived from *pf*14 × wild-type dikaryons was that 16 polypeptides of the group of 17 missing in *pf*14 axonemes would be restored in a radioactive form when flagellar function was restored. The seventeenth component, representing the gene product, would be drawn only from the wild-type pool and would therefore not

be radioactive. When the experiment was carried out with $^{35}SO_4$-labeled
pf14 and unlabeled wild-type gametes, two-dimensional electrophoresis
of the isolated dikaryon axonemes showed restoration in radioactive form
of 16 polypeptides of the 17 originally missing in pf14. The seventeenth
polypeptide, component RS 3 was never found in a radioactive form and
was considered to be the putative gene product.

This conclusion was verified by application of a second technique,
namely, the production by mutagenesis of intragenic revertants of pf14.
The expectation was that among the revertants cases would be found
where changes in polar amino acid composition of the gene product had
occurred. The altered gene products would be identified because of their
different behavior in isoelectric focusing. Several revertants of the mutant
pf14A showed easily detectable changes in the position of component RS
3 in two-dimensional gels (Luck et $al.$, 1977; Huang et $al.$, 1981). The data
obtained by the application of these two different methods indicated that,
for the pf14 locus, radial spoke component 3 is the gene product.

The other radial spoke mutants were also analyzed by these methods.
As seen in Table II, for six out of the seven mutants studied, evidence was
obtained that the mutant gene product represented a component of the
radial spoke. With this knowledge of the gene products and the structural
consequences of the mutations, the molecular phenotypes of the mutants
could be interpreted in terms of the assembly hierarchy of the radial
spoke. A mutation in stalk component 3 (pf14) results in the failure to
assemble the entire radial spoke, while genetic lesions in head components 4 (pf1) or head component 9 (pf17) result in failure of assembly of
the spokehead but do not have any consequence on the assembly of the
radial spokestalk. The absence of stalk component 2 (pf24) results in the
instability or partial assembly of radial spokeheads, while a defect in stalk
component 11 (pf25) has no apparent consequence for radial spoke morphology, but does appear to alter the functional properties of the radial
spokes. In the case of the temperature-sensitive mutant pf26$_{ts}$, the altered
gene product, head component 6, is assembled into the radial spokes at
both the permissive and restrictive temperatures, but has an affect on
radial spoke function only if the polypeptide is assembled at the nonper-
missive temperature. In the remaining radial spoke mutant, pf27, evidence was obtained that the mutant gene product is extrinsic to the radial
spokes and required for the normal phosphorylation of five of the radial
spoke polypeptides (Piperno et $al.$, 1981; Huang et $al.$, 1981).

B. Central Pair Microtubule Complex

The central pair microtubules in $Chlamydomonas$ differ from the outer
doublets in several respects. In contrast to the outer doublet microtu-

bules, which are composed of a composite set of microtubules with 13 and 11 protofilaments, the central pair are singlet microtubules with 13 protofilaments (see Fig. 2 and inset of Fig. 4). The outer doublet microtubules arise as a continuation of the microtubules of the basal bodies, while the central pair microtubules originate distal to the transition zone of the basal body (see Fig. 3) and terminate in specialized structures associated with the flagellar membrane (Dentler and Rosenbaum, 1977). Light and electron microscopic autoradiography of pulse-labeled cells during flagellar regeneration have indicated that the outer doublets are assembled by the distal incorporation of subunits (Rosenbaum et al., 1969; Witman, 1975). There is evidence that the central pair microtubules may grow by the proximal addition of subunits (Dentler and Rosenbaum, 1977). In Chlamydomonas the central pair microtubules have been also shown to be more labile than those of the outer doublets. The central pair microtubules are sensitive to depolymerization by low-ionic-strength dialysis and by exposure to detergents (Jacobs et al., 1969; Witman et al., 1972a).

The two microtubules of the central pair also differ from one another in terms of their morphology and chemical stability (Hopkins, 1970; Witman, 1978; Adams et al., 1981). In Chlamydomonas one of the central pair microtubules bears two projections that measure ~18 nm in length. This microtubule, designated CP 1, has been shown to be more stable than the other central pair microtubule, designated CP 2, which bears two shorter projections of ~8 nm in length (Witman et al., 1972a; Adams et al., 1981; Dutcher et al., 1984).

As we have previously noted, radial spoke interactions with projections from the central pair microtubules have been interpreted as a mechanism by which doublet sliding is converted into flagellar bending. Recent evidence that the central pair microtubules rotate during the bending cycle (Omoto and Kung, 1979; Kamiya, 1982) has led to the proposal that central pair–radial spoke interactions may constitute a "distributor mechanism" controlling activation of dynein arms.

In early studies on a series of paralyzed flagellar mutants of Chlamydomonas, Randall and colleagues identified a group of mutants in which the flagella appeared to be rigidly held out in the form of a "V," with variable amount of twitching at the distal ends (for a review, see Starling and Randall, 1971). When analyzing the ultrastructure of these paralyzed straight flagellar mutants, each was found to be missing normal central pair microtubules (Randall et al., 1964; Warr et al., 1966; McVittie, 1972a). In the mutants the central pair microtubules were replaced by a core of dense material running longitudinally down the center of the axoneme. At the genetic level all of the mutants were found to fall into four complementation groups (pf15, 18, 19, and 20) mapping to different linkage groups (see Table II). In a more recent study additional mutant strains with the paralyzed rigid flagellar phenotype were isolated (Adams *et al.,* 1981). Each of these mutants showed the central pairless phenotype and were found to be mutant alleles for the previously identified *pf*18, 19, and 20 loci.

Witman *et al.* (1976, 1978) and Adams *et al.* (1981) observed that during the course of flagellar isolation and conversion to axonemes by treatment with nonionic detergents the dense core material is lost from the central pair mutants (Fig. 8). Adams *et al.* (1981) and Dutcher *et al.* (1984) isolated flagella and axonemes from representative mutant alleles for the *pf*15, 18, 19, and 20 loci and found that all of the mutants showed deficiencies in isolated axonemes for a common set of 23 ^{35}S-labeled polypeptides normally found in the wild type (some of these components are indicated in Fig. 6 as 5*–23*).

In sea urchin sperm it has been reported that the central pair α- and β-tubulins differ from their outer doublet counterparts in amino acid composition and their tryptic peptides maps (Stephens, 1978). In wild-type *Chlamydomonas* flagella and axonemes the microtubular proteins have been observed to be present as two α-components and one β-component (Lefebvre *et al.,* 1980; Adams *et al.,* 1981). In axonemes isolated from each of the mutants in which the central pair tubules are missing, the same microtubular protein subunit pattern was observed. These results indicate that neither of the two α-tubulins is specifically associated with the central pair microtubules.

At the present time the identity of the *pf*15, 18, 19, and 20 gene products have not been determined. It would appear, however, based on the large numbers of isolated mutant alleles for these four loci, that mutations for only a small number of genes can give rise to the central pairless phenotype.

A further structural analysis of the central pair microtubule complex was afforded when Dutcher *et al.* (1984) identified and analyzed mutations for two unlinked loci, *pf*6 and *pf*16, which appear to specifically affect the structure of the CP 1 microtubule. Three mutant alleles at the *pf*16 locus showed a lability of the CP 1 microtubule in isolated axonemes. This phenotype allowed Dutcher *et al.* (1984) to determine that at least 10 polypeptides (CP 1, 2, 4, 9, 12, 14, and 17–20) of the 23 components previously identified as missing from the axonemes of central pairless mutants are unique to the CP 1 microtubule and associated structures. By analysis of intragenic revertants and dikaryon rescue techniques the gene product of the *pf*16 locus was found to be central pair component 14 with a molecular weight of 57,000. The observation that mutations for the central pair component 14 structural gene result in a chemical instability of the CP 1 microtubule suggests that this polypeptide may have properties related to microtubule-associated proteins shown to influence the stability of microtubules from other systems (for a review, see Vallee *et al.,* 1984).

In addition to the *pf*16 mutants, Dutcher *et al.* (1984) found that mutants for the *pf*6 locus were defective for the assembly of one of the two long projections of the CP 1 microtubule. Biochemical analysis revealed that the *pf*6 mutant axonemes lacked three polypeptide components (CP 9, 12, and 18), a subset of polypeptides missing in isolated CP 1-less *pf*16 axonemes. No evidence for a structural gene product for the *pf*6 locus was obtained using dikaryon rescue analysis; all three polypeptides that were missing returned with label in the isolated *pf*6 × wild-type dikaryon axonemes. There are two alternative explanations of these results, one is that the gene product is a structural component of the axoneme, as yet unidentified or, alternatively, the gene product may be extrinsic to the axoneme and function in regulation of the assembly of the projection onto the CP 1 microtubule.

C. Outer and Inner Arms

In *Chlamydomonas,* as in most motile eukaryotic cilia and flagella, two rows of arms (outer and inner) are found along the A-subfiber of the outer doublet microtubules (see Figs. 4 and 5). Electron microscopic studies have shown that these microtubule arms are complex structures and that those found on the outer and inner rows differ in their morphology (Allen, 1968; Warner, 1970; Goodenough and Heuser, 1982, 1984, 1985). Largely based on the work of Gibbons and co-workers it has been established that the doublet microtubule arms contain the major flagellar ATPases, dyneins, and that the arms are responsible for generating ATP-dependent sliding displacements by forming transient cross bridges with the B-subfiber on adjacent doublets (reviewed by Gibbons, 1981).

In 1979 Huang and colleagues reported on the identification and characterization of paralyzed flagellar mutants showing selective deficiencies for either the outer or inner row of arms. The discovery of these mutants offered new opportunities to analyze their biochemical composition and the localization of different purified dynein-ATPases from *Chlamydomonas* flagella.

1. *Outer Arms*

In their initial report, Huang *et al.* (1979) described four mutants, mapping to two independent genetic loci (*pf*13 and *pf*22), as missing or deficient for the outer arms (Fig. 9). Utilizing two-dimensional axonemal polypeptide mapping, each of the mutants showed a common deficiency for a set of nine axonemal components in the molecular-weight range of 86,000–15,000 (labeled in Fig. 6 as 1♦–9♦). In addition to these components, the outer arm mutants also showed deficiencies for wild-type axonemal components whose molecular weights were too large to be resolved in the two-dimensional gel systems. Other electrophoretic methods were required to resolve these polypeptides. Figure 13 showed the one-dimensional electrophoretic methods used to identify deficiencies in the arm defective mutants for polypeptides in the molecular-weight range of 330,000—300,000 (for detailed methods, see Piperno and Luck, 1979a). In wild-type axonemes, 10 components (numbered I–X) are resolved. In the outer arm mutants (illustrated in Fig. 13), in the lane loaded with axonemes isolated from the mutant *pf*13A, the major polypeptides, I, II, and V, as well as X, were found to be absent or greatly diminished.

From *Chlamydomonas* flagella four different ATPase activities which meet the definition of dynein-ATPases set forth by Gibbons and Rowe (1965) have been isolated (Watanabe and Flavin, 1976; Piperno and Luck, 1979a, 1981; Pfister *et al.*, 1982). Each of these ATPases has a large, but different, sedimentation coefficient of 18, 12, 12.5, or 10–11 S; each specifically hydrolyzes ATP in preference to other nucleotides, is activated by both Ca^{2+} and Mg^{2+}, and contains polypeptide subunits with molecular weights in excess of 300,000.

As seen in Table III, a striking correlation was found to exist between the polypeptides observed to copurify with the isolated 18 and 12 S dyneins (Piperno and Luck, 1979b) and the polypeptide deficiencies of the *pf*13 and *pf*22 outer arm mutants (Huang *et al.*, 1979). The polypeptide components of the 18 and 12 S dyneins accounted completely for the

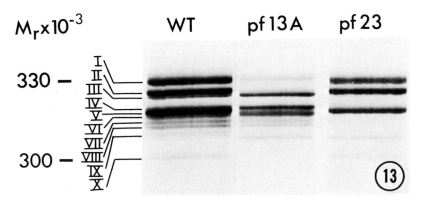

FIG. 13. Autoradiogram of the one-dimensional separation of ³⁵S-labeled axonemal polypeptides in the molecular-weight range of 330,000–300,000 found in the wild type (wt), the outer armless mutant *pf*13A, and the inner armless mutant *pf*23.

TABLE III
CORRELATION OF DYNEIN MOLECULAR COMPOSITION WITH
OUTER AND INNER ARM SIGNATURES

Dyneins[a]	High-molecular-weight polypeptides[b]	Lower molecular-weight polypeptides[b]
18 S	I◆, II◆, X◆	1◆, 2◆, 4◆–9◆ (10[c], 11[c])
12 S	V◆	3◆
12.5 S	VII'	3', 4', 5' (6[c])
10–11 S	VI'	3', 4', 5'

[a] As isolated by Piperno and Luck (1979b, 1981).

[b] ◆, Polypeptides missing or deficient in four different outer arm mutants, *pf*13, 13A, 22, and 22A [Huang *et al.* (1979), see Figs. 6 and 13]. ', Polypeptides missing or deficient in the inner arm mutant, *pf*23 [Huang *et al.* (1979), see Figs. 6 and 13].

[c] Polypeptides copurifying with the 18 and 12.5 S dyneins but not found to be missing from axonemes of the arm mutants (Piperno and Luck, 1979b, 1981).

molecular signature of the outer arm mutants. Fractionation of high-salt axonemal extractions from the different arm mutants confirmed selective deficiencies for these dyneins (Huang *et al.*, 1979). These findings suggested that both the 18 and 12 S dyneins are localized in the outer arms. The observation that single site mutations for two unlinked genetic loci resulted in a deficiency for both activities and associated molecules suggests that the two dyneins are likely to be found in a single arm. More recently, Pfister and Witman (1984) have reported that the 18 S dynein can be further subfractionated into two unique subunits containing ATPase activity. These data, that the outer arms may be composed of three ATPase-containing subunits, are in good agreement with recent work on the fine structure of the outer arm dyneins using scanning transmission electron microscopy (STEM) (Witman *et al.*, 1983) and quick-freeze deep-etch techniques (Goodenough and Heuser, 1984), which indicate that the 12 S dynein is a single globular subunit and that the 18 S dynein contains two globular domains.

The finding that the outer arm has three ATPase-containing subunits raises new and important questions concerning the role of these ATPases in force generation. Elucidation of the mechanism of action of the outer arms will require an understanding of the relationships of the ATPases to the overall architecture of the arm, as well as information on the interactions of the subunits with each other and with the A- and B-subfibers during the mechanochemical cycle of the arm. It should be noted that in *Tetrahymena* cilia (Johnson, 1983; Johnson and Wall, 1983) and sea urchin sperm flagella (Tang *et al.*, 1982) evidence has been obtained that multiple ATPases sites are found also to be associated with the outer arms.

Although analysis of the outer arm mutants for the *pf*13 and *pf*22 loci provided new information on the polypeptide composition and localization of specific dyneins, the paralyzed flagellar phenotypes of the mutants provided no new data on the functional role of the outer arms in flagellar movements.

In studies on sea urchin sperm flagella, Gibbons and Gibbons (1976) observed that extraction of the outer row of arms resulted in a reduced beat frequency which could be restored by recombination of the outer arms. These observations suggested that the outer and inner row of arms might be functionally equivalent in generating sliding displacements. Recently, the isolation of two motile outer armless mutants of *Chlamydomonas* has been reported. These two mutants, *pf*28 (Mitchell and Rosenbaum, 1985) and *oda*38 (Kamiya, 1984), show a total absence of outer dynein arms and a reduced flagellar beat frequency *in vivo* and in isolated axonemes. Although the flagella exhibit wild-type-like waveforms for the normal forward swimming mode, both show apparent defects in the normal beat pattern associated with backward swimming. Each of the mutants has been reported to be deficient for selected high-molecular-weight axonemal polypeptides, and in the case of *pf*28, to lack two ATPases which correspond to the 18 and 12 S outer arm dyneins previously described (Mitchell and Rosenbaum, 1985; Kamiya, 1984).

The motility phenotypes of these two new outer armless mutants suggest that the outer arms are not required for a wide range of normal flagellar motility functions and that the outer and inner arms may share some common functions in the generation of motive force.

In addition to the previously described mutants with defects for the assembly of the outer arms, a mutant, *sup*$_{pf}$1, has been isolated which assembles normal-appearing outer arms, but has been shown to contain a structural deletion mutation for high-molecular-weight component II of the outer arm 18 S dynein (Huang *et al.*, 1982b). This mutant was originally isolated as a second site mutation which suppresses flagellar paralysis associated with radial spoke and central pair mutations (Huang *et al.*, 1982b). The signficance of the suppressor function of this mutation will be discussed in detail later. It should be noted here, however, that in a wild-type background, *sup*$_{pf}$1 has a phenotype of its own. Like the other armless mutants *oda*38 and *pf*28, *sup*$_{pf}$1 shows a reduction in the flagellar beat frequency to about half the normal frequency without alteration of the flagellar bending pattern in the forward swimming mode (Brokaw *et al.*, 1982).

In an analysis of ATPase activities solubilized from $sup_{pf}1$ axonemes by high-salt extraction and fractionation on sucrose gradients, the enzymatic activity corresponding to the 18 S dynein was found to be present with a specific activity similar to that observed for wild type. These observations indicate that the $sup_{pf}1$ lesion does not eliminate the ATPase activity associated with the 18 S dynein. The effect of this mutation on outer arm function appears to be more subtle and may be revealed only through a detailed characterization of the enzymatic properties of the subfractionated 18 S dynein (Pfister and Witman, 1984) or through analysis of its catalytic function assembled into outer arms and coupled to motility.

2. *Inner Arms*

To date, only a single mutation affecting the inner arms in *Chlamydomonas* flagella has been identified. This paralyzed flagellar mutant, *pf*23, shows a marked quantitative deficiency for inner arm structures assembled into the axoneme (Fig. 10) and a deficiency for a set of polypeptides distinct from those missing in the previously described outer arm mutants (Huang *et al.*, 1979). These include five axonemal components in the molecular-weight range 110,000–28,000 resolved by two-dimensional polypeptide mapping (indicated in Fig. 6 as components 1'–5') and five high-molecular-weight components III, IV, VI, VII, and VIII (Fig. 13).

The observation that the outer and inner arm mutants showed deficiencies for different sets of axonemal polypeptides provided the first evidence of the biochemical distinctness of the two rows of arms. The molecular description of the composition of the inner arms could also be correlated with the polypeptide composition of the dynein-ATPases isolated from wild-type axonemes. As seen in Table III, the components of the 12.5 and 10–11 S dyneins isolated by Piperno and Luck (1979b, 1981) define, in part, the molecular deficiencies of the inner arm mutant, *pf*23. Like the outer arms, analysis of *pf*23 suggests that at least two distinct ATPases are associated with the inner arms. Recent studies from quick-freeze deep-etch images of the inner arms in *Chlamydomonas* and other organisms (Goodenough and Heuser, 1985) indicate that there are two different inner arms morphologies which alternate along the length of the doublet microtubules. At the present time, since only a single mutation affecting inner arm assembly has been identified, it is uncertain how the 10–11 and 12.5 S dyneins are organized within the inner arms. It is also unknown how the other high-molecular-weight polypeptides III, IV, and VIII and intermediate-molecular-weight components 1' and 2' which are missing from *pf*23 axonemes but not found to be associated with the 10–11 and 12.5 S dyneins, relate to the structure of the inner arms. An intriguing observation is that inner arm component 3', which is found to be associated with both the 12.5 and 10–11 S dyneins (Piperno and Luck, 1981), has been shown to closely resemble β-actin (Piperno and Luck, 1979a).

D. Beak Projections on the B-Subfiber of Outer Doublets 5 and 6

All of the mutants described thus far are defective for components of the flagellar axoneme which are uniformly distributed along the length of the axonemal microtubules. There are, however, structures which are asymmetrically distributed on certain of the outer doublets and localized to specific regions along the length of the doublets. These specifications include the absence of outer dynein arms on one of the nine outer doublets, outer doublet 1 (Huang et al., 1979; Hoops and Witman, 1983), the presence of a two-part bridge that extends from the proximal region of the A-tubule of the outer armless doublet 1 to the B-tubule of the adjacent doublet (Hoops and Witman, 1983), and the presence of beak-like projections in the proximal region of the B-tubules of the outer armless doublet 1 and the two doublets directly opposite to it, doublets 5 and 6 (Witman et al., 1972a; Hoops and Witman, 1983; Segal et al., 1984).

In a detailed examination of the fine structure of the flagellar axoneme in serial thin sections having a known orientation relative to the basal bodies, Hoops and Witman (1983) have shown that the outer armless doublet 1 of each flagellum faces the other flagellum, indicating that each axoneme has the same rotational orientation relative to the direction of its effective stroke during normal forward swimming. These observations prompted Hoops and Witman (1983) to suggest that these asymmetrically distributed structures might be important in defining the direction of the effective stroke in normal forward swimming.

Direct evidence to support this theory has been obtained through the analysis of a series of mutants that display only backward movement. Two such backward swimming mutants were isolated by Nakamura (1979, 1981) and an additional seven mutants were independently isolated by Segal et al. (1984). In each of these mutants it has been shown that the flagella beat only in the symmetrical "flagellar" bending pattern typical of reverse swimming. This inability to execute the normal asymmetric "ciliary" type of bend pattern associated with forward swimming was found to be a property of the axonemes themselves. Isolated axonemes showed only symmetrical wave forms at concentrations of Ca^{2+} ranging from $<10^{-6}$ to $10^{-3} M$.

In a genetic analysis of seven of these mutants, Segal et al. (1984) found that they fell into three complementation groups, mapping to three unlinked genetic loci, designated mbo (moves backward only) -1, -2, and

−3. In an ultrastructural analysis of the *mbo* mutants, each was found to be specifically missing or deficient for the beak-like projections normally found in the proximal half of the β-tubules of the outer doublets 5 and 6. This structural defect is illustrated in Figs. 11 and 12, which show proximal cross-sectional images of wild-type and *mbo*−1 mutant flagella. The observation that the absence of the beak-like projections on outer doublets 5 and 6 correlates with an inability to execute the ciliary-type stroke gives further indication that these structures play an important function in controlling the normal forward asymmetric stroke.

In an analysis of isolated axonemes from the *mbo* mutants, a set of six polypeptides in the molecular-weight range of 245,000–30,000 regularly present in the wild type were found to be commonly missing in all of the mutants (those resolved in Fig. 6 are labeled 2^+–5^+). Four of these components were observed to be phosphoproteins that in the wild type are highly labeled after a short *in vivo* pulse of $[^{35}P]PO_4$. In a more recent study Segal and Luck (personal communication) have found that the *in vitro* phosphorylation of one of the phosphoproteins missing in the *mbo* mutants, component b4, is selectively stimulated by calcium. Segal and Luck have interpreted these observations to suggest that the calcium-dependent phosphorylation of polypeptide b4 may be one of the early axonemal events associated with the conversion from a ciliary-type stroke to a flagellar-type stroke.

V. Suppressor Analysis Reveals Regulatory Mechanisms in Flagellar Motility

In the previous section we have described how the analysis of mutants defective for the assembly of specific axonemal substructures in *Chlamydomonas* have contributed to our knowledge of their molecular composition, hierarchy of assembly, identity of associated enzymatic activities, and function in flagellar motility. Although each of the different substructures of the flagellar axoneme may have defined roles in the generation of flagellar movements, it is evident that their individual functions must be regulated to produce complex flagellar bending patterns. For example, the direction of sliding generated by the dynein arms has been shown to be polarized (Sale and Satir, 1977). It is clear that regulatory mechanisms must exist for controlling dynein arm activity on the nine different doublets and along the length of a single doublet. Sequential activation and inhibition of cross bridging must occur if sliding is to result in the cyclic formation of bending waves.

During the course of reversion analysis of paralyzed mutants defective

for the radial spokes and central pair microtubule complex, Huang *et al.*
(1982b) identified a series of extragenic mutations that suppressed flagellar paralysis in the original mutant strains without altering their characteristic radial spoke or central pair microtubule defects. Analysis of four of
these suppressor mutations representing independent genetic loci ($sup_{pf}1$–
4) has revealed the existence of a control mechanism that may inhibit
dynein arm activity (Huang *et al.*, 1982b).

Although these suppressor mutations were originally isolated in a background of a specific paralyzed radial spoke or central pair mutant, when
combined with other central pair and/or radial spoke mutants for different
genes they were found to exhibit function rather than gene-specific suppressor activity. Two of the suppressor mutations, $sup_{pf}1$ and $sup_{pf}2$, were
found to rescue some form of flagellar motility to all paralyzed mutants
with defects for either the radial spokes or the central pair microtubule
complex. $sup_{pf}3$ and $sup_{pf}4$ were also observed to act on each of the
different radial spoke mutants, but, in contrast to $sup_{pf}1$ and $sup_{pf}2$, were
not found to restore flagellar activity to the central pair mutants. It should
be noted that none of the sup_{pf} mutations showed the capacity to suppress
the paralyzed phenotype of mutants lacking or deficient for either the
outer of inner row of arms.

Analysis of the axonemal polypeptides of three of the suppressors indicated that the mutations restore flagellar activity to paralyzed radial spoke
or central pair mutants by altering other components of the flagellar axoneme; $sup_{pf}1$, $sup_{pf}3$, and $sup_{pf}4$ each showed a characteristic defect for
axonemal polypeptides (Huang *et al.*, 1982b). The defects were found to
be present in a wild-type background, and in recombinants with different
flagellar mutants where suppressor activity and polypeptide phenotype
were always correlated.

$sup_{pf}1$, as we have previously noted, contains a molecular defect for
high-molecular-weight polypeptide II, a subunit of the outer arm-associ-
ated 18 S dynein-ATPase (Piperno and Luck, 1979b; Pfister *et al.*, 1982;
Pfister and Witman, 1984). $sup_{pf}3$ and $sup_{pf}4$ were observed to be
missing polypeptides subunits of a newly identified functional and/or
structural compartment of the flagellar axoneme (Huang *et al.*, 1982b).
Components of this "spoke-specific suppressor system" are indicated in
Fig. 6 as components 1″–6″.

The existence of these suppressor mutations demonstrated that the
paralyzed flagellar phenotype of radial spoke and central pair defective
mutants represents the operation of a control mechanism that inhibits the
generation of flagellar movements. It is likely that this inhibition occurs at
the level of the principal force-generating processes, i.e., dynein arm
activities and doublet microtubule sliding. The suppressor mutations re-

store flagellar function by interrupting or bypassing this inhibitory mechanism. The molecular phenotypes of $sup_{pf}1$, $sup_{pf}3$, and $sup_{pf}4$ suggest that this control mechanism may be interrupted at at least two different levels of axonemal function: at the level of outer dynein arm activities ($sup_{pf}1$) and at the level of a group of interacting axonemal polypeptides ($sup_{pf}3$ and $sup_{pf}4$), whose localization and functional characteristics have yet to be clearly defined.

Although the function of this regulatory mechanism was revealed through analysis of the suppressive effects of the sup_{pf} mutations on paralyzed radial spoke and central pair mutants, it undoubtedly has a counterpart function in normal flagellar motility. In the mutants where the sup_{pf} mutations have been shown to act, interaction between the radial spokes and the central pair microtubules is globally disrupted. In wild-type flagella, interactions between the radial spokes and the central pair microtubules are likely to be transient and focal. Therefore, the control functions revealed by the suppressor mutations may represent a mechanism by which a basic, dynein arm-driven pattern of flagellar bending may be locally and temporarily influenced by the interactions of the radial spokes and central pair microtubules.

The existence of these suppressor mutations made it possible to study the forms of flagellar bending that can occur in the absence of the radial spoke and central pair structures. Employing computer-assisted image analysis, a description of the flagellar bending patterns exhibited by recombinants of the radial spokehead defective mutant $pf17$ with $sup_{pf}1$ and $sup_{pf}3$ was obtained (Brokaw et al., 1982).

Both $sup_{pf}1$ $pf17$ and $sup_{pf}3$ $pf17$ recombinants showed very similar flagellar bending patterns associated with forward swimming. However, in contrast to the wild-type asymmetric bending pattern a nearly symmetric, large amplitude pattern was recorded in the suppressed $pf17$ strains. This analysis has provided evidence that a functional radial spoke system is not required for the initiation or propagation of bends along the flagellum. The function of the radial spoke system appears to be to convert a relatively simple symmetric flagellar bending pattern into an asymmetric form required for effective forward motility in *Chlamydomonas*.

VI. Conclusion

This article reviews the characteristics of the unicellular alga *C. reinhardtii* that make it uniquely suited for genetic analysis of the structure and motility of eukaryotic flagella. We have described how the use of classical genetic methods combined with structural and biochemical analysis of *Chlamydomonas* flagellar mutants has contributed to dissecting the enormous complexity of the flagellar axoneme. Of the more than 150 different axonemal polypeptides, approximately 75 have been identified through mutational dissections as components of specific axonemal substructures or polypeptides of regulatory function. Most of the studies described here were undertaken during the past decade, and, as documented in this article, were dependent on the development and application of several methodologies to the analysis of flagellar mutants; these included high-resolution polypeptide mapping procedures, methods to visualize flagellar axonemal substructures at the electron microscopic level, techniques to identify mutant gene products, and *in vitro* systems for analyzing different parameters of flagellar movements. In future genetic analyses, the further use and refinement of these techniques coupled with the application of recent advances in such areas as the development of transformation systems in *Chlamydomonas* (Rochaix and VanDillewijn, 1982; Rochaix *et al.*, 1984), the generation of cDNA libraries encoding for flagellar proteins (Schloss *et al.*, 1984), high-resolution electron microscopic methods for analysis of the *in situ* and isolated structure of flagellar components (Goodenough and Heuser, 1982, 1984, 1985; Witman *et al.*, 1983), and the isolation of immunological probes specific for flagellar proteins (King and Witman, 1984) will undoubtedly add greater detail to our understanding of the complex structural and functional interactions involved in flagellar movements.

ACKNOWLEDGMENTS

Appreciation is extended to Brenda Cipriano and Lisa Satterwhite for their assistance in preparing this manuscript.

REFERENCES

Adams, G. M. W., Huang, B., Piperno, G., and Luck, D. J. L. (1981). *J. Cell Biol.* **91,** 69–76.
Adams, G. M. W., Huang, B., and Luck, D. J. L. (1982). *Genetics* **100,** 579–586.
Adams, G. M. W., Wright, R. L., and Jarvik, J. W. (1985). *J. Cell Biol.* **100,** 955–964.
Allen, C., and Borisy, G. G. (1974a). *J. Cell Biol.* **63,** 5a.
Allen, C., and Borisy, G. G. (1974b). *J. Mol. Biol.* **90,** 381–402.
Allen, R. D. (1968). *J. Cell Biol.* **37,** 825–831.
Baker, E. J., Schloss, J. A., and Rosenbaum, J. L. (1984). *J. Cell Biol.* **99,** 2074–2081.
Bean, B. (1977). *J. Protozool.* **24,** 394–401.
Bernstein, E. (1960). *Science* **131,** 1528.

Bessen, M., Fay, R. B., and Witman, G. B. (1980). *J. Cell Biol.* **86**, 446–455.
Bloodgood, R. A. (1977). *J. Cell Biol.* **75**, 983–989.
Bloodgood, R. A. (1981). *Protoplasma* **106**, 183–192.
Brokaw, C. J., and Luck, D. J. L. (1983). *Cell Motil.* **3**, 131–150.
Brokaw, C. J., Luck, D. J. L., and Huang, B. (1982). *J. Cell Biol.* **92**, 722–732.
Brunke, K. J., Young, E. E., Buchbinder, B. U., and Weeks, D. P. (1982). *Nucleic Acids Res.* **10**, 1295–1310.
Cavalier-Smith, T. (1974). *J. Cell Sci.* **16**, 529–556.
Chasey, D. (1974). *Exp. Cell Res.* **84**, 374–380.
Dentler, W. L., and Rosenbaum, J. L. (1977). *J. Cell Biol.* **74**, 747–759.
Dutcher, K., Huang, B., and Luck, D. J. L. (1984). *J. Cell Biol.* **98**, 229–236.
Ebersold, W. T. (1967). *Science* **157**, 447–449.
Ebersold, W. T., Levine, R. P., Levine, E. E., and Olmsted, M. A. (1962). *Genetics* **47**, 531–543.
Feinleib, M. E., and Curry, G. M. (1971). *Physiol. Plant.* **25**, 346–352.
Foster, K. W., and Smyth, R. D. (1980). *Microb. Rev.* **44**, 572–630.
Gibbons, B., and Gibbons, I. R. (1976). *Biochem. Biophys. Res. Commun.* **73**, 1–6.
Gibbons, B. H., and Gibbons, I. R. (1973). *J. Cell Sci.* **13**, 337–357.
Gibbons, B. H., and Gibbons, I. R. (1974). *J. Cell Biol.* **63**, 970–985.
Gibbons, I. R. (1981). *J. Cell Biol.* **91**, 107–124.
Gibbons, I. R., and Rowe, A. J. (1985). *Science* **149**, 424–426.
Gitelman, S. E., and Witman, G. B. (1980). *J. Cell Biol.* **87**, 764–770.
Goodenough, U. W. (1983). *J. Cell Biol.* **96**, 1610–1621.
Goodenough, U. W., and Heuser, J. E. (1982). *J. Cell Biol.* **95**, 798–815.
Goodenough, U. W., and Heuser, J. (1984). *J. Mol. Biol.* **180**, 1083–1118.
Goodenough, U. W., and Heuser, J. E. (1985). *J. Cell Biol.* **100**, 2008–2018.
Goodenough, U. W., and St. Clair, H. (1975). *J. Cell Biol.* **66**, 480–491.
Goodenough, U. W., and Thorner, J. (1983). *In* "Cell Interactions and Development: Molecular Mechanisms" (R. M. Yamada, ed.), pp. 29–75. Wiley, New York.
Gorman, D. S., and Levine, R. P. (1965). *Proc. Natl. Acad. Sci. U.S.A.* **54**, 1665–1669.
Harris, E. H. (1984). *In* "Genetic Maps" (S. J. O'Brien, ed.). National Cancer Institutes, Bethesda, Maryland.
Hirschberg, R., and Stavis, R. (1977). *J. Bacteriol.* **129**, 803–808.
Hoops, H. J., and Witman, G. W. (1983). *J. Cell Biol.* **97**, 902–908.
Hoops, H. J., Wright, R. L., Jarvik, J. W., and Witman, G. B. (1984). *J. Cell Biol.* **98**, 818–824.
Hopkins, J. M. (1970). *J. Cell Sci.* **7**, 823–839.
Huang, B. (1984). *J. Protozool.* **31**, 25–30.
Huang, B., Rifkin, M., and Luck, D. J. L. (1977). *J. Cell Biol.* **72**, 67–85.
Huang, B., Piperno, G., and Luck, D. J. L. (1979). *J. Biol. Chem.* **254**, 3091–3099.
Huang, B., Piperno, G., Ramanis, Z., and Luck, D. J. L. (1981). *J. Cell Biol.* **88**, 80–88.
Huang, B., Ramanis, Z., Dutcher, S., and Luck, D. J. L. (1982a). *Cell* **29**, 745–753.
Huang, B., Ramanis, Z., and Luck, D. J. L. (1982b). *Cell* **28**, 115–124.
Hudock, G. A., and Hudock, M. O. (1973). *J. Protozool.* **20**, 139–140.
Hyams, J. S., and Borisy, G. (1975). *Science* **189**, 891–893.
Hyams, J. S., and Borisy, G. G. (1978). *J. Cell Sci.* **33**, 235–253.
Jacobs, M., Hopkins, J. M., and Randall, J. (1969). *Proc. R. Soc. London Ser. B* **173**, 61–62.
Jarvik, J. W., and Rosenbaum, J. L. (1980). *J. Cell Biol.* **85**, 258–272.
Johnson, K. A. (1983). *J. Biol. Chem.* **258**, 13825–13832.
Johnson, K. A., and Wall, J. S. (1983). *J. Cell Biol.* **96**, 669–678.

Johnson, U. G., and Porter, K. R. (1968). *J. Cell Biol.* **38**, 403–425.
Kamiya, R. (1982). *Cell Motil.* **1** (Suppl.), 169–173.
Kamiya, R. (1984). *In* "International Cell Biology, 1984" (S. Seno and Y. Okada, eds.), pp. 497. The Japan Society for Cell Biology, Academic Press, New York.
Kamiya, R., and Witman, G. B. (1984). *J. Cell Biol.* **98**, 97–107.
Kates, J. R., and Jones, R. F. (1964). *J. Cell Comp. Physiol.* **63**, 157.
Keller, L. R., Schloss, J. A., Silflow, C. D., and Rosenbaum, J. L. (1984). *J. Cell Biol.* **98**, 1138–1143.
King, S. M., and Witman, G. B. (1984). *J. Cell Biol.* **99**, 46a.
Kuchka, M. R., and Jarvik, J. W. (1982). *J. Cell Biol.* **92**, 170–175.
Kuchka, M. R., Adler, S. A., and Jarvik, J. W. (1983). *J. Cell Biol.* **97**, 195a.
L'Hernault, S. W., and Rosenbaum, J. L. (1983). *J. Cell Biol.* **97**, 258–263.
L'Hernault, S. W., and Rosenbaum, J. L. (1985). *J. Cell Biol.* **100**, 457–462.
Lefebvre, P., Nordstrom, S., Moulder, J., and Rosenbaum, J. (1978). *J. Cell Biol.* **78**, 8–27.
Lefebvre, P., Barsel, S., Stuckey, M., and Swartz, L. (1984). *J. Cell Biol.* **99**, 185a.
Lefebvre, P. A., Silflow, C. D., Wieben, E. D., and Rosenbaum, J. L. (1980). *Cell* **20**, 469–477.
Levine, R. P., and Ebersold, W. T. (1960). *Annu. Rev. Microbiol.* **14**, 197–215.
Lewin, R. A. (1952). *Biol. Bull.* **103**, 74–79.
Lewin, R. A. (1954). *J. Gen. Microbiol.* **11**, 358–363.
Lewin, R. A. (1960). *Can. J. Microbiol.* **6**, 21–25.
Luck, D. J. L. (1984). *J. Cell Biol.* **98**, 789–794.
Luck, D. J. L., Piperno, G., Ramanis, Z., and Huang, B. (1977). *Proc. Natl. Acad. Sci. U.S.A.* **74**, 3456–3460.
McVittie, A. (1972a). *J. Gen. Microbiol.* **71**, 525–540.
McVittie, A. (1972b). *Genet. Res.* **9**, 157–164.
May, G. S., and Rosenbaum, J. L. (1983). *J. Cell Biol.* **97**, 195a.
Minami, S. A., Collis, P. S., Young, E. E., and Weeks, D. P. (1981). *Cell* **24**, 89–95.
Mitchell, D. R., and Rosenbaum, J. L. (1985). *J. Cell Biol.* **100**, 1228–1234.
Nakamura, S. (1979). *Exp. Cell Res.* **123**, 441–444.
Nakamura, S. (1981). *Cell Struct. Funct.* **6**, 385–393.
O'Farrell, P. Z., Goodman, H. M., and O'Farrell, P. H. (1977). *Cell* **12**, 1133–1142.
Omoto, C. K., and Kung, C. (1979). *Nature (London)* **279**, 532–534.
Otter, T., and Witman, G. B. (1984). *J. Cell Biol.* **99**, 185a.
Pfister, K. K., and Witman, G. B. (1984). *J. Biol. Chem.* **259**, 12072–12080.
Pfister, K. K., Fay, R. B., and Witman, G. B. (1982). *Cell Motil.* **2**, 525–547.
Piperno, G., and Luck, D. J. L. (1979a). *J. Biol. Chem.* **254**, 2187–2190.
Piperno, G., and Luck, D. J. L. (1979b). *J. Biol. Chem.* **254**, 3084–3090.
Piperno, G., and Luck, D. J. L. (1981). *Cell* **27**, 331–340.
Piperno, G., Huang, B., and Luck, D. J. L. (1977). *Proc. Natl. Acad. Sci. U.S.A.* **74**, 1600–1604.
Piperno, G., Huang, B., Ramanis, Z., and Luck, D. J. L. (1981). *J. Cell Biol.* **88**, 73–79.
Randall, J. (1969). *Proc. R. Soc. London Ser. B* **173**, 31–62.
Randall, J., and Starling, D. (1972). *In* "The Genetics of the Spermatozoon" (R. A. Beatty and S. Gluecksohn-Daelsch, eds.), pp. 13–36. Bogtrykkeriet, Edinburgh.
Randall, J. T., Warr, J. R., Hopkins, J. M., and McVittie, A. (1964). *Nature (London)* **203**, 912–913.
Randall, J. T., Cavalier-Smith, A., McVittie, J., Warr, R., and Hopkins, J. (1967). *Symp. Soc. Dev. Biol.* **1**, 43–83.
Remillard, S. P., and Witman, G. B. (1982). *J. Cell Biol.* **93**, 615–631.

Ringo, D. L. (1967). *J. Cell Biol.* **33**, 543–571.
Rochaix, J. D., and Van Dillewijn, J. (1982). *Nature (London)* **296**, 70–72.
Rochaix, J. D., Van Dillewijn, J., and Mahire, M. (1984). *Cell* **36**, 925–931.
Rosenbaum, J. L., Moulder, J. E., and Ringo, D. L. (1969). *J. Cell Biol.* **41**, 600–619.
Sager, R., and Granick, S. (1953). *Ann. N. Y. Acad. Sci.* **56**, 831–838.
Sale, W., and Satir, P. (1977). *Proc. Natl. Acad. Sci. U.S.A.* **74**, 2045–2049.
Satir, P. (1968). *J. Cell Biol.* **39**, 77–94.
Schleicher, M., Lukas, T. J., and Watterson, D. M. (1984). *Arch. Biochem. Biophys.* **299**, 33–42.
Schloss, J. A., Silflow, C. D., and Rosenbaum, J. L. (1984). *Mol. Cell. Biol.* **4**, 424–434.
Schmidt, J. A., and Eckert, R. (1976). *Nature (London)* **262**, 713–715.
Segal, R. A., Huang, B., Ramanis, Z., and Luck, D. J. L. (1984). *J. Cell Biol.* **98**, 2026–2034.
Silflow, C. D., and Rosenbaum, J. L. (1981). *Cell* **24**, 81–88.
Sleigh, M. A., ed. (1974). "Cilia and Flagella," pp. 79–92. Academic Press, New York.
Smyth, R. D., and Ebersold, W. T. (1970). *Genetics* **64**, 562.
Starling, D., and Randall, J. T. (1971). *Genet. Res.* **18**, 107–113.
Stavis, R. L., and Hirschberg, R. (1973). *J. Cell Biol.* **59**, 367–377.
Stephens, R. E. (1978). *Biochemistry* **17**, 2882–2891.
Summers, K. E., and Gibbons, I. R. (1971). *Proc. Natl. Acad. Sci. U.S.A.* **68**, 3092–3096.
Summers, K. E., and Gibbons, I. R. (1973). *J. Cell Biol.* **58**, 618–629.
Tang, W. J. Y., Bell, C. W., Sale, W. S., and Gibbons, I. R. (1982). *J. Biol. Chem.* **257**, 508–515.
Vallee, R. B., Bloom, G. S., and Theurkauf, W. E. (1984). *J. Cell Biol.* **99**, 38S–44S.
Van Eldik, L., Piperno, G., and Watterson, D. M. (1980). *Proc. Natl. Acad. Sci. U.S.A.* **77**, 4779–4783.
Warner, F. D. (1970). *J. Cell Biol.* **47**, 159–182.
Warner, F. D., and Satir, P. (1974). *J. Cell Biol.* **63**, 35–63.
Warr, J. R. (1968). *J. Gen. Microbiol.* **52**, 243–251.
Warr, J. R., McVittie, A., and Hopkins, J. M. (1966). *Genet. Res.* **7**, 335–351.
Watanabe, T., and Flavin, M. (1973). *Biochem. Biophys. Res. Commun.* **52**, 195–201.
Watanabe, T., and Flavin, M. (1976). *J. Biol. Chem.* **251**, 182–192.
Weiss, R. L. (1984). *J. Cell Sci.* **67**, 133–143.
Witman, G. B. (1975). *Ann. N. Y. Acad. Sci.* **253**, 178–191.
Witman, G. B., Carlson, K., Berliner, J., and Rosenbaum, J. L. (1972a). *J. Cell Biol.* **54**, 507–539.
Witman, G. B., Carlson, K., Berliner, J., and Rosenbaum, J. L. (1972b). *J. Cell Biol.* **54**, 540.
Witman, G. B., Fay, R., and Plummer, J. (1976). *Cell Motil.* Book A–C 1976.
Witman, G. B., Plummer, J., and Sander, G. (1978). *J. Cell Biol.* **76**, 729–747.
Witman, G. B., Johnson, K. A., Pfister, K. K., and Wall, J. S. (1983). *J. Submicrosc. Cytol.* **15**, 193–197.
Wright, R. L., Chojnacki, B., and Jarvik, J. W. (1983). *J. Cell Biol.* **96**, 1697–1707.

INTERNATIONAL REVIEW OF CYTOLOGY, VOL. 99

Genetic, Biochemical, and Molecular Approaches to *Volvox* Development and Evolution

DAVID L. KIRK AND JEFFREY F. HARPER

Department of Biology, Washington University, St. Louis, Missouri

I. Introduction: *Volvox carteri* as a Developmental Model

Volvox possesses a level of complexity that evokes with crystalline clarity one of the most fundamental questions in biology: What molecular and genetic controls account for the regular development of divergent phenotypes by the descendants of a single cell?

All members of the genus *Volvox* possess two, and only two, distinctive cell types: mortal somatic cells and immortal germ cells. But it is in *Volvox carteri* (and its nearest relatives, in the "Section Merillosphaera": Smith, 1944) that this distinction is most clearly established in very early development, and hence that the challenge of elucidating the mechanisms for establishing such a dichotomy is most forcefully presented. Indeed, when J. H. Powers, (re)discovered the species in a sample of pond water collected in 1904, he was so struck with the "early period at which the primary reproductive cells are set apart as distinct from those which are to become the . . . somatic cells of the adult . . . " and with "the conclusion that we have in this simple *Volvox* aggregate a perfect example of the continuity of germ cells," that he proposed (Powers, 1908) naming the species *Volvox weismannia,* in honor of August Weismann, who had championed the concept of a fundamental germ/soma dichotomy in animal development and had recognized that *Volvox* might provide particularly favorable material for investigating both the developmental and evolutionary origins of the dichotomy (Weismann, 1889). [Although the species has since been renamed *V. carteri* in recognition of its earlier description by Carter, the form Powers described is now known as *V. carteri* forma *weismannia* (Starr, 1969).]

Volvox has been known and has captured the imagination of biologists since von Leeuwenhoek's time (Dobell, 1932). And at least since Weismann's time many biologists have clearly recognized its great promise as a potential object of developmental investigations. However, it was not until the 1960s, when conditions for controlling the reproductive activities of various *Volvox* species under axenic culture conditions were worked out in the laboratories of Richard Starr, that this developmental promise finally began to be realized. William Darden first accomplished this for *Volvox aureus*. Using a medium developed by Provasoli and Pintner (1959), Darden not only provided the first detailed descriptions of both asexual and sexual reproduction under controlled conditions for any *Volvox* species, he demonstrated the existence of an activity, in certain culture filtrates, that was capable of inducing sexual development in other cultures (Darden, 1966). Gary Kochert, another student of Starr's, soon followed with a complete description of asexual and sexual development—and evidence for a powerful sexual inducer produced by sexual males—in strains of *V. carteri* f. *weismannia* isolated from Powers' native state of Nebraska (Kochert, 1968). Then Starr himself provided a similar description for another isolate of *V. carteri,* forma *nagariensis* (Starr, 1969).

However, it was the following year, when Starr reported the first developmental–genetic studies to be done in the genus, that the latter forma was moved to the forefront for those interested in the genetic and molecular analysis of development. Specifically, he described a number of aberrations of asexual or sexual development that characterized certain spontaneous variants that had appeared in his cultures and which segregated in crosses as simple Mendelian traits (Starr, 1970). Subsequently, Robert Huskey (not a student of Starr's!) and his associates elaborated on these initial genetic observations of Starr's, demonstrated that a number of additional, simple Mendelian developmental mutants could be induced in forma *nagariensis* by chemical mutagenesis, established a preliminary linkage map, and thereby firmly established the first formal genetic system for the genus (Sessoms and Huskey, 1973; Huskey, 1979; Huskey and Griffin, 1979; Huskey *et al.*, 1979a,b; Callahan and Huskey, 1980). Others have added to the list of developmental variants that have been described (Pall, 1975; Kurn *et al.*, 1978; Kurn and Sela, 1981; Kirk *et al.*, 1982).

Although the complete life cycles of several other species of *Volvox* (and their respective sexual inducers) have been described by Starr and his associates in the intervening years (McCracken and Starr, 1970; Vande Berg and Starr, 1971; Starr, 1972a; Karn *et al.*, 1974; Miller and Starr, 1981), none other has been shown to hold the promise for developmental–genetic analysis that *V. carteri* f. *nagariensis* does. [Curiously, even *V. carteri* f. *weismannia,* which bears strong morphologic similarity to *V. carteri* f. *nagariensis* (although the two formas are neither cross-inducible nor interfertile), has proved to be extremely refractory to mutagenesis and genetic analysis (G. Kochert, personal communication).] Thus, most of the laboratories currently investigating the genetic, biochemical, and molecular regulation of development in *Volvox* are utilizing *V. carteri* f. *nagariensis,* and this review will exhibit a similar bias: All comments not otherwise qualified will refer exclusively to the latter species and forma.

The qualities of *V. carteri* f. *nagariensis* which recommend it as a model for exploring the molecular–genetic control of development include the following. (1) It has two, and only two, extremely different cell types, somatic and reproductive cells, which are set apart very early in embryonic development. (2) These two cell types are normally present in predictable numbers, proportions, and arrangements. (3) The spheroid (individual) bearing these two cell types has an extremely regular and simple symmetry. (4) The morphogenetic processes by which the dichotomy of cell types and organismic symmetry arise are well defined. (5) The life cycle is simple and can be readily synchronized by controlling culture conditions. (6) The two cell types can be separated at most developmental stages and can be subjected to differential analysis. (7) Because the organism is haploid in all active phases of the life cycle, mutants—including those exhibiting aberrations of virtually every aspect of development—are readily isolated. (8) Because reproductive cells develop with considerable autonomy, mutants with a wide range of bizarre deviations from wild-type morphology are fully viable under laboratory conditions. (9) Sexual reproduction can be induced at will, and hence formal genetic analysis is possible. (10) In addition to permitting genetic analysis, the readily induced sexual pathway provides an additional dimension for developmental exploration.

The existence of two entirely different and interdependent cell types—somatic cells specialized for motility and germ cells specialized for reproduction, neither of which could succeed by itself in perpetuating the germ plasm in nature—raises *Volvox* to the level of a multicellular organism with complete division of labor. In this it is sharply distinguished from its well-known unicellular and colonial relatives in the order Volvocales (such as *Chlamydomonas, Gonium, Pandorina,* and *Eudorina;* see Kochert, 1973). In all of these other genera, the two sets of cellular functions that are separately assigned to members of two distinct lineages in *Volvox* (i.e., vegetative and reproductive) are fulfilled sequentially by all cells. Therefore, to contemplate *Volvox* ontogeny is, in a more obvious way than for many organisms, to contemplate its phylogeny. Consequently, once we have reviewed what is known (and what remains distressingly unknown) about the control of *Volvox* development, we will engage in speculation about what our current, fragmentary developmental insights may be telling us about the evolutionary origins of multicellularity within the order Volvocales.

II. Asexual Development

Two asexual spheroids (individuals) of *V. carteri,* strain HK 10 [a strain described by Starr (1969), and now employed as the standard female strain for many laboratory investigations] are shown in Fig. 1. The two are at different stages of asexual development and are presented to illustrate the "Chinese-box-within-a-box" arrangement that characterizes the asexual phase and frequently perplexes the first-time observer. The spheroid on the left is a young adult, comprising about 2000 biflagellate somatic cells (similar to *Chlamydomonas* in their morphology) that are arranged in a monolayer at the surface, plus 16 much larger asexual reproductive cells, or "gonidia," which lie just below the somatic cells in a very regular pattern, toward one end of the spheroid (the posterior end, as it swims). The cells are separated from one another and yet held together by an intricate glycoprotein-rich extracellular matrix (ECM) which is completely transparent in life. The spheroid on the right is a somewhat older individual in whom the gonidia have given rise to juveniles, each of which is a miniature of the adult and contains the full complement of somatic cells and gonidia. Thus, within the single spheroid on the right three generations can be seen: The somatic cells of the adult constitute

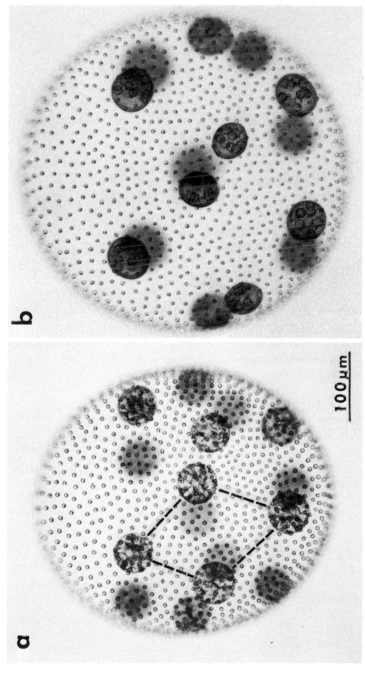

FIG. 1. Two spheroids of *V. carteri* f. *nagariensis*, strain HK 10, at different developmental stages. (a) An adult spheroid containing about 2000 somatic cells and 16 nearly mature gonidia. The dashed lines connect the four gonidia of one quadrant to emphasize the tipped rhomboidal arrangement each such set of four exhibits. (b) A spheroid about 15 hours older than the one in Fig. 1a. The spheroid has expanded slightly and the gonidia have undergone asexual embryogenesis to generate 16 juveniles that are miniatures of the adult form. Although the 2000 somatic cells of the juveniles cannot be resolved at this magnification, many of the young gonidia can.

what remains of generation 1, the juveniles constitute generation 2, and their little gonidia constitute (at least prospectively) generation 3. Eventually, after the juveniles have escaped from the parental spheroid and become free-living, this latter potential will be realized when the gonidia, having matured, will give rise to a new generation of daughters asexually.

The asexual life cycle is conveniently subdivided into two phases: During embryogenesis juveniles containing all of the cells and cellular arrangements that will be present in the adult are generated; then during spheroid maturation their cells differentiate, the spheroid expands by deposition of extracellular matrix, the juveniles hatch, and eventually the somatic cells of the parental spheroid senesce and die. Each cycle is completed when the mature gonidia of the juveniles initiate a new round of embryogenesis. A diagram indicating when these various events occur in the course of an asexual life cycle that has been synchronized and entrained by a 48-hour light–dark cycle (Kirk and Kirk, 1976) is shown in Fig. 2.

A. EMBRYOGENESIS

In this section we will first describe in broad outline the visible events of embryonic development and will then review, in a series of separate sections, what has been learned from experimental and genetic analysis concerning the control of various important component steps.

1. The Morphogenetic Events

a. *Cleavage.* The process by which a mature gonidium is subdivided by about 11 rapid clevage divisions, in the absence of growth, to generate a hollow, spherical, cellular monolayer has been thoroughly described (Starr, 1969, 1970; Green and Kirk, 1981, 1982); what follows has been abstracted from those sources.

Prior to cleavage, the previously spherical gonidium flattens markedly, particularly at its anterior pole (the pole closest to the surface of the maternal spheroid), but also at its posterior pole, to assume a rather discoidal shape. As viewed from the anterior pole, the first two cleavage furrows are nearly perpendicular to one another and divide the embryo into four quadrants lying in a cloverleaf arrangement. Because these four quadrants exhibit parallel and synchronous behavior through the rest of cleavage, discussion is simplified by consideration of what happens within each quadrant.

Prior to the third cleavage each quadrant extends a cytoplasmic lobe toward the anterior pole. The third cleavage furrows pass obliquely to cut off these extending lobes as a tier of anterior blastomeres which overlap

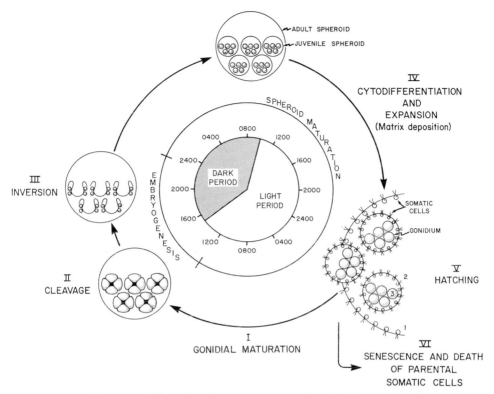

FIG. 2.   The asexual life cycle of *V. carteri,* as synchronized by a 48-hour light–dark cycle. For clarity, somatic cells have been indicated only in the spheroids in the lower right, where arabic numerals are used to indicate the three generations that can be seen within an adult spheroid prior to release of the juveniles. The roman numerals used to idenitify the major developmental stages are those employed by Gilles *et al.* (1984), except that stage VI has been added.

their more posterior sister cells. The fourth cleavage is also oblique and subdivides each quadrant into four overlapping cells, arranged along an arc running from the anterior to posterior pole. Thus, the embryo as a whole by now consists of four overlapping tiers of four cells each. Gonidia will be derived from cells of the two more anterior tiers.

Because the cellular extension begun prior to third cleavage continues until the four quadrants come into contact with one another at the anterior pole, by the 16-cell stage the cells have enclosed a central space, and the embryo has a hollow, spherical configuration superficially resembling a blastula. The analogy with a blastula is imperfect, however, because where the four quadrants establish contact at the anterior pole a cross-shaped slit, called a "phialopore," remains. This phialopore will figure importantly in a later stage of embryogenesis.

The fifth cleavage plans are nearly equatorial in orientation; they separate each of the cells into an anterior–posterior pair. Because the first five cleavages are all equal divisions, by the end of the fifth cleavage the embryo has become a hollow ball containing 32 cells of approximately equal size. It is at the sixth division, which is unequal in one-half of the cells, that the somatic and gonidial lineages of the next generation are visibly set apart. At this division all 16 (or under suboptimum conditions only a portion) of the cells, which were derived from the two anterior tiers of the 16-cell embryo, divide unequally to yield a small-anterior/large-posterior pair of sister cells. The larger member of each pair becomes a gonidial initial; the smaller a somatic initial. Meanwhile, the posterior cells all divide equally to generate additional somatic initials. The gonidial initials subsequently divide unequally about two more times—throwing off additional somatic initials each time—and then cease dividing. Somatic initials divide three or four times more than the gonidial initials do, for a total of 11 or 12 cycles altogether. Under conditions where one complete asexual life cycle is completed every 48 hours, cleavage is completed in about 6–7 hours.

By the end of cleavage, all of the cells to be found in the adult are present in a hollow sphere, but their arrangement differs in one significant aspect from that found in the adult: Whereas in the adult the gonidia are on the inside and somatic cells have their flagellar ends toward the exterior and their chloroplast ends toward the interior of the spheroid, the relationships are precisely the reverse of this at the end of cleavage. In other words, the fully cleaved embryo is inside out! This predicament is corrected in the next phase of embryogenesis.

b. *Inversion.* The dramatic process by which the embryo turns itself completely right-side out through its phialopore, to achieve the adult configuration, is called inversion (although "eversion" would have been a more appropriate term). Like cleavage, inversion has been thoroughly described (Starr, 1969, 1970; Viamontes and Kirk, 1977); the following summary is abstracted from those descriptions.

Once cleavage has been completed, there is a brief pause in visible activity. Then a wave of contraction sweeps over the embryo, the phialopore opens wide and the four lips of cells adjacent to the phialopore curl backward over the surface of the embryo. The zone of curvature accompanying this backward bending of the lips now moves in a wavelike fashion, involving cells further and further from the phialopore and bringing progressively more of the anterior hemisphere to the exterior. Suddenly, when the region of maximum curvature has just passed the equator, the posterior hemisphere "pops through" the opening (rather like the fingers of a glove being blown inside out). Now all that remains is to bring the phialopore lips back together, and a juvenile spheroid, with gonidia in the interior and all cells in the adult configuration, has been created. Under optimum conditions inversion takes about 45 minutes. When a 48-hour light–dark cycle is used to synchronize development, cleavage is initiated near the end of the illumination period and inversion occurs early in the dark period (Fig. 2).

(It should be pointed out that by long-standing convention the anterior–posterior axis of the spheroid is considered to be reversed when the embryo inverts. The anterior pole of the preinversion embryo is defined as that which lies closer to the surface of the maternal spheroid, whereas the anterior pole of the mature spheroid is defined by the direction of swimming. Therefore, the gonidia which are formed by unequal division of anterior blastomeres during cleavage come to lie in the posterior end of the mature spheroid, even though their relationships to neighboring somatic cells have not changed.)

2. Analytical Studies of Cleavage and Inversion

A number of challenging questions inevitably force themselves upon an experimental biologist watching any embryo step through a stereotyped morphogenetic program: What are the cytological forces at work which account for the predictable sequence of divisions and movements that each cell executes? How are these forces regulated within cells and integrated among cells to generate the emerging species-specific morphology of the embryo? What portion of the morphogenetic program is encoded in the genome? And how is it encoded? And how is it decoded? Although the relative simplicity of the *Volvox* morphogenetic program raises hope that ultimately it may be possible to obtain meaningful answers to some or all of those questions, and some progress in that direction has already been made, the ultimate goal is far from accomplished as yet.

a. *Cytology and Biochemistry of Morphogenesis.* Throughout embryogenesis polypeptide synthesis varies in a highly stage-specific manner (Kirk and Kirk, 1983), and inhibition of protein synthesis with antibiotics leads to a virtually immediate cessation of morphogenetic activities at all stages (Weinheimer, 1973, 1983; Kirk and Kirk, 1983, and unpublished observations). A number of inhibitors of nucleic acid synthesis have also been shown to inhibit cleavage and/or inversion in a number of *Volvox* species (Tucker and Darden, 1972; Weinheimer, 1973, 1983; Ireland and Hawkins, 1980). Of these studies, the effects Weinheimer (1973, 1983) observed on embryogenesis in *V. carteri* f.*weismannia* as a consequence of treatment with extremely low levels of actinomycin D (0.5 μg/ml) seem most instructive: Initiation of cleavage and inversion were both blocked. However, if the drug was added after cleavage had begun, a number of cleavages could be completed in its presence; this number increased from early to midcleavage and then declined again. But, most interestingly, if the drug was added shortly before the sixth cleavage, unequal cleavage division did not occur, and the resulting embryo lacked gonidial initials! Taken altogether, these studies appear to suggest that morphogenesis is under much more immediate nuclear control in *Volvox* than in many other organisms.

The first part of the morphogenetic program to be analyzed cytologically was inversion. Kelland (1964, 1977), working predominantly with *V. aureus,* observed that embryos during inversion contained cells of a variety of interesting and unusual shapes. He then showed that living cells released from inverting embryos elongated and assumed a variety of the shapes he had described in the intact embryo. Therefore he postulated that changes in cell shape were the cause, not an effect, of inversion. But because the temporal–spatial sequence of shape changes was not worked out in detail, a more comprehensive theory was not generated.

Viamontes and Kirk (1977) reinvestigated inversion in *V. carteri,* confirmed that all cells of the inverting embryo go through a highly regularized sequence of shape changes, and reiterated the view that these shape changes—in conjunction with a relocation of cytoplasmic bridges that link all cells throughout embryogenesis—provided the driving force for inversion. The following description is abstracted from that report.

Although other shape changes are seen at earlier and later stages, the most important one appears to be that associated with generating the curvature that bends the anterior portion over the surface of the more posterior portion. Prior to the curling back of the phialopore lips, all cells are "spindle shaped," i.e., they are elongated and tapered, particularly at their outer ends. At this stage, each cell is linked to its neighbors, as it has been throughout cleavage, by numerous cytoplasmic bridges at its widest point about midway along its long axis. At the time that the phialopore lips curl back, two important changes occur in the cells in the region of maximum curvature: Their tapered outer ends elongate substantially more, converting them to "flask cells" (of the sort that have been seen regularly in metazoan embryos at points where gastrulation or other morphogenetic movements are occurring; Holtfreter, 1943); simultaneously, cytoplasmic bridges are moved to the outermost tips of these now greatly elongated cells. As the region of maximum curvature moves away from the phialopore, more distal cells progressively undergo the spindle-to-flask transition and bridge relocation, while cells that had been on the proximal edge of the region shorten to assume a simple columnar shape (remaining linked at their originally outermost, but now innermost ends). As this process is reiterated, the region of maximum curvature moves in a wavelike manner from the neighborhood of the phialopore toward the opposite pole. Viamontes and Kirk (1977) postulated that it was the elongation of the outer ends of the cells combined with movement of attachment points (bridges) from the widest, central region of the cells to their narrowest, outermost regions that provided the driving force for inversion.

This hypothesis was reinforced by additional studies of a more quantitative nature (Viamontes *et al.*, 1979): Geometric analysis demonstrated that the shapes of the cells, plus location of attachment points, were sufficient to account for the shape of the embryo at all stages of inversion. Furthermore, pharmacological studies implicated cytoskeletal rearrangements as the mediators of cell shape changes and bridge relocation and showed that those changes were sufficient and necessary to account for inversion. Antimicrotubule drugs, such as colchicine, caused rapid disappearance of the girdle of cortical microtubules that normally runs from the basal body region to the opposite end of the cell; such drugs prevented normal cellular elongation and cytoplasmic bridge relocation, and they blocked inversion. Similarly, cooling midinversion embryos to 5°C caused immediate dissolution of the microtubular girdle, collapse of cells to a globular shape, and cessation of inversion. On rewarming, microtubules reformed, cells reassumed the shapes they had prior to cooling, and inversion resumed where it had left off. Perhaps one of the most instructive aspects of those studies was the selective inhibition of bridge relocation observed following treatment of embryos with cytochalasin D. In the presence of cytochalasin, cells throughout the embryo elongated relatively normally, assumed the characteristic flask shape, and then arrested; neither relocation of cytoplasmic bridges nor inversion occurred. When the drug was washed out, cells went through the sequential bridge-relocation process, and inversion occurred.

Ireland and Hawkins (1981) demonstrated that concanavalin A (Con A) blocks inversion in *Volvox tertius* in a manner that bears strong resemblance to the cytochalasin effect just described: In the presence of the Con A, embryos went through the initial sequence of shape changes and formed "long pseudopodia" at their chloroplast ends (i.e., became flask cells). But outward bending of the phialopore lips never occurred. Embryos could be held in this "strained state" for hours, but when washed free of Con A (or treated with an α-mannoside to inactivate the drug), they inverted rapidly and violently, often tearing themselves into fragments in the process. The simplest interpretation is that Con A, like cytochalasin, blocks cytoplasmic bridge relocation specifically and reversibly.

What is the mechanism by which bridges become relocated by this cytochalasin-sensitive, Con A-sensitive process? It turns out that it is not the bridges that are moved relative to the cells on which they occur, but the cells that move relative to the bridges linking them to their neighbors (Green *et al.*, 1981). This concept was inferred originally on purely structural grounds: The demonstration that the cytoplasmic bridges of all cells are linked into a continuous network, called "the cytoplasmic bridge system," at all stages. More direct evidence that it is the cells that move, and not the bridges, came from observations of live cell clusters isolated from inverting embryos: In a cell pair isolated at the time when the two cells were joined at their midpoints, one cell simultaneously underwent the spindle-to-flask transition and moved so that it came to be joined by its tip to the midpoint of its neighbor; then the companion cell elongated and moved until the two were joined only at their tips. The current working hypothesis is that it is the movement of individual cells relative to the cytoplasmic bridge system (apparently by a cytochalasin- and Con A-sensitive motile system) that provides the driving force of inversion (Kirk *et al.*, 1982).

The cytoplasmic bridge system has a characteristic ultrastructure throughout cleavage and inversion, which is shared with other members of the genus and the family Volvocaceae. Pickett-Heaps (1970) described a system of parallel (concentric or spiral) cortical striations in the cytoplasmic bridges of *V. tertius*. Marchant (1976, 1977) observed similar concentric striations in the bridges of the related colonial species, *Eudorina elegans,* and described them as being embedded in a uniform, electron-dense coating of the plasmalemma in the bridge region. Both specializations are present in the bridge region of *V. carteri* embryos: Striations embedded in electron-dense material and spaced about 25 nm apart ring the entire bridge proper and extend out as concentric rings under the adjacent plasmalemma of the cell body to abut similar rings encircling neighboring bridges (Green *et al.*, 1981). The latter authors also reported what they interpreted to be a third cortical specialization of bridges, an "annulet" of "vaguely fibrillar material" in the very center of each bridge. They postulated that all of these specializations may act as reinforcements, providing the bridges with tensile strength to withstand the stresses of inversion. [In time-lapse movies of mutant embryos inverting from two centers at once, bridges of cells caught in a tug-of-war between the two centers have been observed to stretch many times their original length before breaking (J. L. Bryant and D. L. Kirk, unpublished observations).]

How are the movement-generating events of inversion initiated and coordinated? Again the first insight came from the work of Kelland (1964, 1977), who showed that fragments of preinversion embryos inverted from all free edges and that a surgically generated, adventitious slit acted as a second site for the initiation of inversion (i.e., as a "pseudophialopore"). Sessoms and Huskey (1973) reported that the latter was also true for adventitious slits appearing in morphological mutants. Kirk *et al.* (1982) also observed this and reported that in a mutant that develops numerous adventitious slits during cleavage, inversion was initiated at each one of these slits at the same time it was initiated at the true phialopore. They also showed that in cell clusters released by fragmentation of a preinversion embryo, cells all around the edge not only initiated inversion movements (as Kelland had described), they did so before inversion was initiated in intact sibling embryos, suggesting that they had been released from inhibition by the fragmentation. It was postulated that the timing and coordination of inversion are regulated by an inhibitor of inversion movements that is lost preferentially by—and therefore falls to subinhibitory levels first in—cells adjacent to a free edge and then falls to subinhibitory levels sequentially in cells progressively further from the edge (Kirk *et al.*, 1982). Since the phialopore constitutes the only free edge in the intact, wild-type embryo, this model could account for the fact that inversion is initiated in the phialopore region and then involves cells progressively further from it.

Earlier it had been shown (Viamontes *et al.*, 1979) that exogenous cyclic GMP or isobutylmethylxanthine (an inhibitor of cyclic nucleotide phosphodiesterase) both inhibited inversion initiation. Whether endogenous cyclic GMP may act as the putative inhibitor of inversion in the normal embryo remains unresolved, however.

The cytoplasmic bridge system which plays such a crucial role in inversion arises during cleavage as a result of incomplete cytokinesis. This has been known since Janet's classical analysis of cleavage in *V. aureus* (Janet, 1923) and reiterated by more contemporary ultrastructural analysis (Bisalputra and Stein, 1966). It was when cleaving *V. carteri* embryos were studied by scanning electron microscopy (Green and Kirk, 1981), however, that the full extent and regularity of the cytoplasmic bridge system was revealed. Rather than the one bridge per cleavage furrow that Janet suggested, the latter authors found great numbers: In the first furrow, more than 200 bridges are formed, hexagonally packed into a narrow bridge band (about 4–5 bridges wide) and linking sister blastomeres at the subnuclear level. Each successive cleavage results in formation of a new cohort of bridges—in bands that are perfectly aligned and continuous with those of the preceding furrow—and subdivision of existing ones between sister cells. Thus, the number of bridges per embryo rises progressively, even though the number per cell falls from the fourth cleavage onward. By the end of cleavage, more than 40,000 bridges are present, and each cell is linked to its neighbors by more than 20 bridges.

How are these regular bridge bands formed during cytokinesis? A definitive answer is not available, but discussion of current ideas requires a digression to describe the mechanism(s) of cytokinesis in this and related species.

Cytokinesis in various members of the order Volvocales has been shown to involve an ingressive (or constricting) furrow reminiscent of the furrows typically seen in animal cells (Bisalputra and Stein, 1966; Johnson and Porter, 1968; Deason et al., 1969; Marchant, 1977; Fulton, 1978; Birchem and Kochert, 1979b). The plane in which this furrow bisects the cell is defined by two parallel arrays of microtubules that emanate from the basal bodies, pass between the two sister nuclei that have just completed karyokinesis, and extend into the chloroplast end of the cell. These have been termed the "cleavage microtubules" (Johnson and Porter, 1968) and are the major feature of what Pickett-Heaps (1975) has termed the "phycoplast" to distinguish it from the "phragmoplast" system of microtubules found in certain other algae and the higher plants. But although ingression of the furrow in the space between these two parallel arrays of cleavage microtubules accounts for cytokinesis of the anterior (basal body) end of the cell, it does not account for cytokinesis in the bridge-forming region of V. carteri. Here furrow formation seems to involve alignment and partial fusion of vesicles; in some cases this aspect of cytokinesis appears to be completed before partitioning of the anterior end of the cell by the ingressive furrow has begun (Green et al., 1981). Cleavage of the posterior, or chloroplast end of the cell is initiated by punctate depressions of the plasmalemma and appears to involve yet a third mechanism.

The conclusion drawn from morphological studies that three different cytokinetic mechanisms are involved in Volvox cleavage was reinforced by differential drug effects: Cytochalasin blocks cytokinesis at the chloroplast end selectively, Con A blocks cytokinesis of the anterior end selectively, and cytokinesis in the central bridge-forming region is resistant to the action of both drugs (Green, 1982). Ireland and Hawkins (1981) also reported that Con A blocked cleavage, but the demonstration that this inhibition is regionally specific is a refinement that is consistent with the observation of Jaenicke and Gilles (1982) that FITC–Con A binds selectively to the anterior region of the gonidium and very young embryo.

The central importance of microtubules in establishing proper orientation of cleavage furrows was indicated by abnormalities observed in colchicine- or colcemide-treated embryos. Cleavage is much less sensitive to these antimicrotubule drugs than inversion is. Whereas inversion is rapidly interrupted by 25 μM colchicine (Viamontes *et al.*, 1979), karyokinesis is unaffected and cytokinesis continues for at least 2–3 cycles in colchicine concentrations as high as 1 mM (Green, 1982). At that concentration, however, Green reported that the number of phycoplast microtubules appeared markedly reduced and the orientation of cleavage furrows was highly aberrant: Not only was the regular pattern of oblique cleavages completely lost, highly asymmetric partitioning of cells sometimes occurred, generating sister blastomeres of vastly different sizes.

Using cytochalasin, Green (1982) also provided insight into the mechanism by which cellular elongation occurs between the second and fourth cleavage divisions to generate the spherical configuration and form the phialopore. She first showed that embryos beyond the first cleavage would continue cleaving partially for up to three cycles in cytochalasin D. As mentioned above, the chloroplast ends of the cells did not divide under these conditions, but they did so rapidly after the drug was washed out. And, with one type of exception, embryos treated with the drug for two cycles and then washed yielded perfectly normal adults. The exceptions occurred when the drug was present during the period (the third and fourth cleavage cycles) when cellular elongation and closing of the phialopore normally takes place. Embryos exposed to the drug during this interval continued cleaving, but as a flattened cup with a wide gap at the anterior pole. When the drug was washed out, the cells rapidly elongated in the anterior direction, enclosed a central cavity, and closed off the phialopore in a superficially normal pattern. However, a hidden abnormality became apparent during inversion: Such embryos exhibited failure of phialopore closure at the end of inversion and generated misshapen adults. The assumption is that formation of normal cellular relationships in the phialopore region can only occur if the cells have undergone the (cytochalasin-sensitive) movements necessary to bring them into proper alignment at the time the bridges of the third and fourth cleavage furrows are formed.

In summary, all these studies suggest that normal morphogenesis requires a carefully orchestrated, stage-specific sequence of cytoskeletal rearrangements.

b. *Genetic Analysis of Morphogenesis.* To what extent and in what manner are such cytoskeletal rearrangements preprogrammed in the genome? The facts that adult form is species-specific and heritable and that in many species simple Mendelian mutations leading to gross malformations have been identified both argue that cytoskeleton-mediated events of embryogenesis must be genetically preprogrammed to some extent in all embryos. Potential advantages of *Volvox* for analyzing such preprogramming have been discussed in Section I.

In the first paper reporting induced mutations in *Volvox* (Sessoms and Huskey, 1973), 4 of the 12 phenotypes described exhibited severe inversion defects: "Inversionless" (*inv*) strains failed to initiate inversion. In "quasi-inverter" (*q-inv*) strains, inversion was initiated, but because the phialopore was extraordinarily large, the tension required for "pop-through" of the posterior hemisphere apparently could not be generated, and only the anterior hemisphere inverted. In "double posterior" (*post*), inversion was initiated, but did not proceed beyond the initial stages. In "doughnut" (*do*), inversion was initiated and proceeded from both poles; when the two sets of phialopore lips met near the equator, they fused, resulting in the toroidal shape from which the name of the strain was derived. One of the *inv* strains [of the 15 originally isolated (Sessoms and Huskey, 1973)] and the *do* strain were subsequently analyzed genetically, found to be the result of single Mendelian mutations, and assigned to linkage groups (Huskey *et al.*, 1979a). The *post* phenotype, however, turned out to be the result of multiple mutations acting in concert (R. J. Huskey, personal communication).

Subsequently, more than 50 independent clones exhibiting a wide range of inversion defects have been isolated and maintained in this laboratory. Included are strains resembling the four categories described above, plus a number that appear to arrest at various intermediate stages of inversion (Kirk *et al.*, 1982). Preliminary phenotypic analysis (Bryant *et al.*, 1980; K. J. Green, unpublished observations) led to the conclusion that in a clear majority of these strains the gross defects that became apparent during inversion were actually a consequence of somewhat more subtle aberrations that had developed during cleavage. [From their descriptions, summarized above, we presume that the *q-inv, post,* and *do* strains of Sessoms and Huskey (1973) fall into this category also.] Exceptions to this generality include several *inv* strains which appear to cleave normally, but never initiate inversion, plus what has been termed an "unco-ordinated inverter," or *unc,* strain (Kirk *et al.*, 1982), which also appears to cleave normally, but initiates inversion simultaneously from numerous locations and fragments itself in the resulting tug-of-war between inversion centers. One possibility is that mutants of the former type possess an excess and those of the latter type a deficiency of the putative inversion inhibitor.

Formal genetic analysis of such mutants has been slow because although they reproduce asexually at normal rates, their morphological defects often interfere with mating. The first step of mating involves attachment of a sperm packet to a fertile female via a flagellar interaction between the partners (see Section III,C), and many of the strains of interest bear their flagella in inaccessible locations. Furthermore, phenotypic analysis is complicated by the fact that detailed study usually reveals a range of pleiotropic defects, and determination of which defect is primary is problematic.

An obvious approach to minimizing both kinds of difficulties is to use temperature-sensitive mutants. This should allow genetic analysis under permissive conditions where wild-type morphology and mating behavior is exhibited, and determination of the critical period for gene product action (via temperature-shift experiments) should aid in focusing attention on the time period when the primary defect is being expressed. Of the 19 strains Sessoms and Huskey isolated in the four phenotypic categories described above, three were reported to be temperature sensitive. Ten of the mutants we have isolated also exhibit temperature sensitivity.

One such temperature-sensitive mutant (strain S-16) has now been analyzed in some detail (Green, 1982; Mattson, 1985; Mattson *et al.*, 1985). This strain possess a single Mendelian defect, unlinked to mating type. At 24°C it inverts to generate a sphere that resembles wild type, but at 32°C it tears itself apart during inversion to generate a bizarre assortment of multilobed and fragmented progeny, as previously described (Kirk *et al.*, 1982). Fragmentation during inversion is the consequence of tensions generated when each of the adventitious slits and gaps that develop during cleavage acts as a site for inversion initiation, as previously described (Section II,A,2,a). These slits and gaps result from a marked deficiency of cytoplasmic bridges in some cleavage furrows (although the bridges that do form appear to have relatively normal morphology). Other severe cleavage defects are seen: cleavage furrows oriented in inappropriate directions, highly asymmetric cleavages that yield sister blastomeres of vastly different sizes, and blastomeres that cease cytokinesis altogether while continuing to undergo karyokinesis, yielding giant, multinucleate cells. S-16 appears to be temperature sensitive throughout cleavage, and the earlier defects set in, the greater the portion of the resulting adult that exhibits such abnormalities. All these abnormalities now appear to be secondary consequences of a primary cytological defect in the basal body complex of S-16.

In wild-type embryos, basal bodies separate at the time, in the direction, and to the extent that nuclei elongate in preparation for mitosis. [In the Volvocales, the spindle is intranuclear, and the nucleus and spindle curve so that poles are brought into the neighborhood of the basal bodies, which themselves remain firmly attached to the plasmalemma (Coss, 1974).] As a consequence, the cleavage microtubules that fan out from the basal bodies subsequent to karyokinesis run perpendicular to the spindle axis and provide the pathway for the furrow to bisect the cell midway between the sister nuclei. In S-16, however, both the direction and extent of basal body separation may be entirely abnormal, resulting in wholly inappropriate cytokinetic patterns. In some cases, the separation of sister basal bodies is so abnormal that the cleavage microtubules they give rise to run in almost perpendicular directions—whereas in wild-type the cleavage microtubules from sister basal bodies run in parallel. In these cases cytoplasmic bridge-forming elements apparently are not properly guided into position in the furrow, and defective connections between sister blastomers result. The concept that the primary cytological defect lies somewhere in the basal body complex was strengthened by ultrastructural examination of the large, multinucleate, nondividing blastomeres referred to in the preceding paragraph: In such cells the basal bodies are found to have dissociated from the plasmalemma entirely and moved deep into the cytoplasm, where they are associated with abortive intracellular axonemes of bizarre configuration (Mattson, 1985; Mattson et al., 1985).

Strain S-16 has focused attention on the basal body complex as one of the most important links between the genotype and normal adult morphology. To explore this relationship further, we are seeking more temperature-sensitive mutants of this type; at least one additonal one has been found. Although there are undoubtedly many structural genes which code for essential components of the basal body complex, selection for mutants which lead to aberrant cleavage patterns (as opposed to absence of cleavage) under restrictive conditions should permit identification of those loci which are important in this one function of the basal body complex.

c. *Molecular Analysis of the Cytoskeleton.* The preceding sections illustrate the importance of cytoskeletal rearrangements in *Volvox* development. As a first approach to understanding the control of these morphogenetic events at the molecular level, this laboratory has initiated a two-pronged effort to characterize the individual components of the cytoskeleton that are involved, by use of monoclonal antibody and molecular-cloning techniques. Both of these by-now conventional approaches exploit both the close relationship between *Volvox* and *Chlamydomonas* and the progress that has already been made in such studies using the latter organism. A comparison of the *Volvox* and *Chlamydomonas* cytoskeletons should provide a unique opportunity to assess the molecular basis for evolving (from the basic components essential for life as a unicellular eukaryote) all the dynamic cytoskeletal interactions required for morphogenesis of a multicellular organism with division and integration of cellular labor. Since the basic architecture of vegetative cells of the two genera are virtually indistinguishable (Kochert and Olson, 1970a; Pickett-Heaps, 1975), molecular differences detected in the cytoskeletons of the two species will lead, we are hopeful, to the key features related to multicellularity. Because of the unparalleled opportunities which we believe such a comparison holds, we will briefly review work in progress, even though it is at an extremely early stage of development.

Evidence that the *Chlamydomonas–Volvox* relationship has considerable practical (as well as theoretical) importance is derived from our attempts to use antibodies to heterologous proteins to study protein synthesis and cellular organization in *Volvox*. Over many years we tested antibodies from a wide variety of sources (including our own efforts) that were directed toward "highly conserved" eukaryotic proteins from more distantly related species; none has ever been reactive with *Volvox* material. In contrast, antibodies raised against *Chlamydomonas* proteins have proved extremely useful· Polyclonal antibodies to five *Chlamydomonas* proteins (including the tubulins) permitted us to quantitate relative rates of synthesis of the cognate polypeptides in *Volvox* (Kirk and Kirk, 1985b); a monoclonal antibody to a *Chlamydomonas* tubulin has permitted us to analyze the organization of microtubules and the separation of basal bodies in a cleavage mutant (Mattson, 1985; Mattson *et al.*, 1985).

As our first attempt to exploit this antigenic relationship in a more systematic way, we have initiated a collaborative effort to develop a library of monoclonal antibodies to components of *Chlamydomonas* basal body complexes (bbcs). Interest in bbcs grows out of work with cleavage mutants summarized in the preceding section. But whereas isolation of these organelles from *Volvox* is technically difficult, Robin Wright and Jon Jarvik have established a simple and effective procedure for isolating them from *Chlamydomonas*—with the nucleus still attached, as it apparently is *in vivo* (Wright and Jarvik, 1985). Using bbcs provided by Wright and Jarvik as both immunogen and test antigen, we have isolated more than a dozen ELISA-positive clones; the specificities of these clones are now being examined on Western blots and in immunocytochemical assays on cells from both species.

The cloning of *Volvox* cytoskeletal genes has focused initially on tubulin because of the elegant work already completed on the *Chlamydomonas* genes, which provides complete gene sequences for detailed comparisons (Brunke *et al.*, 1982, 1984; Youngbloom *et al.*, 1984). *Volvox* genomic clones with homology to *Chlamydonomas* α- and β-tubulin cDNAs (courtesy of Carolyn Silflow and associates) have been isolated in this laboratory; independently and simultaneously, another laboratory has isolated clones homologous to *Chlamydomonas* α-tubulin, using a probe provided by D. Weeks and associates (R. Schmitt, Regensburg University, personal communication). (Details of cloning procedures will be given in Section V.) Between the two laboratories, multiple clones have been isolated corresponding to two α- and two β-tubulin genes. Southern blot analysis (J. F. Harper, unpublished; M. Salbaum and R. Schmitt, unpublished) confirms that *Volvox* has two α and two β genes. Northern blot analysis of total RNA has resolved two transcripts homologous to *Chlamydomonas* β-tubulin cDNA (J. F. Harper, unpublished); by analogy to *Chlamydomonas* (Brunke *et al.*, 1982), these are believed to correspond to unique transcripts from each β gene.

The two β-tubulin genes of *Chlamydomonas* differ in sequence, but code for identical polypeptides (Youngbloom *et al.*, 1984). The same may also be true for the α genes (J. Schloss, personal communication). When the restriction maps of the *Volvox* β-genomic clones are compared to the published maps for the *Chlamydomonas* equivalents, differences are seen both in the regions that are common and those that differ in the two *Chlamydomonas* genes. In addition, differences in intron organization are indicated. Whether these restriction site changes reflect divergence of one or both of the protein sequences awaits completion of DNA sequencing, which is under way.

If the amino acid sequences of *Volvox* tubulins are found to be different from one another, the question will arise whether these differences have functional significance. Fulton and Simpson (1976) have postulated organelle-specific tubulins in their "multitubulin hypothesis." The only organelle-specific subunits yet detected in *Chlamydomonas* are a result of posttranslational acetylation and deacetylation of α-tubulin (L'Hernault and Rosenbaum, 1983, 1985a,b). However, genetic diversity of tubulins is widespread, and although the validity of the multitubulin hypothesis can now be ruled out in some specific cases, it cannot yet be ruled out in all (Cowan and Dudley, 1983). Whether *Volvox* will be found to utilize different primary sequences, posttranslational modifications, or microtubule-associated proteins in order to mobilize its microtubules differently for a variety of morphogenetic tasks not seen in its simpler cousin remains an open and intriguing question.

The Schmitt laboratory has also isolated and begun to characterize several genomic clones homologous to a chicken β-actin cDNA. Further information on these clones is not yet available, however (R. Schmitt, personal communication).

3. *Segregation of Somatic and Germ-Line Potentials*

Insight into the mechanism(s) by which a germ/soma dichotomy may be established is one of the most fundamental developmental problems that *V. carteri* may help us to understand. In this section, we will review the pattern in which gonidia (germ cells) arise in wild type, variations of that pattern which are seen in mutants, and current models for the mechanism of gonidial specification.

a. *Wild-Type Cell Lineages.* As discussed in Section II,A,1,a, under optimum conditions, approximately one-half of the cells in the 32-cell embryo normally divide unequally in the sixth division cycle to yield 16 large gonidial initials and 16 smaller somatic initials. Because the first two cleavages divide the embryo into four quadrants which subsequently behave similarly, discussion of cell lineages is simplified by considering only a single quadrant at a time. Figure 3 identifies the asymmetrically dividing cells within a quadrant in two ways. Figure 3a traces the cellular pedigrees of the presumptive gonidal cells, whereas Fig. 3b is a stylistic representation of the spatial distribution of these cells within a quadrant (Starr, 1969, 1970; Green and Kirk, 1981, 1982). [Note, however (as discussed in Section II,A,1,b), that the anterior–posterior axis is reversed during inversion and hence the cells which are most anterior in the cleaving embryo are most posterior in the adult.]

Before discussing these patterns in more detail, it is important to point out that although 16 gonidia per spheroid is both an "ideal" number and the modal number seen in a wild-type population under optimum culture conditions, it is not in any way an absolute. Under deteriorating culture conditions, this number declines in a nonlinear manner to be discussed later. But even under optimum conditions, variation is seen, and within a clone, individuals with greater or lesser numbers of gonidia are produced. In one survey of several hundred adults selected at random over a 2-week period from cultures growing under our standard conditions (Kirk and Kirk, 1983), 72% of the population had the ideal number of gonidia (16); but we observed a range of gonidial numbers between 12 and 20 (with numbers smaller than 16 five times more frequent than numbers larger than 16) and obtained a mean and standard error of 15.50 ± 1.05 for the population (D. Kirk, unpublished observations). As Gilles and Jaenicke (1982) have properly pointed out, accounting for the lineages of the "extra" or "missing" gonidia in such cases will ultimately be as important as accounting for the lineages of the ideal set of 16; hence, after discussing details of the ideal pattern, we will return to a discussion of the pattern variations that are seen under both optimum and suboptimum culture conditions.

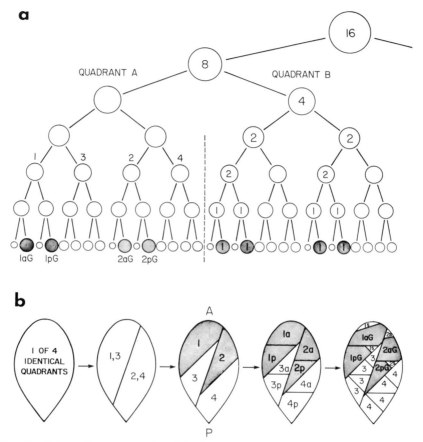

FIG. 3. Schematic representation of the lineages of the gonidia, through the sixth cleavage division, in the ideal case in which 16 gonidia are formed in the embryo as a whole. Because the four quadrants derived from the first four blastomeres cleave in a similar fashion, behavior of only a single quandrant need be considered. (a) The cellular pedigrees; shaded circles indicate gonidial initials. In quadrant A the cells are given the identifying numbers and letters suggested by Green and Kirk (1981). In quadrant B the numbers within the circles indicate the number of gonidial progeny the corresponding cells normally generate. (b) A stylistic representation of the spatial arrangement of the gonidial precursors (shaded) within a quadrant.

The idealized cellular pedigree shown in Fig. 3a does not lend itself to any simple formulation. At the top and right-hand side of this figure, numbers within circles indicate the way in which the potential for forming the ideal number of 16 gonidia is distributed among daughter cells at each of the first six cleavage divisions. At each of the first three cleavages, this potential is divided symmetrically so that each of the first eight blastomeres normally has the potential to give rise to two gonidia. At the fourth division, however, this potential is divided asymmetrically, all passing to the more anterior daughter of each dividing cell—the "tier 1" and "tier 2" cells of each quadrant (see left side of Fig. 3a and b). At the fifth cleavage, gonidia-forming potential normally is divided symmetrically again. The second asymmetrical division is the sixth one (which is the first visibly "unequal" division at which daughters of different size are produced). But in the sixth division, in contrast to the fourth one, gonidial potential is passed to the more posterior of the sister cells. Not shown in this figure is the fact that the gonidial initials created at the sixth cleavage normally divide asymmetrically three more times, passing gonidial potential to the more posterior daughter at each of these divisions (Green and Kirk, 1981, 1982, and unpublished observations). Thus, the sequence in which gonidial potential is normally distributed to the daughters in the first nine division cycles can be summarized as follows: equal, equal, equal, anterior, equal, posterior, posterior, posterior, posterior. This is the basic pattern for which any satisfactory model of gonidial specification must account.

But, as outlined above, a successful model of germ cell specification should also be able to account for the deviations from this idealized cell lineage that are observed under both optimum and suboptimum conditions. In a simple but elegant analysis, Gilles and Jaenicke (1982) not only provided clear evidence that such deviations are distinctly nonrandom, they elucidated the rules that are most frequently followed in such cases. The rules they established (restated, and modified slightly) are as follows: (1) Gonidial numbers do not follow a simple binomial distribution; they tend to be clustered either in the range of 12–16 or in the range of 8–12, and gonidial numbers outside the modal range are significantly rarer than would be predicted on the basis of a so-called normal distribution. (2) When spheroids containing 12–15 gonidia are examined, the "missing" gonidia are almost always those which "should have been" derived from 2a blastomeres (Fig. 3). The 2a tier of gonidia tends to drop out entirely before any gonidia are lost from other tiers; thus, an individual with 12 gonidia most frequently has gonidia in all of the locations appropriate for the 1a, 1p, and 2p cells, but none in the locations usually occupied by 2a cells. (3) As cultures become overcrowded, gonidial numbers tend to drop from the higher (12–16) to the lower interval (8–12). (4) The second tier in which gonidia are preferentially lost, in individuals with 8–12 gonidia, is the 1a tier. Thus, an 8-gonidiate individual most frequently has four gonidia in positions expected for cells of the 1p and 2p tiers and none corresponding to the other two tiers. (Data on the third tier to drop out, in individuals with fewer than 8 gonidia, are not available, but by extrapola-

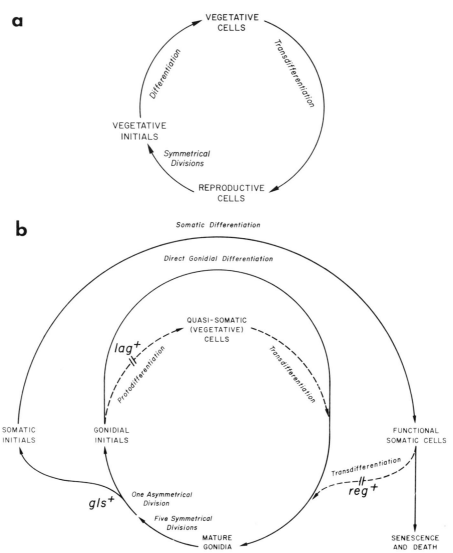

FIG. 4.   Schematic representation of Volvocalean life cycles. (a) The cycle characteristic of unicellular and colonial members of the order, in which all cells function first as vegetative and then as reproductive cells. (b) The life cycle of *V. carteri* and the proposed actions of the three genes proposed to account for the complete division of labor between somatic cells and reproductive cells. It is proposed that the *gls*⁺ gene is responsible for the asymmetric division that sets the two cell types apart, that the *lag*⁺ gene causes the reproductive initials to be shunted directly into the gonidial pathway of development (bypassing the vegetative stage), and that the *reg*⁺ gene functions to prevent the somatic cells from returning to the reproductive pathway.

tion from the preceding, one would predict it would be the 2p tier.) (5) When two gonidia are missing from a tier, losses from neighboring and opposite quadrants are equally frequent. (6) When "extra" gonidia are present, they are always the result of "twinning" of one of the normal gonidial cells—i.e., division of a gonidial initial after the sixth cleavage to give two smaller than usual gonidia in the area where only one is normally found. Twinning is most frequent in the 1a tier, but may be seen in all layers. (7) Gonidial patterns that do not follow these rules (e.g., loss of one or more gonidia from other tiers in individuals which still have one or more gonidia in the 2a tier) do occur, but at nearly 10-fold lower frequencies.

The one possible exception to these rules which we have detected appears noteworthy. Working with a population (described above) that had a higher mean and modal gonidial number than that studied by Gilles and Jaenicke (1982), we observed more than twice the frequency of individuals with more than 16 gonidia. Twinning clearly accounted for some of these extra gonidia, but it did not appear to account for all of them. We occasionally observed extra gonidia that were of normal size (as opposed to the reduced size of twin gonidia) and in locations which we interpreted to mean that they had been derived from the 3a and/or 3p cells. Based on these observations, the one generalization made by Gilles and Jaenicke (1982) on which we would reserve judgment is that "the mechanism of variation is unable to increase the basic gonidial number of 16." Whether our differing observations are a result of strain differences or differences in culture conditions remains to be determined.

The mechanisms by which these deviations may occur and their implications for various models of gonidial specification will be considered after discussing other types of pattern variations seen in mutant strains.

b. *Cell Lineages in Pattern Mutants.* Huskey *et al.* (1979a) have defined four loci, termed the *mul* (for multiple gonidia) loci, at which the only apparent effect of mutation is to affect numbers and locations of gonidia. In addition, they and others have described other types of mutations which result in abnormal gonidial number and distribution in combination with other developmental anomalies.

Mutation at the *do* (doughnut) locus (Huskey *et al.*, 1979a) falls in the latter category. It results in an embryo that inverts from both ends (to yield the toroidal adult shape from which the name is derived) and that possesses twice the usual number of gonidia, uniformly distributed. These features cosegregate [although certain other features of the phenotype originally described by Sessoms and Huskey (1973) do not], and the current interpretation is that this mutant possesses a mirror duplication of the posterior hemisphere (the anterior, or gonidium-forming hemisphere of the embryo), formally analogous to the "bicaudal" mutation of *Drosophila* (R. J. Huskey, personal communication).

Mutation at *mulA* affects gonidial number only. However, this locus appears not to control pattern formation per se; rather, it appears to be involved in control of gonidial twinning. In the *mulA* mutant, clusters of two, three, or four gonidia of reduced size are frequently found in each of the areas where one gonidium is normally located in wild type (Huskey *et al.*, 1979a). These locations and sizes are consistent with the hypothesis that gonidial initials are originally specified in the wild-type pattern, but then may divide equally one or two times to yield daughter gonidia of one-half or one-fourth the usual volume. This is the phenotype earlier described by Starr (1970) and called "multi."

The *mulB* mutation, in contrast, was reported to result in an "increased number of gonidia in a random pattern over most of the spheroid" (Huskey *et al.*, 1979a). However, apparently no detailed analysis has been performed to show that gonidial distribution truly is random—as opposed to following some other (and possibly more complicated) set of patterning rules than in wild type. Indeed, the possibility has not been ruled out that the *mulB* mutant merely expresses a variant of the reproductive cell specification pattern characteristic of sexual females (which will be described in Section III,B,1) (R. J. Huskey, personal communication). In any case, it would appear that this locus is a candidate for one involved in some fundamental way in the control of reproductive cell patterning.

Similarly, *mulC* and *mulD*, two unlinked loci of apparently similar function, appear to be of central importance for this process. One of these loci (*mulC*) is linked to mating type; the other is not. Three mutations at each locus were analyzed genetically and described as yielding spheroids with "fewer cells and variable number of gonidia" (Huskey *et al.*, 1979a). Further examination (R. J. Huskey, personal communication) has revealed that under optimum conditions the cleavage pattern observed in both *mulC* and *D* mutants actually resembles the pattern seen in the sexual male, namely, the unequal cleavage is delayed two or more divisions, at which time nearly all cells divide unequally, and then cleavage ceases. This cleavage pattern in the mutant asexual female results, as in the sexual male, in a smaller spheroid, with fewer total cells, essentially a 1 : 1 ratio of somatic to reproductive cells, and a retention of all reproductive cells at the surface of the spheroid (Starr, 1969; see also Section III,B,2). Because mutation at either locus appears to result in the specialized cell lineage pattern normally seen only in sexual males, a reasonable interpretation of the role of these genes would seem to be that they produce negative regulators, both of which must be present to prevent emergence of the male-specific pattern in other phases of the life cycle. But such a model leads to the unexpected conclusion that the sexual male pattern might be the basic cleavage pattern of the species, which appears "spontaneously" if not repressed. While not impossible, there was no prior basis for suspecting that such would be the case.

Pall (1975) described three mutants (similar in phenotype to one another), which he termed *pcd,* for "premature cessation of division," and which he suggested might provide a genuine insight into the basis for gonidial specification. Embryos of these strains divided only 5–7 times, rather than the usual 11-12 times, resulting in greatly reduced cell numbers and increased cell sizes. The resulting spheroids always had increased ratios of gonidia to somatic cells, and frequently absolute increases in gonidial numbers were observed. Unlike the *mul* mutants (which affect only the asexual phase), *pcd* mutations had the same effect in sexually induced males and females as in asexual embryos: They ceased cleaving prematurely and exhibited both larger than normal average cell size and a pronounced increase in the ratio of reproductive to somatic cells. Pall interpreted these results to mean that "some cell parameter closely related to cell size is critical in determining reproductive cell versus somatic cell differentiation." He pointed out, quite correctly, that this idea is consistent with the normal processes of reproductive cell development. In wild type the first visible step in development of either asexual or sexual reproductive cells is always an unequal division that sets aside larger cells as the presumptive reproductive cell lineage.

Many of the cleavage mutants in our collection, although they have very different phenotypes from Pall's strains, behave in a manner that is consistent with his hypothesis. For example, the S-16 strain (described in Section II,A,2,b) divides very asymmetrically throughout cleavage to generate a wider assortment of cell sizes than is normally produced; it also has an elevated ratio of gonidial to somatic cells and an increased number of gonidia. Although the question has not been examined systematically, our impression is that among our many cleavage mutants, such an elevation of the gonidial to somatic cell ratio may be more the rule than the exception.

c. *Theories of Gonidial Specification.* The first theory of gonidial specification that needs to be considered, then, is Pall's—that "some cell parameter closely related to cell size is critical in determining reproductive cell versus somatic cell differentiation." As mentioned immediately above, many observations that have been made on both wild-type and mutant strains are consistent with this hypothesis.

However, clear evidence indicating that large size is not an essential prerequisite for establishing reproductive potential comes from study of two categories of mutants: "late gonidial" (or *lag*) mutants and "somatic regenerator" (or *reg*) mutants. In *lag* strains, presumptive gonidia cannot be distinguished on the basis of size at the end of cleavage. But by the time the juvenile spheroids hatch, it is apparent that gonidial specification has occurred, because by then cells in the locations appropriate for gonidia have enlarged sufficiently that they are perceptibly larger than the neighboring somatic cells. These cells continue enlarging until they approximate the size of mature, wild-type gonidia, whereupon they cleave to yield another generation—which likewise lacks obvious gonidia at first (R. J. Huskey, personal communication; D. L. Kirk, unpublished observations). In *reg* mutants, gonidial cleavage is normal and generates juvenile spheroids that possess the normal complement of presumptive gonidia and somatic cells, which differ in size in the usual manner. The somatic cells differentiate normally and carry out the usual vegetative functions for a time. But then, in the most extreme form of the mutation, all somatic cells enlarge and redifferentiate (or "transdifferentiate") to become functional gonidia that cleave and give rise to a new generation of like phenotype (Starr, 1970; Huskey and Griffin, 1979). In both these categories of mutants, the capacity to differentiate as gonidia appears not to be related to size at the end of cleavage.

Pall (1975) dealt with the problem of the *reg* mutants by suggesting that somatic cells in this mutant "have lost their normal growth control and, after a period of growth, the enlarged somatic cells differentiate into reproductive cells." Presumably he would interpret the *lag* mutants in a parallel manner. Extended in this manner, it becomes clear that the Pall hypothesis speaks to the problem of "reproductive cell versus somatic cell differentiation" only in the narrow sense of the word "differentiation" (development of mature phenotype) and does not speak to the question of the initial process by which differential commitment occurs. It is clear that even if some minimal size were shown to be crucial for establishing reproductive potential, the key question would still be how the pattern of cells possessing critical size (or the potential to reach critical size) is specified in the cleaving embryo.

Even in this narrow sense, the Pall hypothesis apparently is not valid, however, as detailed study of *reg* mutants indicates. In his original description of such mutants, Starr (1970) reported that in the presence of sexual inducer, the somatic cells of asexual males could divide, without prior enlargement, to form small sperm packets. Huskey and Griffin (1979) subsequently demonstrated that such division without prior enlargement can also occur in the asexual phase, in *reg* mutants that exhibit less than complete regeneration. More recently, Baran (1984) has shown that even in mutants with the most extreme *reg* phenotype, somatic cells begin to make gonidia-specific proteins before they have enlarged appreciably. Thus, it seems clear that although large cell size is frequently associated with reproductive behavior, it is not an essential prerequisite.

The first explicit attempt to account for the gonidial determination process in *V. carteri* was made by Kochert and Yates (1970), based on experiments performed with forma *weismannia*. Their strains usually cleaved unequally at the 16-cell stage, so that adults normally possessed eight gonidia in two tiers of four gonidia each. However, when adult spheroids containing mature, uncleaved gonidia were immobilized in random orientation and ultraviolet (UV) irradiated, progeny derived from the gonidia present at the time of irradiation frequently exhibited missing gonidia. Although the experimental design did not permit correlation between sites irradiated and locations of later defects, the authors concluded—by explicit analogy with metazoan embryos that had been shown to have a UV-sensitive "germ plasm" responsible for germ cell specification (see Smith and Williams, 1975)—that "these experiments seem to indicate that mature, but undivided gonidia of *V. carteri* contain a morphogenetic substance (or substances) localized in their cytoplasm which controls the formation of gonidia in the daughter . . . ." Subsequently, Kochert (1975) also reported that "centrifuged gonidia cleaved to produce spheroids which were viable, but which had altered distribution of gonidia." Again a parallel was drawn with germ plasm-containing metazoan embryos.

Appealing as these parallels with animal embryos are, their validity has not been clearly established. The centrifugation studies are difficult to assess because details of experimental methods and results have not yet been published. And results of several subsequent studies have suggested that perhaps the UV irradiation results also bear reexamination. Results from Kochert's laboratory (Kochert and Crump, 1979) and others (Zagris and Kirk, 1980; Baran, 1984) have shown that UV irradiation can affect reproductive behavior of *V. carteri* cells in a number of ways that cannot easily be explained on the basis of destruction of localized morphogenetic substances. Furthermore, although one of the most compelling aspects of the irradiation study was the nonrandom nature of the gonidial deficiencies produced, Gilles and Jaenicke (1982) have shown that other unfavorable conditions lead to nonrandom gonidial deficiencies. Indeed, the specific results obtained by Kochert and Yates (nearly five times as many gonidia missing from the "anterior" tier of the adult as from the "posterior" tier—which correspond to tiers 2 and 1, respectively, in the terminology used above) are not necessarily the results one might predict from physical considerations, but they are consistent with the patterns of gonidial deficiencies seen by Gilles and Jaenicke as gonidial numbers decreased for other reasons (Section II,A,3,a). Until studies are done that would, for example, permit correlating the site of irradiation with the location of subsequent defects, it would appear prudent to suspend judgment concerning the existence of localized morphogenetic substances as the source of gonidial specification. And even if such localized determinants were shown to be a reality, we would still face the problem of how those determinants were normally localized, why missing determinants do not exhibit a random pattern, what goes on differently in pattern mutants to generate a different distribution of determinants, etc.

The first attempt to account for gonidial specification with a specific molecular model capable of generating a spatially defined pattern was made by Sumper (1979). He postulated that determination is mediated by the chemical nature of the "cell contacts" each cell shares with its neighbors; that during development cell contact proteins are synthesized in a hierarchical sequence such that when one type of contact protein is used up in establishing contacts, the next member of the hierarchy is produced; that cell contacts are not subdivided symmetrically, but pass unilaterally to one daughter at each cleavage; and that "the presence (or absence) of a defined contact-forming protein species signals gonidial differentiation." One of the most attractive features of this model was its apparent ability to account for both the temporal and spatial aspects of reproductive cell specification in both asexual and sexual spheroids with a single molecular mechanism.

Data that seemed both to support this attractive hypothesis and to provide identification of the postulated hierarchical family of contact proteins came from studies by the Sumper laboratory of the $^{35}SO_4^{2-}$ pulse-labeling patterns of intact spheroids (Sumper and Wenzl, 1980; Wenzl and Sumper, 1979, 1981, 1982). They observed rapid incorporation of sulfate, as such, into a number of high-molecular-weight extracellular glycoproteins. One of these, which they named "SSG 185" (for sulfated surface glycoprotein, 185 kDa), showed a labeling pattern that had many of the properties proposed by the model: Enzyme sensitivity studies showed that SSG 185 was external to the cell membrane. Incorporation into SSG 185 rose 10-fold at the time that the gonidia began to cleave. During early cleavage, incorporation into SSG 185 remained high, but the position of the band on SDS gels shifted downward, suggesting that perhaps SSG 185 was "a family of structurally related cell-surface glycoproteins" whose "synthesis . . . is controlled in a hierarchical manner" (Sumper, 1984), rather than a single entity. Prior to the differentiative cleavage, the position of the band on gels stabilized, but incorporation into it dropped at this time, then rose again after the differentiative cleavage and remained elevated throughout the remainder of embryogenesis. Adding to the attractiveness of this model and its candidate molecule(s) was the fact that in sexually induced male and female spheroids the time of minimal SSG labeling was shifted, just as would be predicted on the basis of the known shifting of the asymmetric cleavage divisions.

Subsequent work, however, has indicated that as interesting as these labeling patterns are, they (and SSG 185) appear to have no causal relationship to gonidial specification in the cleaving embryo. The labeling patterns described were obtained with intact spheroids containing cleaving embryos. But when the somatic cells and cleaving embryos were separated, either before or after exposure to radioisotope, labeled SSG 185 was observed in the somatic cell fraction, but not in the embryo fraction (Kirk and Kirk, 1983; Wenzl *et al.*, 1984). Furthermore, the molecule is not a cell-surface molecule, as originally postulated; it is a precursor of the extracellular matrix surrounding the somatic cells, where it becomes insolubilized shortly after its synthesis (Wenzl *et al.*, 1984). Any idea that SSG 185 might exert some critical influence on gonidial specification in the cleaving embryo from this distant vantage point is ruled out by the knowledge that gonidia, isolated from the somatic cells before this programmed synthesis has begun, undergo perfectly normal cleavage and gonidial specification (Starr, 1969, 1970). One interesting possibility is that information might be flowing in the other direction in the intact spheroid, that substances emanating from the embryo at different cleavage stages might influence the biosynthetic activities of somatic cells. This possibility could be tested by studying the time course of SSG 185 labeling in isolated somatic cells.

Although SSG 185 does not have the properties predicted by the original Sumper (1979) model, is the model itself sufficiently compelling to warrant continued consideration? Probably not, at least in its original form. Electron microcopy has revealed that *Volvox* embryos simply do not possess "cell contacts" that have the properties required by the Sumper model: Cells are linked only by cytoplasmic bridges during cleavage; these connections do not appear to be qualitatively different in different parts of the embryo, and they are divided equally between daughter cells at each division, not unequally as the Sumper model postulated (Green and Kirk, 1981). A more recent variant of the model which shifts the emphasis to a spatially differentiated "extracellular matrix space" within which "embryonic cells become embedded in differently composed microenvironments" (Sumper, 1984) appears to have no firmer empirical basis than the earlier, cell contact-based version: Scanning electron microscopy of cleaving embryos gives no evidence of any extracellular matrix in the vicinity of embryonic cells (Green and Kirk, 1981). More recent studies using quick-freeze, deep-etch analysis have confirmed that this is the one stage of the life cycle during which *Volvox* cells appear to be totally devoid of any extracellular matrix (U. W. Goodenough, J. Heuser, and D. L. Kirk, unpublished observations).

Jaenicke and Gilles (1982) proposed a variant of the Sumper model that attributed to two different membrane proteins separate control over the time and the sites of unequal cleavage divisions: a D-protein responsible for *d*ifferentiating the cells destined to cleave unequally, and a C-protein responsible for *c*ounting divisions and specifying when such differentiated cells should actually execute an unequal division. The concentration of the D-protein was postulated to vary in concentration spatially, as a consequence of insertion into the membrane of the uncleaved gonidium at the anterior pole followed by radial diffusion, and a candidate protein with such distribution was identified by FITC–Con A labeling (Jaenicke and Gilles, 1982). The model proposed that cells having more than some threshold concentration of D-protein would be competent to cleave unequally. The C-protein was postulated to be spatially homogeneous, but to fall in concentration stepwise at each cell division (as a consequence of dilution by new membrane at each cleavage); below some threshold concentration of C-protein, cells competent to cleave unequally would do so. Modifications of the concentrations and/or distribution were proposed to account for patterns of reproductive cell specification in sexual males and females.

The potential applicability of the C–D model appears to hinge on the mechanism by which pattern variations arise in embryos producing fewer than 16 gonidial initials. Gilles and Jaenicke (1982) assumed that gonidial deficiences arose as a consequence of precocious asymmetrical division in one or more of the cells with gonidium-forming potential. If, for example, tier 2 cells were to occasionally divide asymmetrically at the fifth (instead of the sixth) division—with the more posterior sister cell becoming a gonidial initial and the more anterior a somatic initial—the preferential loss of gonidia from the 2a tier would be accounted for. And this phenomenon could readily be accomodated within the C–D model by simply assuming that in such cases the concentration of the C-protein was sufficiently marginal that one or more tier 2 cells happened to pass threshold, and cleave asymmetrically, one division early. With yet lower concentrations of C-protein, tier 1 cells would be expected to undergo premature asymmetric division also, accounting for loss of gonidia from the 1a tier before loss of any from the 2p tier. Thus, if precocious asymmetric division proved to be the mechanism by which variations in gonidial pattern arise, the C–D model would have great explanatory power.

If, however, it were found that missing gonidia were not the result of this sort of *temporal* variability in asymmetric division, but were due to the failure of a *spatially defined subset* of presumptive gonidial precursors to divide asymmetrically at the sixth division, the C–D model, in its original form, would appear to have been invalidated. This is because the spatial variation in behavior of tier 2 cells that would be required to generate the gonidial deficiencies observed by Gilles and Jaenicke (1982) is exactly the opposite of that which their model (Jaenicke and Gilles, 1982) predicts: The model predicts that the 2a cells should have a stronger gonidium-forming potential than their 2p sister cells, because they are more anterior and hence at a higher position on the postulated D-protein gradient; but in individuals with 12–15 gonidia, it is the 2a cells, not the 2p cells, that most frequently fail to leave gonidial progeny.

Thus the crucial question is whether embryos that possess fewer than 16 gonidia do so because of temporal or spatial deviations from the ideal pattern of unequal cleavage divisions.

The reasoning that Gilles and Jaenicke used to support their assumption that premature unequal division of some cells could occur was indirect. The only observation provided in support of the idea was one Starr (1969) has also made, namely, that under extremely crowded culture conditions all asymmetric division occurs at the fifth cleavage, rather than the sixth, to yield progeny with only eight gonidia. From this, they extrapolated that "it is a reasonable assumption that the intermediate numbers of gonidia result from . . . asynchronous differentiation." However, Starr (1969), having observed that unequal division could occur one cycle prematurely in *all* cells under adverse conditions, specifically examined the question of whether premature unequal division ever occurred in *some subset of cells* under more favorable conditions. His observations led him to conclude quite unambiguously that it did not. Observations of this laboratory are in accord with those of Starr. Thus, until or unless contrary evidence is obtained, the explanatory power of the current C–D model is very much in question.

Where does this leave the problem? No model yet proposed appears to adequately explain all the observed facts of gonidial specification.

But as various models have been proposed and subjected to scrutiny, at least the specific features of gonidial patterning that any satisfactory model must account for have been more and more precisely defined. One important advance in this regard has been the description by Gilles and Jaenicke (1982) of the regular deviations from ideality that are observed. As they have put it, any model which seeks to explain the gonidial specification process must now "also explain the variability of the gonidial pattern; and, more important the kind of variation should be correlatable to the mechanism of differentiation involved." A second important advance has been the way in which the Jaenicke and Gilles (1982) C–D model has focused attention on the importance of both temporal and spatial aspects of the gonidial specification process and has raised the possibility that these two aspects are under separate control. Incorporation of pattern-forming mutants, of the sort described by Huskey *et al.* (1979a), into experimental explorations of this problem would appear to be an obvious and potentially productive approach for the future.

B. Spheroid Maturation

By the end of embryogenesis a miniature spheroid has been produced that contains, in proper orientation, all the cells that will be present in the adult. The remainder of the asexual cycle involves completion of cytodifferentiation, expansion of the spheroid by deposition of extracellular matrix, hatching, and preparation for a new cycle of embryogenesis. And as one generation begins to give rise to the next, somatic cells of the preceding generation undergo senescence.

1. *Cellular Differentiation*

Somatic cells and gonidia of adult *Volvox* are more than 1000-fold different in volume. They are nearly as different in cellular architecture, biosynthetic activities, and physiology as they are in size. Somatic cells are postmitotic cells specialized for motility, while the gonidia are nonmotile cells specialized for mitotic activity. Mature somatic cells are organized in a highly polarized fashion: Paired flagella, basal bodies, and contractile vacuoles, surrounded by mitochondria, are at the extreme anterior end. One large, cup-shaped chloroplast (with an eyespot in the anterior-most portion of one cusp and a single, basal pyrenoid) fills the posterior end of the cell. The nucleus, fringed by endoplasmic reticulum and Golgi elements, lies in between, nestled in the hollow of the chloroplast. Mature gonidia, in contrast, are organized with almost perfect spherical symmetry: A large nucleus with prominent nucleolus is suspended at the center by radiating strands of cytoplasm that pass from nucleus to periphery between the many radially arrayed vacuoles. The central cytoplasm is dominated by elements of the endoplasmic reticulum and Golgi systems, the peripheral region by chloroplast(s) and mitochondria, and the radial strands of cytoplasm contain a graded mixture of these organelles. Instead of the one pyrenoid and two contractile vacuoles characteristic of somatic cells, gonidia have numerous organelles of each type, distributed liberally through the cell periphery. The spherical symmetry of the mature gonidium is broken only at the point closest to the surface of the maternal spheroid, where the basal body complex lies attached to the plasmalemma, marking the future anterior pole of the embryo to be. Although flagellar outgrowth is initiated from these basal bodies at the end of inversion, the flagella are resorbed before contributing to motility. [For more details of cytological features see Kochert and Olson (1970a), Pickett-Heaps (1975), and Dauwalder *et al.* (1980).]

Development of the specialized features of the two cell types occurs predominantly after the completion of morphogenesis. At the end of inversion presumptive somatic cells and gonidia are cytologically indistinguishable (except in size). Although both have a polarized organization, they lack many of the features of mature somatic cells; cytologically they are undifferentiated embryonic cells. At this stage—possibly as a consequence of the fact that they remain interconnected by numerous cytoplasmic bridges—the two cell types make virtually identical polypeptides (Kirk and Kirk, 1983).

Several different light–dark regimens have been used to synchronize development (Yates and Kochert, 1972; Yates *et al.*, 1975; Kirk and Kirk, 1976; Wenzl and Sumper, 1979; Pomerville and Kochert, 1981; Baran, 1984; J. A. Zeikus, personal communication). In the light-dark regimens we have tested, morphogenesis is completed in the dark, but little cytodifferentiation occurs as long as darkness is maintained—even if it is extended by many hours. Within just a few hours after the lights come on, however, cytoplasmic bridges between the two cell types are broken, the cells assume their differing appearances, and they are found to be making extremely different patterns of polypeptides (Kirk and Kirk, 1983). Thus, the first question to be posed in this section is how does light influence cytodifferentiation?

a. *Differential Protein Synthesis.* The earliest change that has been detected after the lights come on, to end the dark period in which embryogenesis was completed, is a sweeping revision of the pattern of major polypeptides being synthesized by all cells. This change is extremely rapid, fully reversible (if the lights are turned off again), and sufficiently comprehensive that labeling patterns in light and darkness exhibit more differences than similarities (Kirk and Kirk, 1983). As would be expected, the polypeptides exhibiting selectively enhanced synthesis in light include chloroplast proteins made both on chloroplast ribosomes (the large subunit of RUBPcarboxylase) and cytosol ribosomes (the small subunit of RUBPcarboxylase and the chlorophyll *a/b* binding protein); however, they also include polypeptides (e.g., the tubulins) which are not destined for the chloroplast (Kirk and Kirk, 1985b).

Two types of studies indicated that this light-induced transition in protein synthetic activity is mediated almost entirely at the translational level: (1) Actinomycin D, at a level that is adequate to block heat shock-induced protein synthetic changes,[1] had very little effect upon the light-induced changes in protein synthesis in the same cells at the same time; (2) RNA fractions isolated from dark-adapted and illuminated cultures yielded extremely similar polypeptide products in heterologous *in vitro* translation systems (Kirk and Kirk, 1985b). By both methods some qualitative differences were detected due to differential transcription in the light; these are of considerable interest in their own right, but contribute in only a minor way to the overall change in protein synthetic pattern seen as a consequence of illumination.

The action spectrum for this light-mediated change in translational activity (Kirk and Kirk, 1985b) resembles that for phototaxis, which has recently been shown to be mediated by rhodopsin in *Volvox* and its relatives (Foster *et al.*, 1984). Therefore, studies are under way to test the hypothesis that rhodopsin also mediates light-induced translational regulation.

The evidence that this initial burst of protein synthesis in the light is causally related to the subsequent onset of overt cytodifferentiation in the presumptive somatic cells and gonidia is purely circumstantial, based on the temporal relationship between the two events. However, although many different dark–light regimens have been devised that successfully synchronize cytodifferentiation, and hence the life cycle, no alternative method of synchronization has yet been reported. (Growth is even more rapid, but asynchronous, in continuous light; heterotrophic growth in continuous darkness has not yet been reported for any member of the genus.)

A parallel effect of light on one specific aspect of cytodifferentiation has been detected by Coggin and Kochert. In addition to studying flagellar regeneration in *Volvox* (which exhibits numerous parallels with that of *Chlamydomonas*), they have been examining the curious dynamics of flagellar outgrowth by differentiating somatic cells. During inversion, both flagella of each presumptive somatic cell are about 2 μm long. As Starr (1970) reported, in the juvenile spheroid only one flagellum of each cell elongates at first; then after a delay of many hours the second elongates. Coggin and Kochert showed that it is always the right-hand flagellum of each cell (as viewed from the posterior pole) that elongates first, at a rate of 0.06 μm/min, reaching 12 μm in 3 hours. Some 6 hours later, the

---

[1] As in many other organisms (Schlesinger *et al.*, 1982) a sublethal heat shock (e.g., 42.5°C) induces a transcription-dependent synthesis of a number of protein species of defined MWs; however, in contrast with some other species, it does not interfere with other protein synthetic activities of the cell (Kirk and Kirk, 1985b). In addition to providing thermoprotection, heat shock in *Volvox* has a number of other interesting biological effects that are highly stage specific, including "autoinduction" of sexuality (Kirk and Kirk, 1985a; Section III,A,2).

left-hand flagellum begins its elongation at the same rate its partner had exhibited earlier; when it reaches 12 μm the two elongate together at half that rate until they eventually reach about 25 μm. In the light, elongation of the first flagellum begins 2 hours postinversion. But if individuals with either two short or one long and one short flagellum are put back in the dark, no flagellar outgrowth occurs. Cycloheximide applied at the same stages mimics these effects of darkness. Studies of stage-specific tubulin synthesis are under way (S. J. Coggin and G. Kochert, 1980, and personal communication).

From early in the light period through the remainder of the life cycle, the major newly synthesized polypeptides of somatic cells and gonidia, examined on one-dimensional SDS–PAGE gels, are strikingly dissimilar (Kirk and Kirk, 1983). However, when extant (as opposed to newly synthesized) polypeptides are examined on the same gel system, far fewer differences between the two cell types are apparent (D. L. Kirk and M. M. Kirk, unpublished observations). The probable basis of this apparent paradox has been elucidated in another study, which will now be reviewed.

Baran (1984) examined both the accumulated (silver-stained) and newly synthesized (^{35}S-labeled) polypeptides of gonidia and somatic cells, isolated at various points in the life cycle, on parallel two-dimensional gels. Regarding accumulated proteins of newly hatched (young adult) spheroids, he observed that "the majority of proteins are found in both cell types and displayed only a slight variation in their relative concentration." Differences in accumulated proteins were primarily (but not exclusively) in the form of specific additions to the common repertoire in somatic cells, which he attributed to the more "specialized" nature of those cells. When the newly synthesized polypeptides of cells at the same stage were examined, however, many more differences were detected: He estimated that about 20% of the labeled components were cell-type specific.

Somatic cells exhibited a more complex pattern than gonidia in terms of biosynthesis, just as they had in stained gels. As expected, of the components that had been classified as somatic cell specific on the basis of stained gels, those that exhibited label incorporation did so only in somatic cells. In addition, a number of somatic cell-specific components were detected on labeled gels that were not seen on stained gels. Chief among the latter was a family of intensively labeled components of very high molecular weight and heterodisperse charge; these probably correspond to sulfated precursors of extracellular matrix which Wenzl *et al.* (1984) have shown are rapidly incorporated into the maxtrix in an insoluble form shortly after their synthesis. As spheroids of increasing maturity were examined, somatic cells exhibited more progressive changes in both stained gel and labeling patterns than gonidia did. (Patterns observed with senescent somatic cells will be discussed in Section II,B,5.)

Certain aspects of gonidial labeling patterns paralleled those of somatic cells. For example, of the components that had been classified by Baran as gonidia specific on the basis of stained gels, any that exhibited label incorporation at all did so only in gonidia. But there were some surprises also. One unanticipated observation was that a number of the proteins that had been identified as "shared" proteins, on the basis of stained gels, exhibited label incorporation only in gonidia. As progressively more mature spheroids were examined, synthesis of these shared proteins continued to be observed in gonidia—but not in somatic cells. Baran interprets these data to mean that gonidia accumulate these proteins throughout their development and then pass them on (via cleavage) to their progeny somatic cells as an endowment, sparing those cells (which will never divide again) the need to make such proteins for themselves.

Even more remarkable, perhaps, was Baran's observation that certain polypeptides that had been identified as major gonidia-specific components on stained gels at all stages, but were not detected on labeled gels derived from gonidia at any stage of maturation, were labeled during early cleavage. The implication of these data seems to be that these components are selectively segregated into the presumptive gonidia during cleavage. If so, questions about their possible causal relationship to the gonidial specification process inevitably arise.

Unfortunately, technical problems in separating sufficient quantities of young somatic cells and gonidia prevented Baran from being able to analyze either accumulated proteins or protein labeling patterns in the critical period between the completion of embryogenesis and hatching (approximately one-third of the life cycle under his growth conditions). Hence, his provocative ideas about inheritence and sorting of proteins during cleavage could not be subjected to direct test by examining stained and labeled gels of proteins derived from the two cell types of very young spheroids.

Since preparation of monoclonal antibodies to single spots from two-dimensional gels is now feasible, a productive area for future investigation would appear to be the production and use of such an antibody to further explore the behavior of the gonidia-specific components Baran concluded were synthesized only during cleavage and (by inference) selectively segregated into presumptive gonidia.

b. *Differential Nucleic Acid Synthesis.* Two methods have been used to estimate the relative amounts of chloroplast and nuclear DNA in somatic cells and gonidia; both indicate that DNA is accumulated during gonidial maturation, but result in very different estimates of the amount and type of DNA accumulated. Kochert (1975) and his associates used a

chemical assay to estimate total DNA content and analytical ultracentrifugation to estimate the ratio of plastid to nuclear DNA in each cell type
of forma *weismannia*. They found that nearly mature gonidia contained 25
times the DNA content of somatic cells, of which 30% was plastid DNA
(as contrasted to 6.5% plastid DNA in somatic cells). These data indicate
that gonidia possess about 19 times the nuclear DNA and more than 100
times the plastid DNA of somatic cells. Coleman and Maguire (1982) used
microspectrofluorometry to estimate nuclear and plastid DNA contents of
individual cells of forma *nagariensis* at various stages of development.
They estimated that 5.5% of somatic cell DNA was cytoplasmic and
observed that somatic cells had an average of 2.3 plastid "nucleoids,"
with no significant change in numbers or fluorescence intensity during
maturation. (A somewhat surprising 5% lacked any detectable nucleoids.)
They observed that gonidial fluorescence varied uniformly with gonidial
size and observed about a 50-fold difference in fluorescence between the
largest and smallest gonidia. However, they attribute all of this difference
to accumulation of plastid DNA, since they found no significant difference in nuclear fluorescence between somatic cells and gonidia.

Labeling studies have not clearly resolved this question of whether
gonidia do or do not accumulate nuclear DNA during maturation. Yates
and Kochert (1976) observed that incorporation into total DNA rose prior
to cleavage, peaked during cleavage, and then declined during expansion,
but for technical reasons a complete picture of incorporation into the two
DNA fractions of the two cell types was not obtained. Margolis-Kazan
and Blamire (1976) concluded that "while nuclear DNA is made to some
extent throughout the life cycle, cytoplasmic DNA synthesis appears to
occur at discrete intervals," but the labeling periods they used (10–50
hours) greatly complicate interpretation of their data.

Yates and Kochert (1976) showed that both cytoplasmic and chloroplast rRNA are made (in somewhat varying proportions) in both cell types
over most of the life cycle, but with a perceptible decline during cleavage.
Kochert (1975) suggests that gonidia contain enough rRNA "to supply all
the cells of the embryo with ribosomes during cleavage." Weinheimer
(1973, 1983) has shown many stages of development to be exquisitely
sensitive to actinomycin inhibition (see Section II,A,2,a), and some preliminary reports bearing on messenger RNA have appeared (Caplan and
Blamire, 1980), but detailed studies of transcriptional activities throughout the life cycle remain to be reported.

c. *Genetic Control of Cytodifferentiation.* Unquestionably there must
be many genes involved in the differentiation of somatic cells and gonidia.
But ever since Starr (1970) described the bizarre developmental behavior
of Mendelian mutants with "fertile somatic cells" (now known as "so-

matic regenerator,'' or *reg* mutants), the locus responsible for such behavior has held particular fascination for students of *Volvox* development.
As outlined in Section II,A,2,b, gonidial development in *reg* mutants is
normal and cleavage generates spheroids with the normal complement of
cell types. But after apparently differentiating normally (and executing
vegetative functions normally for a time) somatic cells redifferentiate as
fully functional reproductive cells.

Baran (1984) studied protein synthetic patterns of regenerating cells.
He found that in newly released adults such cells synthesize a mixture of
polypeptides that he had identified as somatic cell specific and gonidia
specific in wild type, as if the regenerating cells were expressing both
developmental programs at once. Unfortunately, however, he was unable
to study such cells in younger individuals to determine how early this
pattern of mixed gene expression began.

Huskey and Griffin (1979) described 39 *reg* mutants that fell into four
phenotypic classes and explored their genetic relationships: Class I (33
isolates, presumed to be equivalent to null mutants) were characterized
by regeneration of all somatic cells, class II (3 isolates) by regeneration of
cells only in the posterior $\frac{3}{4}$ (the gonidial end) of the spheroid, class III (2
isolates) by a random intermixture of regenerating and nonregenerating
cells, and class IV (1 isolate) by a pattern similar to that in III, but with
''regeneration'' occurring earlier and ''nonregenerating cells'' failing to
ever complete somatic cell differentiation.

In 57 two-factor crosses involving all 38 members of classes I, II, and
III, in different pairwise combinations, only one wild-type recombinant
was detected in more than 9000 progeny. Therefore, these mutations were
tentatively ascribed to allelic variants of a single locus, called *regA*. The
validity of this assignment was enhanced by the demonstration that phenotypes equivalent to all three classes could be displayed by a single cold-
sensitive mutant under different temperature-shift regimens. The class IV
mutation segregated independently of the other three and was assigned to
a second locus, called *regB*. But because of the entirely different cellular
phenotype of the class IV mutant (see above), inclusion of class IV in the
reg category is questionable at best (Baran, 1984). All the cases in which
regeneration occurs in cells that appear to have first differentiated as
somatic cells appear to be a result of mutation at a single locus: *regA*.

Huskey and Griffin (1979) concluded that the function of the *regA* product was necessary ''to establish and/or maintain the terminally differentiated state'' of somatic cells. This implies a positive action of the *regA*
product. However, an alternative interpretation is possible. Under conditions of asexual reproduction, regenerating cells like gonidia give rise to
asexual spheroids whose somatic cells repeat the cycle. But in the pres-

ence of the sexual inducer, regenerating cells give rise to sexual spheroids that produce either eggs or sperm (depending on the mating type of the mutant individual) plus somatic cells. The "somatic" cells of the sexual *reg* spheroids then give rise to more gametes directly. Indeed, following exposure to the sexual inducer, regenerating cells of asexual *reg* males occasionally divide without first enlarging, to form sperm packets directly (Starr, 1970). Thus, it appears that in *reg* mutants, somatic cells, which normally are excluded from reproducing, have all pathways of reproductive behavior opened up to them. Our working hypothesis, therefore, is that the *regA* product acts as a negative regulator to repress all pathways of reproductive development in the somatic cells. In either case, this locus is clearly of great interest as a developmentally significant "control gene."

When, and in what cell type, is the *regA* gene expressed? Conventional thinking would suggest that it must be expressed in the somatic cells, since the behavior of these cells is dramatically modified in *reg* mutants, although the behavior of gonidia is not perceptibly different than in wild type. However, in an important paper, Kurn *et al.* (1978) provided suggestive evidence to the contrary. They showed that the frequency with which *reg* mutants appeared spontaneously in a wild-type population (2.5×10^{-6}) was not significantly different from that observed for appearance of spontaneous BUdR-resistant mutants (3.4×10^{-6}). Because wild-type somatic cells cannot reproduce, most spontaneous mutations that occurred in them would not be detected. However, Kurn *et al.* argued, a mutation that conveyed reproductive capacity on somatic cells should be an exception to this rule, and because there are 100 times as many somatic cells as gonidia in a spheroid, one should expect a higher observable mutation rate for *reg* than for other loci—if such mutations were capable of affecting the behavior of the somatic cells in which they arose. The fact that the observed mutation rate for *reg* was not elevated was interpreted to mean that mutations that arose in somatic cells could not be expressed. This, in turn, was taken to mean that the gene must be expressed during, or prior to, embryogenesis. There are two possible flaws in this argument. The first is that it was never clearly established that the small sector of a clone which would exhibit the mutant phenotype if one somatic cell had mutated and reproduced could have been reliably detected by the screening method used (naked-eye inspection of plates). The second (and possibly more serious) problem is that there is no independent way of establishing the mutation rate in somatic cells. The failure to recover more mutations might only mean that the spontaneous mutation rate in somatic cells (which are nondividing) is two or more orders of magnitude lower than in dividing embryonic cells. Such a differential seems not unreasonable, since chemical mutagens are clearly more effective in producing recoverable mutations if applied during cleavage than at any other time (Huskey *et al.*, 1979a). Therefore, suggestive as this experiment was, it appears wise to suspend judgment on the validity of the conclusions drawn from it.

Temperature-shift experiments performed by Huskey and Griffin (1979) with their cold-sensitive mutant indicated that the critical period for action of the *regA* gene product was after cleavage of the embryo and before hatching of the resultant young adult spheroid (i.e., during somatic cell differentiation). More recent studies are said to have narrowed the critical period further to a 2- to 4-hour peroid during and immediately after the completion of inversion (Baran, 1984). These results are not necessarily at odds with those of Kurn *et al.* (1978), however. It is conceivable that the time of gene "expression" (i.e., transcription) could occur well in advance of the critical period for action of the gene product. Other studies, to be discussed later, suggest that this may in fact be the case.

Zagris and Kirk (1980) discovered that UV irradiation of spheroids 3 hours after inversion (i.e., approximately the stage when the above studies suggested that the *regA* gene product is acting in somatic cells) led to selective generation of *reg* mutants with extremely high efficiency. The UV exposure employed had little if any mutagenic effect at most other periods of the life cycle. The window of sensitivity was brief, less than 1 hour. The effects of UV could be completely reversed by immediate exposure to visible light, but were fixed within 3 hours in the dark. The effects were exerted entirely on the gonidia; no effects on somatic cells could be detected. Gonidia irradiated under optimum conditions subsequently exhibited one of three behaviors: About two-thirds failed to exhibit any gonidial differentiation, but instead developed as large somatic cells, complete with prominent eyespots and flagella. The remaining third developed as gonidia and cleaved, but only about two-thirds of these gave wild-type progeny. The remainder—10% of all presumptive gonidia in the target population!—yielded *reg* progeny. No mutations of other types could be detected above spontaneous mutation rates.

Of the many hundred such UV-induced mutants identified, many have been observed over numerous generations, and one has now been followed for over 1000 generations and repeatedly subjected to mutagenesis. No spontaneous or induced revertants to wild type have yet been detected. The absence of reversion suggests the possibility of a deletion. In crosses of UV-induced *reg* to the temperature-sensitive *regA* mutant described by Huskey and Griffin (1979), conditional and nonconditional mutants segregate in a 1:1 ratio with no wild-type recombinants (K. S. Huson and D. L. Kirk, unpublished observations). Thus, these UV-induced lesions appear to be at the *regA* locus.

If (as results summarized earlier have suggested) the *regA* gene is normally expressed during embryogenesis and its function is to prevent reproductive development, it would obviously have to be turned off in the presumptive gonidia before they could develop as reproductive cells. The fact that UV sensitivity occurs in gonidia, at about the time the gene product is apparently acting in somatic cells, led us to propose that the mutability of the gene is related to the mechanism by which the gene is normally turned off. Specifically, we postulated (1) that shortly after completion of embryogenesis a DNA rearrangement normally occurs in gonidia at the *regA* locus to inactivate it, (2) that UV damage induces an error-prone repair system (Haseltine, 1983) that may either interfere with completion of the inactivating rearrangement (resulting in continued expression of the gene and transformation of gonidia into somatic cells) or cause a deletion of the rearranging sequence (resulting in a nonrevertible *regA* mutation).

This model led to two testable predictions: (1) There should be a second UV-sensitive period before or at the beginning of cleavage reflecting "reactivation" of the gene, in preparation for its expression. (2) Other agents which either induce error-prone repair or interfere with normal DNA repair processes should selectively and efficiently induce *reg* mutations at one or both of the UV-sensitive periods. Both of these predictions have been tested in two laboratories with interesting results (Baran, 1984; Baran *et al.*, 1986; Kirk *et al.*, 1986).

Both laboratories have detected the predicted second UV-sensitive period at or just before cleavage (differences in timing may reflect differences in growth conditions). Both laboratories have detected stage-specific, high-frequency, selective induction of *reg* mutants during the postinversion UV-sensitive period by agents with known effects on DNA. The two most effective, and hence best studied of these agents are bleomycin (Kirk laboratory) and novobiocin (Baran); the former is a chromosome-breaking agent (Vig and Lewis, 1978), and the latter inhibits repair of UV-induced DNA damage (Collins and Johnson, 1979). More than 20 other mutagens, carcinogens, intercalating agents, etc., have been tested; most of these exerted little or no effect under the conditions tested.

One important observation has come out of these studies that was unanticipated on the basis of the initial UV studies, or precedents from other systems, and has caused a complete reexamination of the working hypothesis. In the initial studies (Zagris and Kirk, 1980), all the mutants detected were first-generation progeny of the irradiated (presumptive) gonidia. In addition to—or, in the case of novobiocin, instead of—this pattern, the chemical agents gave rise to a high frequency of stable mutations that did not appear until one generation later, and then in only a portion of the second-generation progeny. Baran (1984) has seen a similar kind of delayed appearance of mutants among second-generation progeny of UV-irradiated spheroids also, particularly when the irradiation is somewhat later than the UV-sensitive window originally detected by Zagris and Kirk.

Attempts to fit all these observations together into a credible, comprehensive hypothesis about what is going on at the *regA* locus at these critical stages of development continue. The only thing that is abundantly clear by now is that the more this locus is studied, the more interesting its behavior seems to be.

2. *Postdifferentiative Events*

a. *Expansion: Matrix Deposition.* Most of the enlargement of a *Volvox* spheroid that occurs after embryogenesis can be attributed to deposition of extracellular matrix (ECM); and more than 90% of the volume of a mature spheroid is ECM. Synthesis of ECM components undoubtedly constitutes one of the major activities upon which the cells expend energy. Recent studies (Section III,B,1) suggest that in addition to its obvious role in maintaining spheroid integrity, the ECM plays a crucial role in developmental regulation.

Morphological aspects of the ECM structure, and its secretion, have been described for several *Volvox* species (Kochert and Olsen, 1970a; Burr and McCracken, 1973; Sessoms, 1974; McCracken and Barcellona, 1976, 1981; Birchem, 1977b; Dauwalder *et al.,* 1980; Kirk *et al.,* 1985). Distinctive fibrous layers surround the entire spheroid (except where penetrated by flagella), envelop each cell, and surround cells at some distance to form contiguous cellular compartments. One of these layers, a "tripartite layer" near the surface of the spheroid, is a highly conserved feature, seen in every Volvocalean species yet examined. Spaces enclosed by these various fibrous layers are filled with less well-organized, gelatinous, or mucilaginous materials.

Lamport (1974) has shown that, like the plant cell wall and the ECM of animals, the *Volvox* ECM contains hydroxyproline-rich glycoprotein (HPRGs). A subsequent study from his laboratory (Mitchell, 1980) reported that nearly half the total spheroid hydroxyproline could be extracted with SDS–mercaptoethanol, leaving the spheroids morphologically intact. After mechanical disruption, extracted spheroids could then be separated by centrifugation into insoluble (fibrous) and soluble (mucilaginous) fractions that accounted for about 40 and 20% of the total spheroid hydroxyproline and had amino acid compositions of 40 and 25% hydroxyproline, respectively. Only the SDS-extracted HPRG was more extensively analyzed. After purification, HF-deglycosylation, limit digestion with trypsin and pronase, and further purification, most of the hydroxyproline was recovered in a fraction that was more than 90% hydroxyproline and had an apparent MW >100K. Sequencing data indicated a string of 750–1000 hydroxyproline residues interspersed with only traces of other amino acids. Whether this polyhydroxyproline exists as such in life or is merely the protease-resistant core of some larger and more complex protein is presently indeterminant.

One set of HPRGs clearly is located in the tripartite layer near the surface of the ECM. Roberts (1974) demonstrated that in *Chlamydomonas* this structure was composed of HPRG in a crystalline organization and that the crystalline lattice in the equivalent region of *Volvox* ECM was virtually indistinguishable. Disassembly and *in vitro* reassembly of this crystalline structure from *Chlamydomonas* has been described (Hills *et al.*, 1975; Catt *et al.*, 1978), and the crystals have been shown to consist of three HPRGs of distinctive morphology in stoichiometric proportions and distinctive arrangements (Goodenough and Heuser, 1985). More recently, the *Volvox* equivalent has been recrystallized and shown to be composed of three similar HPRGs, albeit of distinctly different sizes (U. W. Goodenough and J. Heuser, unpublished data).

Another of the HPRGs is the sulfated matrix component SSG 185 identified by the Sumper group and discussed in Section II,A,3,c; it has been shown to be 34% hydroxyproline (Wenzl *et al.*, 1984). Although the localization of this component in the matrix is as yet uncertain, it is unlikely to be one of the components of the crystalline layer, since in the work just cited it was shown to remain with the insoluble matrix under extraction conditions that solubilize the crystalline layer.

In this laboratory, a number of monoclonal antibodies have been produced which react with specific regions of the ECM and specific glycoproteins on Western blots (Hoffman, 1984; Hoffman and Kirk, 1986); these should aid in further correlations of particular glycoproteins with aspects of ECM architecture.

The primary sugars found in the ECM by Mitchell (1980) were glucose, mannose, galactose, and arabinose; Jaenicke and Waffenschmidt (1981) reported predominantly arabinose, xylose, and mannose. A tunicamycin-sensitive, dolichyl phosphate-based glycosylation system in membrane fractions of *Volvox* cells has been described (Muller *et al.*, 1981). *In vitro* studies show that a chitobiose-(mannose)$_5$ chain is built up on membrane-bound dolichyl diphosphate, much as in animal cells (Bause *et al.*, 1983). Although transfer of the oligosaccharide to mammalian proteins has apparently been accomplished (Jaenicke, 1982), the nature of the *Volvox* protein receptor(s) has not been reported. *In vivo*, tunicamycin has a marked effect on protein glycosylation and secretion (as revealed by

PAGE analysis of proteins from cells and medium); it also interferes with gonidial maturation, cleavage patterns, and hatching of juvenile spheroids (Kurn and Duskin, 1982). The extent to which this dolichyl-P system is involved in biogenesis of ECM components is as yet unclear; but since expansion is not reported to be tunicamycin sensitive, it may be that its principle role is in constructing cell-surface glycoproteins.

In addition to glycoproteins, the ECM contains ill-defined polysaccharides. Contrary to some assertions in the early literature, there is no evidence that cellulose is present in any member of the Volvocales; but Mitchell (1980) did detect a noncellulosic glucan in *Volvox* ECM. Jaenicke and Waffenschmidt (1981) reported the existence of a lysozyme-sensitive polysaccharide fraction which was mainly composed of mannose and arabinose. Further definition of the ECM polysaccharides is awaited.

Many ECM components, apparently both glycoproteins and polysaccharides, are sulfated (Burr and McCracken, 1973; Wenzl *et al.*, 1984). Willadsen and Sumper (1982) have described both a sulfokinase and a sulfotransferase system in a particulate fraction of *Volvox* homogenates that are capable, respectively, of synthesizing phosphoadenosylphosphosulfate (PAPS) *in vitro* and using PAPS as a donor to sulfate a protein that probably is the "SSG 185" described in Section II,A,3,c. Details of the sulfation sites on glycoproteins and polysaccharides have not been reported.

Based on the report of a polyhydroxyproline as a major component of the ECM, and with the realization that all CCX codons code for proline, we probed the *Volvox* genomic bank with synthetic poly(dG)·poly(dC), searching for putative polyproline-coding sequences. So many clones hybridized that not all of them could reasonably be expected to code for ECM components; but we suspect some of them do. (The possible significance of certain others is discussed in Sections V,C and VI.) One such clone has been partially sequenced and found to contain runs of tandemly repeated C-rich simple sequences. If translated, these C-rich regions would code for extremely proline-rich (hence potentially hydroxyproline-rich) peptides. Probes from two of these cloned sequences detect several discrete bands on a Northern blot of RNA isolated from cells at the peak of matrix deposition (Harper, 1986). Further details are given in Section V.

b. *Release: Matrix Degradation.* Midway through expansion the juveniles leave the parental spheroid. Prior to their release, cells overlying each juvenile begin to rotate and then tumble free of the spheroid; each juvenile then swims away through a hole that is precisely its own diameter. After the juveniles have departed, the holes do not enlarge further; a day later the parental hulk retains holes of virtually the same size, marking the former locations of its progeny.

Release is presumed to involve localized enzymatic degradation of the parental matrix. Jaenicke and Waffenschmidt (1979, 1981) have purified a trypsin-like serine protease that accumulates in the culture medium 2–6 hours before release. Although the purified enzyme is capable of degrading formalin-fixed (or to a lesser extent heat-killed) spheroids, no activity was detected using live spheroids. Nonetheless, Jaenicke and Waffenschmidt infer that this is the enzyme responsible for release. If so, the mechanism by which it acts *in vivo* to yield such controlled, localized lesions in the parental wall, with no effect on juveniles, remains mysterious.

3. *Senescence and Death of the Somatic Cells*

In each life cycle, more than 99% of the cells in a *Volvox* spheroid—all of the somatic cells—undergo programmed senescence and death. Pommerville and Kochert (1981, 1982) have examined these processes in detail; the following summary is abstracted from their reports.

During senescence, a number of degradative cytological changes— shrinkage of the cytoplast, disorganization of organelles, accumulation of lipid droplets—were seen to develop progressively. Viability of cells of a particular strain (whether measured by dye exclusion or loss of chloroplast autofluorescence) was lost on a predictable time course under constant conditions. When culture conditions, such as temperature or pH, were varied, senescence was accelerated or delayed, but only in proportion to the effects observed on other aspects of the life cycle, such as gonidial maturation. However, under the same conditions two strains exhibited very different life expectancies; whereas in cells of forma *naguriensis* an abrupt drop in viability began 96 hours after the last cell division, cells of forma *weismannia* survived 48 hours longer. Together these data suggest that senescence is as much part of the genetically encoded developmental program as reproduction.

The decline in viability was preceded by a decline in photosynthetic assimilation of $^{14}CO_2$ and accompanied by a precipitous loss of cellular protein. But when photosynthesis was interrupted artificially by placing middle-aged cultures in the dark for 24 hours, senescence was not induced prematurely; instead it was delayed 24 hours. Similarly, when protein synthesis was partially inhibited for 72 hours, the life-span of the cells was extended 48 hours. Both these results lead to the important conclusion that somatic cell senescence is a metabolically active process and that the losses of photosynthetic activity and protein normally seen are consequences, rather than causes, of senescence.

Two studies have examined protein synthetic patterns in senescent cells (Hagen and Kochert, 1980; Baran, 1984). Both observed a progressive loss of the capacity to take up and incorporate precursors and a progressive simplification of the patterns of newly synthesized polypeptides. Neither detected synthesis of novel polypeptides during this period that might account for the delay in senescence observed when protein synthesis is inhibited. cDNA cloning is now being used in an attempt to identify such senescence-specific molecules (J. C. Pommerville, personal communication).

III. Sexual Development

In *V. carteri,* in distinction to its better known and simpler relative, *Chlamydomonas,* sexuality cannot be elicited by deprivation of nutrients. One reason for this may be that the gonidia of *V. carteri* never differentiate directly into gametes, but instead go through a complete round of asexual reproduction after sexual induction, producing progeny that contain gametes (Kochert, 1968; Starr, 1969). This developmental pattern would hardly be compatible with a signal based on nutrient deprivation.

A. INDUCTION OF SEXUALITY

1. *Nature of the Sexual Inducer*

The sexual inducer of forma *nagariensis* is a single-chain glycoprotein of about 30 kDa; the protein portion is not unusual in composition (no hydroxyproline) and the carbohydrate portion (about 45% of the total) contains seven sugars—pentoses, hexoses, and hexoseamines (Starr and Jaenicke, 1974). It is one of the most potent biological effector molecules known: It exhibits full effectiveness at 6×10^{-17} M, and one mature sexual male produces enough inducer to convert many millions of other males and females to the sexual pathway (Starr, 1970; Gilles *et al.,* 1984). The sexual inducer of forma *weismannia* is superficially similar. It is a glycoprotein of about 32 kDa containing a similar array of sugars; both have a pI above 10 (Kochert and Yates, 1974; Kochert, 1981). But these inducers are as specific as they are potent: The two formas show no cross-inducibility. More striking is the fact that protease digests give little if any evidence of peptide homologies between the two (Kochert, 1981)!

2. *Production of the Inducer*

Sexual inducer is released by mature males at about the time they release sperm packets (Starr, 1970). The kinetics of inducer production by sexual males has been studied by Gilles *et al.* (1981). Pulse–chase labeling experiments indicate that the peptide backbone of the sexual inducer begins to be produced in the presumptive sperm-forming cells more than

30 hours before any active inducer can be detected in the medium. It is apparently stored in an inactive form, however, because its presence can neither be detected by bioassay of cell lysates (for inducer activity) nor by immunofluorescence (using antibody raised against the active inducer). Just as the mature sperm packets are to be released, inducer becomes detectable by immunofluorescence in the interstices of the sperm packet. Then, after the sperm packets are released, the inducer content of the medium (measured by bioassay) rises exponentially for 10 hours. Tunicamycin added at the time of sperm packet release blocks appearance of inducer in the medium. This is interpreted to mean that the stored inducer is glycosylated at the last moment and that glycosylation is essential for both release and biological activity of the inducer.

How does the sexual cycle get started? In response to inducer produced by sexual males, both asexual males and asexual females may be diverted to the sexual pathway. When so diverted, males make more inducer. But where does the first male come from? Starr (1972b) postulated that "first males" might all arise by mutation at a spontaneous sexuality (now known as constitutive sexuality, or sex^c) locus. He went on to suggest that "it would be interesing from an evolutionary point of view to contemplate the relative survival value of a sexual system that can be initiated only by a spontaneous gene mutation rather than by some combination of environmental factors."

As a first test of this hypothesis, Callahan and Huskey (1980) monitored uninduced cultures for spontaneously sexual individuals. In the female strain, sexual individuals arose at a rate of about 10^{-5}, which is equivalent to the spontaneous mutation rate measured for other loci—when the existence of three sex^c loci (see Section III,B,3) is taken into consideration. All spontaneous females that were mated yielded sex^c progeny in Mendelian proportions, indicating that in females all cases of spontaneous sexuality can be accounted for by mutation. In male cultures, the incidence of spontaneously sexual individuals was nearly 10-fold higher than in females. But because not all such first males yielded sex^c progeny when crossed to wild-type females, Callahan and Huskey concluded that "the occurrence of spontaneously sexual males . . . must be accounted for by some other mechanism in addition to mutational events." (They also interpreted certain epistatic interactions observed among loci affecting sexuality to indicate the existence of an "inducer-independent" pathway of sexual development; see Section III,B,3.) Turning their argument around, however, the fact that about half of the first males did yield sex^c offspring (R. J. Huskey, personal communication) indicates that mutation is an important—if not the only—source of first males. Furthermore, the rate at which bona fide mutant males arose in their study appeared to be

·significantly higher than the rate at which mutations are detected at other loci.

Recently, Weisshaar *et al.* (1984) have reexamined this problem. In contrast to Callahan and Huskey, they observed the same high frequency of spontaneous sexuality (about 10^{-4}) in male and female cultures. The evidence that these individuals were all mutants was indirect and came from analysis of the patterns in which sexual individuals occurred in the parental spheroids bearing them. In no case were all the progeny within a spheroid sexual; in most cases only one was. But parents bearing either 2, 4, or 8 sexual progeny were seen, and in all such cases the sexual progeny were arranged in a pattern that was consistent with known cell lineages— and hence consistent with the hypothesis that sexual specification had segregated at one particular cleavage division in the development of the parental spheroid. Although suggestive, this evidence does not appear to rule out unambiguously the possibility that it was an epigenetic rather than a genetic feature which segregated during cleavage. An unspecified high frequency of reversion to wild type (which has not previously been reported for *sex^c* mutants) was also observed in this study. Despite the absence of genetic evidence, considered with the data of Callahan and Huskey, this study certainly raises the distinct possibility that some mechanism may exist for selectively elevating the mutation rate at one or more of the *sex^c* loci. If so, elucidation of that mechanism would appear to be a most interesting avenue for future research.

Mutation is not the only route to initiating sexuality, however. It has recently been discovered that at one stage of the life cycle heat shock triggers immediate production and release of inducer in both asexual females and asexual males, resulting in a phenomenon that has been termed "autoinduction" of sexuality[2] (Kirk and Kirk, 1985a). This is clearly not a case of inducer-independent sexuality, since inducer is detected in the medium shortly after heat shock, and if Con A or anti-inducer antibody is added to the medium to inactivate it, no sexual progeny are produced by the heat-shocked individuals. The amounts of inducer produced are small compared to those produced by sexual males, but they suffice to turn the entire population sexy.

---

[2] This observation may provide an explanation of the fact that Powers (1908) reported that he was unsuccessful over several years in obtaining samples of *V. carteri* with sexually reproducing individuals until a collection was made in a pond that was "decidedly warm to the touch." In this paper he also said, "in the full blaze of Nebraska sunlight *Volvox* is able to appear, multiply, and riot in sexual reproduction in pools of rain water of scarcely a fortnight's duration."

B. Response to the Inducer

1. *Mechanism of Inducer Action*

The ultimate target of inducer action obviously is the gonidium of the asexual spheroid, which responds by cleaving to produce a sexual rather than an asexual offspring. Hence it has generally been assumed that the gonidium is also the site of initial inducer action. But studies of where and how—if at all—the inducer interacts with the gonidium have been complicated by several factors. The first is the exquisite sensitivity of the system. As noted earlier, the inducer is active at $6 \times 10^{-17} M$. Pall (1973) studied the concentration dependence of inducer action, found it to be second order (indicating a bimolecular rate-limiting reaction), and concluded that "only two molecules of inducer are required to induce a gonidium . . . ." Detecting the site of critical interaction, if only two molecules were involved, would tax even the most powerful of modern methods. Second is the fact that gonidia in isolation apparently will not respond to inducer, even though they remain healthy and capable of perfectly normal asexual development. Complicating the matter further has been the repeated observation (Kochert, 1968; Kochert and Crump, 1979; Callahan and Huskey, 1980; Gilles *et al.*, 1984) that as potent as the inducer appears to be from a concentration standpoint, its action is extremely slow. Inducer must be present for many hours during gonidial development in order to be effective; hence an immediate and irreversible effect of inducer addition is not to be anticipated.

Kochert (1981) labeled the inducer of forma *weismannia* with ^{125}I and performed binding tests with whole spheroids. Binding was neither saturable nor species-specific. Studies with isolated gonidia did indicate saturable as well as nonsaturable components of the binding reaction; but the species specificity of this gonidial binding was apparently not determined. By explicit analogy to animal systems involving glycoprotein hormones that regulate reproductive cell behavior (i.e., gonadotropins), Kochert (1981) proposed that the binding of inducer to the gonidial membrane might activate an intracellular cyclic nucleotide/protein kinase cascade. He detected cAMP and phosphodiesterase activity in gonidia, but was unable to affect gonidial behavior with analogs, inhibitors, or activators that are effective in other cyclic nucleotide-based systems.

Gilles *et al.* (1983) reported a transient, rapid decrease in phosphorylation of a high-molecular-weight matrix glycoprotein following exposure to inducer. Stimulated by this observation, these investigators reopened the question of where the inducer is bound and where it exerts its primary effect; they have concluded that the answer to both questions is the matrix (Gilles *et al.*, 1984). Attempts to localize the inducer within the reacting spheroid by use of fluorescent antibody were unsuccessful; so they took a different approach. First they titrated medium containing a marginal concentration of inducer by adding increasing numbers of spheroids. From the steep decline in inductive effectiveness observed, they not only concluded that the spheroids were binding the inducer competitively, but also calculated that at limiting concentration the inducer is concentrated about 300-fold by the spheroids, and that more than 300 molecules per spheroid (about 20 molecules per gonidium) were essential to induce any sexual development. They then performed a similar experiment in which they added isolated somatic cells, gonidia, or matrix (from equivalent numbers of spheroids) to a system containing a small number of target spheroids and a marginal amount of inducer; matrix was most effective in preventing induction of sexuality in the test organisms, so they concluded that the matrix was the primary site of inducer binding. (Gonidia were also quite effective, however, even after pronase digestion to remove matrix.) In what was perhaps the most provocative experiment of this series, they then showed that isolated gonidia, which do not normally respond to inducer, do respond to some extent (20%) if exposed to inducer in the presence of isolated matrix.

In a further series of experiments aimed at discerning possible connecting links between inducer binding and gonidial response, these authors then tested a series of reagents for inhibitory activity. As previously reported (Kurn, 1981), Con A blocked induction, but monovalent (succinyl) Con A did not. Most of the inhibitory effects they observed (e.g., with actinomycin, anisomycin, tunicamycin, monensin, A23184, and polycations) are difficult to interpret unambiguously because of accompanying nonspecific side effects. However, the fact that IBMX (a phosphodiesterase inhibitor) could act, either alone or synergistically with cAMP or cGMP, to prevent sexual induction took on added significance when it was demonstrated that beef heart 3′,5′-cyclic nucleotide phosphodiesterase could induce sexuality! This is the first published report of another molecule capable of mimicking the action of the inducer. Controls to demonstrate that the effect was not due to inadvertent contamination by inducer included the demonstration that the effect could not be blocked with Con A, as inducer activity is. Other types of phosphoesterases were ineffectual.

It is unlikely that either the cyclic nucleotides (which were underivatized) or the beef phosphodiesterase penetrated the cells to affect intracellular metabolism directly. This, combined with the earlier observation of changes in matrix–protein phosphorylation at the time of inducer addition, plus (unpublished) evidence for a phosphodiesterase in the matrix, led the authors to propose that a complete *extracellular* cyclase/diesterase/kinase/phosphatase system exists in the spheroid and is responsible for somehow amplifying and transducing the signal provided by the inducer into one effective in redirecting the activities of the gonidium. Since that time, analytical data have been obtained which demonstrate a difference in cAMP concentrations of matrix, cells, and medium of induced versus uninduced cultures (Gilles *et al.*, 1985). Precedent for an extracellular cyclic nucleotide regulating entry into a new developmental pathway has, of course, been well established in studies of the cellular slime molds. Nevertheless, the concept that this may be the mechanism by which the sexual inducer acts in *Volvox* is a novel departure and certainly warrants further attention.

One other set of experiments reported by Gilles *et al.* (1984) provides the clearest information to date on the time course of inducer action and its reversal. A series of experiments were performed to determine percentage of induction as a function of time of inducer addition and removal, as had been done previously by others (Kochert, 1968; Kochert and Crump, 1979; Callahan and Huskey, 1980). But Gilles *et al.* noted a significant relationship that had not been noted by previous investigators—and which is not intuitively obvious. This was that if the number of hours between inducer removal and gonidial cleavage were subtracted from the number of hours the spheroids had been exposed to inducer, and the resulting "corrected pulse time" was plotted as a function of percentage of sexual induction, a linear relationship was obtained! The corrected pulse time required for 50% induction was just under 5 hours. But this could be equally well obtained, for example, by exposure to inducer for the last 5 hours before cleavage or by exposing for the period −15 to −5 hours. What this appears to mean is that the "induced state" builds up gradually in the presence of inducer (with a half time of about 5 hours) and then decays at the same rate after inducer is removed. This appears to be a powerful observation that should prove extremely useful in guiding future research and model building in this fascinating area.

Not to be overlooked in such studies, however, is the report by Kochert and Crump (1979) that (in forma *weismannia*, at least) no matter how painstakingly the induced state has been built up over how many hours, it can be abolished by a few seconds of UV irradiation that has no deleterious effects on embryogenesis per se or the capacity to develop sexually in the next generation!

2. *Development of Sexual Spheroids*

As has been alluded to several times above, the first visible evidence that the inducer has acted is obtained during cleavage, at which time the induced gonidium cleaves in a different pattern to initiate development of a sexual spheroid. In the asexual cycle of both females and males the first unequal cleavage occurs (ideally) in 16 cells of the 32-cell embryo to generate 16 gonidial initials. In the induced gonidia of females, however, the first unequal division is delayed one cycle and affects a larger portion of the blastomeres, resulting in up to 48 reproductive initials, which will go on to form eggs (Starr, 1970). In contrast, the first unequal cleavage division seen in induced males is usually at the 256-cell stage (occasionally the 128-cell stage) and affects all cells, so that a 1:1 ratio of somatic cells and "androgonidia" is produced; division then ceases until the androgonidia have matured, at which time they initiate a new round of cleavage divisions to produce packets of 64 or 128 sperm (Starr, 1970). Inversion and expansion proceed as in asexual spheroids. The fine structure of developing male (and to a lesser extent female) gametes of forma *weismannia* has been described (Birchem, 1977a, 1978; Birchem and Kochert, 1979a,b).

Cytologically, eggs differ from gonidia primarily by their smaller size and less vacuolated cytoplasm. Analysis of protein synthesis in developing eggs and gonidia, by one-dimensional electrophoretic comparisons of polypeptide-labeling patterns, reveals a number of quantitative changes, but few major qualitative differences (Hoffman, 1983; Hoffman and Kirk, 1986. This is perhaps not as surprising as it appears at first glance, since if the eggs go unfertilized, they will enlarge and revert to asexual reproduction as gonidia.

As mentioned in Section II,A,3,c, modified patterns of high-molecular-weight sulfated glycoprotein biosynthesis by parental somatic cells are seen in both induced females and males and are temporally correlated with the modified cleavage patterns seen in the two types of sexual embryos (Wenzl and Sumper, 1981; Wenzl *et al.*, 1984). As in the asexual cycle, however, these activities appear to play no essential role in development of the embryos, since sexual embryos separated from somatic cells at the beginning of cleavage develop normally. Hoffman (1984) has also detected a novel sulfoglycoprotein of modest size (46 kDa) that is made by the somatic cells of juvenile sexual female spheroids, at the end of the dark period during which embryogenesis is completed. At this time this is the primary entity into which sexual females incorporate label; but a comparable entity is not seen either in asexual spheroids or sexual males. The protein remains with the somatic cell layer throughout development, is not associated with the flagella (the site of mating recognition), and does not appear to be required for mating, since it can be selectively removed proteolytically with no effect on fecundity (Hoffman and Kirk, 1986).

It is clear that somatic cells of sexual females have a much longer life expectancy than those of their asexual sisters (Pommerville and Kochert, 1982); male somatic cells, in contrast, appear to have a shortened lifespan. One as yet untested hypothesis concerning the sex-specific sulfated glycoproteins made by somatic cells is that they are in some way related to the modified senescence patterns of these cells.

3. *Genetic Control of Sexual Development*

In an extensive analysis of mutations affecting sexual development, Callahan and Huskey (1980) provided useful insights into both the entry into the sexual cycle and the modified cleavage patterns of induced gonidia. Three unlinked loci were defined at which mutation leads to constitutive sexuality: the *sex^{c}A, B,* and *C* loci. One locus, *sicA,* was defined at which mutation resulted in "sickling" and developmental arrest of gonidia in the presence of inducer; upon removal of inducer, normal asexual development ensued. Genetic analysis of *sicA* was made possible, despite its sterile phenotype, by isolation of a *sicA,sex^{c}A* double mutant in which the *sicA* phenotype was suppressed. The *sex^{c}C* locus was so tightly linked to the mating-type female locus that no male recombinants were recoverable; effects of the other three mutations were studied in both sexes.

The concept that two genetically specified pathways for entry into the sexual cycle exist (an inducer-dependent and an inducer-independent pathway) hinged on the distinctive behavior of *sex^{c}C* mutants. The other three loci, in various combinations, all showed epistatic relationships that were consistent with the hypothesis that they possessed lesions at different points in a common pathway. This was deduced to be the inducer-dependent pathway, since one of them, *sicA,* clearly responds to inducer, but has a lesion that prevents it from completing the response. *Sex^{c}A* and *B* were assumed to possess modifications that permitted them to bypass the initial (inducer-dependent) steps. Both of the latter mutants produced 100% sexual spheroids in both sexes under all conditions. In contrast, in the absence of inducer the *sex^{c}C* strains produced variable numbers of sexual females under differing conditions; but when inducer was added, they responded like wild type, by producing 100% sexual progeny. This environmental sensitivity and responsiveness to inducer, when combined with the fact that *sex^{c}C* did not demonstrate epistatic interactions with the other loci in double mutants, led Callahan and Huskey to the conclusion that *sex^{c}C* strains had lesions in an inducer-independent pathway of sexual development. Furthermore, the extremely tight linkage to mating type exhibited by the *sex^{c}C* mutations led the authors to propose that this locus (which might actually be part of the *mt* locus) could be responsible for establishing different patterns of environment-dependent, inducer-independent sexuality in male strains. Further tests of this interesting proposal have not been reported, but we have observed extreme differences in the ease of maintaining different male strains in the asexual phase under identical environmental conditions (unpublished). It would be interesting to know the extent to which such differences might be traceable to sex^cC alleles.[3]

In the same study, Callahan and Huskey examined a range of pattern-forming mutants and clearly demonstrated one salient fact that must be taken into account in any model that attempts to explain reproductive cell specification in the species. Specifically, they showed that, although in wild-type strains particular cleavage patterns are tightly correlated with formation of different types of reproductive cells (gonidia, eggs, sperm), *pattern and cell type have no necessary relationship to one another.* Among the mutants studied in this paper and earlier (Huskey *et al.,* 1979a), most of the possible combinations of pattern and reproductive cell type have been observed: gonidia in both sexual male and female patterns, eggs in the sexual male pattern, and sperm in the sexual female pattern. Some of the pattern mutations observed affected patterns in all three types of spheroids, whereas others affected only one type of pattern. This is taken to indicate a hierarchy of controls over pattern formation. Furthermore, one mutant was found in which a wild-type patterning response to the sexual inducer occurred without the corresponding shift of cell type. This was the *pan* (parthenogenetic androgonidia) strain in which males respond to inducer by producing asexual gonidia in the pattern usually reserved for sperm packets. In summary, reproductive cell type and reproductive cell pattern must be thought of as two independent features that both happen to be modified by inducer in wild-type strains.

C. Fertilization, Zygote Formation, and Germination

Shortly after their formation, sperm packets are released from the male spheroid and swim as coherent units in a random, spiralling fashion, flagella forward. Tests for chemoattractants released by females were negative (Coggin *et al.,* 1979). Sperm packets attach briefly to spheroids of all types, but Coggin *et al.* (1979) provided convincing evidence that in forma *weismannia* (as in the *Chlamydomonas* mating reaction) specific recognition and attachment between the mating types occur by flagellar interaction. Females deflagellated by pH shock lost the capacity to bind

---

[3] It should be noted, however, that Zeikus and Starr (1980) have shown that some strains exhibit markedly different degrees of responsiveness to inducer in different environmental conditions, indicating that environmental sensitivity of sexual development is not of itself diagnostic of inducer-independent sexuality.

sperm; within 30 minutes after flagellar regeneration was complete the capacity to bind sperm had returned. Trypsinization or protease treatment of either females or sperm packets abolished the recognition reaction. The females recovered binding capacity within 24 hours, but the sperm packets, being ephemeral, did not. Con A had no regular effect upon the flagellar interaction.

Huskey *et al.* (1979a) reported successful matings between "flagellaless" mutant females and wild-type males in forma *nagariensis*. This might suggest that the sexual recognition mechanism is different in the two formas. But ultrastructural analysis was not performed on the mutants to assure that they were indeed completely without any flagellar stubs (R. J. Huskey, personal communication). The same mutation in males does cause sterility because the sperm packets are immotile.

Subsequent to attachment to the fertile female, the sperm packets dissociate into individual sperm while digesting a hole, or fertilization pore, in the surface of the female spheroid; the individual sperm then wiggle, more than swim, in a spiralling manner through the interior of the spheroid until they encounter an egg cell (Kochert, 1968; Starr, 1969). Upon reaching an egg, sperm have been observed to sweep the egg surface with an extended "proboscis" prior to penetrating it (Birchem and Kochert, 1979a). Within a few days, the zygote resulting from gamete fusion develops into a bright red resistant zygospore with a thick, crenulated sport coat (Kochert, 1975).

Dissociation of the sperm packets and production of the enzyme used for penetration are both triggered by the interaction between the flagella of opposite mating types, since the reaction can be triggered by exposing the sperm packets to flagella isolated from fertile (but not from asexual) females; the enzyme is apparently trypsin-like, since trypsin inhibitor blocks formation of fertilization pores (Coggin *et al.*, 1979). Production of fertilization pores is also blocked by addition of either protein synthesis inhibitors or actinomycin immediately before the mating types are mixed; this is interpreted to mean that release of the penetration enzyme requires that both transcription and translation occur subsequent to the flagellar interaction (Hutt and Kochert, 1971).

Maturation of the zygote requires at least 2 weeks, after which germination may be induced by washing with fresh medium. Meiosis occurs prior to germination to yield one viable haploid germling and three aborted polar bodies (Starr, 1975). The germling undergoes a modified form of asexual embryogenesis to yield a miniature spheroid (red at first, but then turning green) with four or eight gonidia; these gonidia subsequently undergo normal embryogenesis to yield asexual or sexual progeny, depending on the absence or presence of inducer (Starr, 1969). Interestingly, the

TABLE I

DEFINED LOCI AND LINKAGE GROUPS OF *V. carteri* f. *nagariensis*[a]

Linkage group	Locus	Mutant description[b]
I	*mtf*	Mating-type female; spheroids develop as sexual females in the presence of inducer (HK 10 of Starr)
	mtm	Mating-type male; spheroids develop as sexual males in the presence of inducer (69–1b of Starr)
	sex^cC	Sexual constitutive; spheroids always in the sexual pathway in the absence of inducer (1)
	eyeB	Eyespot; randomly oriented eyespots (2)
	expD	Nonexpander; adult spheroids remain smaller than normal due to an apparent defect in matrix synthesis or assembly (3)
	invA	Inversionless; embryo does not invert (1)
	mulC	Multiple gonidia; spheroids have fewer cells and variable number of gonidia (same as *s-reg* of Sessoms and Huskey, 1973) (3)
II	*regA*	Regenerator; somatic cells act as reproductive cells (40)
III	*mesA*	Methionine sulfoximine resistant (3)
	d-disA	Delayed dissolver; spheroids disassociate 2–4 days after inversion (1)
	l^{ts}B	Lethal (ts); grows at 24°C, lethal at 35°C (1)
	flgA	Flagella; no flagella, eyespots randomly oriented (1)
IV	*expA*	Nonexpander; spheroids small and dark with little extracellular matrix (1)
	expB	Nonexpander; spheroids have less matrix than wild type, but phenotype is variable (1)
	relA	Release; daughter spheroids fail to release at normal time (1)
	l^{ts}A	Lethal (ts); grows at 24°C; lethal at 35°C (1)
V	*l^{ts}D*	Lethal (ts); grows at 24°C; lethal at 35°C (1)
	megA	Multiple egg; sexual female strains have increased number of eggs (5)
	u-expA	Unequal expander; anterior of spheroid expands but posterior does not (1)
VI	*doA*	Doughnut; spheroids have increased number of gonidia and embryos invert from both ends (1)
	mulD	Multiple gonidia; spheroids have fewer cells and variable number of gonidia (same as *s-reg* of Sessoms and Huskey, 1973) (3)
VII	*expC*	Nonexpander; similar to *expB* (1)
	flgC	Flagella; no flagella when grown at 35°C; flagella present or regrow at 24°C (1)
VIII	*mulA*	Multiple gonidia; clusters of 2–4 gonidia in wild-type pattern (similar to *multi* of Starr) (1)
	mulB	Multiple gonidia; increased number of gonidia in a random pattern over most of the spheroid (1)
IX	*nitA*	Ammonia requiring; does not grow in presence of nitrate, but will grow in presence of ammonia (15)
	nitB	Ammonia requiring; similar to *nitA* (1)

TABLE I (*Continued*)

Linkage group	Locus	Mutant description[b]
X	*eyc*A	Eyespot; randomly oriented eyespots (1)
	*fr*A	Fruity; variable, abnormal spheroid morphology due to abnormal cleavage (1)
XI	*nit*C	Ammonia requiring; similar to *nit*A (5)
XII	*rot*A	Rotator; spheroids rotate in opposite direction to wild type (1)
XIII	*l*^{ts}*C*	Lethal (ts); grows at 24°C, lethal at 35°C (1)
XIV	*sex*^cA	Sexual constitutive; spheroids always in the sexual pathway in absence of inducer (1)
?	*sex*^cB	Sexual constitutive; similar to *sex*^cA (1)
?	*sic*A	Sickle; gonidia have "sickled" appearance in presence of inducer and do not cleave (1)

[a] Adapted, with permission, from Huskey *et al.* (1979a).

[b] Independent mutant loci of similar phenotype are designated by a letter abbreviation. The number of isolates at each locus is given in parentheses.

germling generation appears either to have a developmental–genetic program all its own or to be preprogrammed in the diploid zygote, since few if any of the mutations that will affect its asexual progeny are expressed in the germling itself (R. J. Huskey, personal communication).

IV. Formal Genetics

Huskey *et al.* (1979) used over 100 strains in 19 phenotypic categories to define 33 loci that were tentatively placed into 14 linkage groups. Their data are reproduced as Table I. No mutations with non-Mendelian segregation were detected. Because the organism is haploid in all active phases of the life cycle, tests for complementation and dominance relationships are not possible. Allelism is therefore defined by recombination. Based on parallels with *Chlamydomonas* in which intragenic recombination has been detected at a frequency of 1.6% (Matagne, 1978), mutations showing less than 1% recombination were defined as allelic. Linkage was defined by x^2 analysis of observed deviations from segregation patterns to be expected if assortment were independent; a deviation significant at the 0.05 level or less was taken as evidence of linkage. The number of linkage groups listed in Table I, 14, happens to correspond to the number of chromosomes, but it would not be surprising if this correspondence turned out to be only fortuitous as more loci are mapped.

The data set on which Table I was based was a fundamentally important contribution to the field. It established for the first time that extensive formal genetic analysis of *Volvox* variants was feasible and provided the methodological guidelines for future genetic work. In addition, it provided useful information about the organization of the genome, such as the fact that genes of related function do not exhibit any significant degree of clustering (e.g., see distribution of *exp, sex^c,* and *nit* genes).

Unfortunately, however, no additions to the map have since been published. One feature that has slowed its expansion is that many of the new mutants one would like to place on it, as well as many of the original markers, have morphological properties that tend to interfere with the mating performance in one or both sexes, and hence exhibit low fecundity. Fertility problems are compounded if one attempts to generate tester strains bearing several such markers. Thus, performing all of the crosses necessary for complete linkage analysis of a novel mutation is often a formidable undertaking. Further complicating the analysis is the fact that certain of the marked strains that would be most useful for linkage studies are apparently no longer available (R. J. Huskey, personal communication).

In an attempt to circumvent these difficulties, we have undertaken the construction of tester strains with normal morphology and fecundity that bear markers in many (ideally all) linkage groups and that could be used to facilitate linkage analysis (K. S. Huson and J. F. Harper, unpublished). Two types of markers are being emphasized in this program: drug-resistant mutations and restriction fragment-length polymorphisms (RFLPs).

Three of the loci defined in Table I were isolated as drug-resistant markers: *mesA* (methionine sulfoximine resistance) and *nitA* and *C* [nitrate reductase mutants isolated on the basis of chlorate resistance (Huskey *et al.,* 1979b)]. Kurn has described two other categories of (unmapped) drug-resistant mutants: BUdR resistance (Kurn *et al.,* 1978) and fluphenazine resistance (Kurn and Sela, 1981; Kurn, 1982). As a preliminary to expanding the list of available resistance markers, we have determined lethal levels for many drugs of potential interest under conditions favorable for selection of resistant strains (Table II). Using X-ray and chemical mutagenesis, so far we have induced resistance mutations for nine of these drugs, as noted in Table II; the drugs we have focused on at the outset are all ones that are known to interfere with developmental processes of specific interest to us, such as DNA synthesis and/or repair (8-azaguanine, novobiocin), cleavage (benzimidazole), inversion (imidazole), nitrogen metabolism (chlorate), or ones for which the specific target gene is known and has been cloned in other organisms (fluoroacetamide).

More recently we have commenced development of a strain bearing

TABLE II
Drug Sensitivities of *V. carteri* f. *nagariensis*

Drugs[a]	LLDT[b]	HNLDT[c]	Solvent[d]
DNA synthesis inhibitors			
Aphidicolin	25	5	DMSO
5-Azacytosine	200	100	susp.
8-Azaguanine	50	25	DMSO
Bromodeoxyuridine	1500	300	DMSO
Cytosine arabinoside	—	275	DMSO
Hydroxyurea[e]	550	75	SVM
Methotrexate	50	10	DMSO
Nalidixic acid	250	100	DMSO
Novobiocin[e]	600	400	DMSO
Oxolinic acid	—	1000	susp.
6-Thioguanine	200	100	DMSO
Protein synthesis inhibitors/antibiotics			
Amphotericin	0.25	0.025	DMSO
Anisomycin	1.3	0.6	EtOH
Chloramphenicol	50	25	DMSO
Cycloheximide	1	0.25	DMSO
Erythromycin[e]	15	1.5	EtOH
Spectinomycin	20	5	SVM
Streptomycin	50	1	SVM
G418 sulfate	5	1	SVM
Anticytoskeletal drugs			
Benzimidazole[e]	250	100	DMSO
Benomyl	100	10	DMSO
Colchicine	600	200	DMSO
Griseofulvin	120	60	DMSO
Isopropyl-N(3-chlorophenyl)carbamate (CIPC)[e]	0.32	0.16	DMSO
Isopropyl-N-phenyl carbamate (IPC)	—	1.8	DMSO
Nacodazole	—	100	DMSO
Podophyllotoxin	250	125	DMSO
Nitrogen antimetabolites			
Chlorate[e]	1000	100	SVM
Fluoroacetamide[e]	1.2	0.12	SVM
Methylamine	300	100	DMSO
Other antimetabolites			
3-Acetylypyridine	5	1	DMSO
β-aminopropionitrile	2500	250	SVM
Canavanine[e]	25	2.5	SVM
Dehydroproline	100	75	DMSO
3-(3,4-Dichlorophenyl)-1,1-dimethylurea (DCMU)	0.02	0.01	DMSO
Fluoroacetate	7.5	2.5	SVM
Fluoroorotic acid	5000	1000	susp.
Fluphenazine	0.1	0.05	DMSO

(*continued*)

TABLE II (*continued*)

Drugs[a]	LLDT[b]	HNLDT[c]	Solvent[d]
Glyphosate	1300	260	SVM
Homoarginine	—	20	SVM
Imidazole	400	200[f]	DMSO
Indole	10	10	DMSO
Methionine sulfoximine[e]	4.5	1	SVM
Methyl bis(guanyl hydrazone)	260	26	SVM
Sulfanilamide	500	100	DMSO

[a] All drugs except chlorate were tested in solidified medium in petri plates. HK 10 (Eve) juveniles 24 hours postcleavage were mechanically released from parental spheroids, suspended in 5-ml SVM containing 0.375% Seaplaque low-melting agarose at 37°C, and poured into a plate containing the test drug at 2× final concentration, dissolved or suspended in 5 ml of (solidified) 1% Bacto agar in SVM. Chlorate toxicity was tested in liquid medium, using cultures that had been derepressed for nitrate reductase by removing reduced nitrogen sources from the medium (Huskey *et al.*, 1979b). Some morphological mutants tested (e.g., *reg*) showed different sensitivity to some of the drugs.

[b] Lowest lethal dose detected in μg/ml; dashes indicate no lethal dose yet detected.

[c] Highest nonlethal dose detected in μg/ml; intermediate doses either have not been tested or gave variable results.

[d] DMSO, dimethylsulfoxide; EtOH, ethanol; SVM, standard *Volvox* medium (Kirk and Kirk, 1983); susp. indicates no nontoxic solvent was identified in which the drug could be kept in solution at the concentrations tested, so it was used in suspension in the agar. Final DMSO and EtOH concentrations were 1% or less in all cases, which had no effects on survival by themselves.

[e] Resistant strains have been recovered following mutagenesis.

[f] Imidazole at concentrations as low as 100 μg/ml completely blocked inversion in all subsequent generations without affecting survival.

multiple resistance markers. New resistances are being added stepwise, selecting after each round of mutagenesis for mutants in which resistance to the new selective agent is present in combination with all the previously established resistances, plus wild-type morphology and fertility. To date, a strain bearing five resistance markers has been produced.

Incorporation of RFLPs in the proposed tester strains was an outgrowth of molecular analysis of the *V. carteri* genome (Harper, 1982, 1984, and unpublished work, reviewed in Section V). In early stages of this work, it was discovered that when DNAs from the standard wild-type female and male strains now in general use (HK 10, and 69-1b, respectively) were digested in parallel with any one of a number of restriction enzymes and then hybridized on Southern blots with any one of a number of repetitive-element probes, numerous polymorphisms were detected.

Further study revealed two important features of such polymorphisms: (1) Different laboratory populations of the same strain (which had been reproducing asexually as separate stocks for several years) exhibited low levels of polymorphism, indicating that the sequences being detected were changing at a slow, but perceptible rate under laboratory conditions. (2) In the great majority of cases, such polymorphisms segregated in a simple Mendelian fashion, indicating their potential utility as unselected markers (Harper, 1984).

Three steps have been taken to capitalize on those observations. First, to minimize polymorphism within strains, one spheroid was selected at random from each of the standard strains (HK 10 and 69-1b) to establish clonally derived female and male stocks for use in all subsequent genetic work within the laboratory. These have been given the mnemonic designations "Eve" and "Adm," respectively. The Eve strain has been used as the progenitor in the drug-resistance protocols described above. Second, for use in later stages of genetic analysis, we have established "congenic" female lines bearing most of the RFLP pattern of Adm and males bearing most of the Eve pattern (except for polymorphisms tightly linked to mating type) by 10 generations of reciprocal backcrosses. These strains have been given the mnemonic designations "Tom" (Tom-boy) and "Drg" (Drag), respectively. Third, we have begun to construct a linkage map of the polymorphisms with Mendelian segregation patterns that we have observed so far by analysis of F_1 progeny from Adm × Eve crosses. To date 93 such RFLPs have been studied. So far, their distribution does not appear to be completely random. They fall into 38 clusters, within which less than 10% recombination is observed. One cluster, containing eight RFLPs, is tightly linked to the mating-type locus, but no polymorphisms have yet been identified that lie within 10 map units of either of the two drug-resistant markers that have been tested. Of greatest immediate interest to us are a pair of RFLPs, detected with two different middle repetitive probes, that appear to lie within 10 map units of the *regA* locus (see Section II,B,1,c). These are the first markers of any kind for which linkage to the *regA* locus has been reported (Huskey *et al.,* 1979a), and they provide possible starting places for a "chromosome walk" designed to clone this interesting control gene.

Recently a new dimension was added to this analysis when Starr provided us with additional male and female strains that had been isolated from Poona, India. The Poona strains, when compared to the original strains (isolated from Japan), define yet more polymorphisms. The Indian and Japanese strains are interfertile, and although there is sometimes high germling lethality, progeny can be obtained with a fair degree of ease.

V. Molecular Aspects of Genome Organization

In the belief that a molecular-level understanding of *Volvox* development will ultimately require a comprehensive characterization of the genome, one of us (Harper, 1982, 1984, 1986) has taken the initial steps of such characterization. Simultaneously and independently, parallel work has been initiated in another laboratory (R. Schmitt, personal communication).

A. GENERAL PROPERTIES OF THE GENOME

Kochert (1975) has estimated the DNA content of somatic cells of forma *weismannia* to be 1.14×10^{-13} g; analysis of forma *nagariensis* yielded a slightly larger value of 1.27×10^{-13} g (M. M. Kirk, unpublished). Since about 93% of somatic cell DNA is nuclear (Kochert, 1975; Coleman and Maguire, 1982) and the cells are genetically haploid, these values indicate that the haploid complement of *Volvox* is in the same range as *Chlamydomonas* and *Drosophila* and contains about 1.2×10^8 bp. With 14 chromosomes per genome (R. C. Starr, personal communication), this means the average chromosome should contain about 8500 kb of DNA, making the electrophoretic separation of the chromosomes a future possibility (Schwartz and Cantor, 1984). Coleman and Maguire (1982) have estimated that the size of the chloroplast genome is about 2×10^{-16} g and that an average somatic cell contains about 50 such genome equivalents, clustered in 2–3 "nucleoids." Gonidia accumulate more chloroplast (and possibly more nuclear) DNA during maturation (see Section II,B,1,b).

Analytical ultracentrifugation of total DNA reveals two major species with buoyant densities of about 1.715 and 1.705 g/cm^3, which are assumed on the basis of relative abundance to be the nuclear and chloroplast species, respectively (Kochert, 1975; Margolis-Kazan and Blamire, 1976). Based on those values and thermal denaturation studies, Margolis-Kazan and Blamire (1976) have estimated the G + C content of nuclear and chloroplast DNAs to be about 55 and 45%, respectively. One strain of forma *weismannia* that had previously been shown to contain a bacterial endosymbiont (Kochert and Olson, 1970b) was shown to contain a third DNA peak with a buoyant density of 1.693 (Margolis-Kazan and Blamire, 1976). Other strains are not suspected to have such bacterial endosymbionts.[4]

[4] It should be noted, however, that in preparative CsCl gradients of *Volvox* nucleic acids, we routinely observe a discrete band of high-molecular-weight nonribosomal RNA; this material has some of the properties to be expected of the genome of an RNA virus (J. F. Harper, unpublished). Virus-like particles have frequently been seen in thin-sectioned specimens in the EM (Dauwalder *et al.*, 1980; D. L. Kirk, G. Olson, and G. M. Veith, unpublished).

B. DNA Isolation and Cloning

Routine isolation of high-molecular-weight *Volvox* DNA capable of being modified by restriction and ligation enzymes has proved to be extraordinarily difficult—as has isolation of RNA capable of being translated *in vitro* (Kirk and Kirk, 1985a). In both cases, the difficulty appears to result from the presence in the matrix of highly sulfated glycoproteins and/or polysaccharides that tend to copurify with nucleic acids and inhibit enzymes of nucleic acid metabolism. Use of the proteolytic enzymes often included in nucleic acid purification protocols (e.g., proteinase K) intensifies these problems, apparently by releasing more polyanionic contaminants from the matrix.

The methods independently developed in this laboratory and in Regensburg for successful DNA isolation are quite similar and avoid the use of such enzymes. Our current protocol involves freezing cells in liquid nitrogen, grinding with dry ice in a Moulinex coffee grinder, lysis in ethylenediaminetetraacetate (EDTA)-sarcosyl-mercaptoethanol at 42°C, extraction with cetyltrimethylammonium bromide (CTAB) in butanol, equilibrium centrifugation in a CsCl gradient containing sarcosyl and ethidium bromide, followed by spermine precipitation. Salbaum and Schmitt have isolated DNA by a modification of the method of Bendich *et al.* (1979), which involves grinding liquid nitrogen-frozen cells (30 minutes) in a mortar and pestle, differential centrifugation to obtain a crude nuclear pellet, sarcosyl lysis, CsCl banding, and phenol extraction of the banded DNA.

The *Volvox* genomic library made in St. Louis was prepared from a partial *Bam*HI digest of total DNA cloned into λ Charon 30; the Regensburg genomic library was prepared from a *Sau*3A partial digest cloned into λ EMBL 3. In both cases, banks representing several genome equivalents were generated. Donald Strauss (Massachusetts General Hospital), in collaboration with this laboratory, is currently establishing cDNA banks from wild-type and mutant strains to be screened for developmentally important genes by subtractive methods.

In addition to the tubulin and actin clones mentioned in Section II,A,2,d, the Regensburg group has isolated and begun analysis of four 18 S rDNA clones homologous to the rRNA of the yeast small ribosomal subunit (these will be sequenced in collaboration with Carl Woese) and a histone H4 clone that has been isolated and mapped using homologous probes from wheat and *Xenopus*. After further mapping and subcloning of these genes and the tubulin and actin genes, the intent is to examine expression of each at selected stages in the asexual life cycle of synchronized cultures (R. Schmitt, personal communication).

In addition to the tubulin clones mentioned in Section II,A,2,d, the attention of this laboratory has been focused on a variety of repetitive elements. The remainder of this section will be devoted to a summary of the present status of those investigations.

C. POLYMORPHIC MIDDLE REPETITIVE ELEMENTS

Six kinds of probes that hybridize with distinct families of middle repetitive DNA sequences have been used to probe Southern blots of genomic DNA in a search for polymorphisms (RFLPs) that might be useful for the sort of genetic analysis described in Section IV (J. F. Harper, unpublished). These were as follows: (1) Synthetic poly(dGdT)·poly(dCdA). This probe, hereafter called "GT/CA," detects repetitive sequences in a wide range of eukaryotes; such sequences have a propensity to form Z-DNA (Hamada et al., 1982); (2) Synthetic poly(dG)·poly(dC). As discussed in Section II,B,2,a, this probe (hereafter called "GG/CC") was initially used in the hope of detecting putative polyproline-coding sequences; (3) A subcloned portion of an intron from one of the β-tubulin genes, which turned out to be a repetitive element; (4 and 5) Two non-cross-hybridizing genomic clones exhibiting homology to the fold-back fraction of the genome (which in turn was isolated as the S_1 nuclease-resistant, double-stranded DNA remaining after boiling and rapid cooling on ice); and (6) A cDNA clone derived from an uncharacterized abundant transcript homologous to a repetitive genomic sequence. RFLPs were detected with all six probes, but the present discussion will be restricted to those detected with the two synthetic probes (GT/CA and GG/CC), which in many ways appear to be the most interesting.

Of the six families of repetitive elements examined in our clonally derived male and female strains (Adm and Eve), the highest frequency of RFLPs was detected in the family (containing more than 1000 dispersed genomic sequences) homologous to the putative Z-DNA probe, GT/CA. When Adm × Eve F_1 progeny are examined, nearly all of these RFLPs segregate in a simple Mendelian pattern, and many allelic pairs can be identified. This pattern of heritability, combined with the fact that simple repetitive sequences in general and GT/CA sequences in particular have been suspected to be hot spots for recombination (Nikaido et al., 1981; Stringer, 1982), leads us to believe that these RFLPs result from structural rearrangements of the DNA. However, we cannot currently rule out the possibility that they result from stable strain-specific differences in methylation of restriction sites, resulting in differential cutting. In either case, these GT/CA-rich sequences are clearly associated with genomic hypervariability. Not only are numerous RFLPs seen between Adm and Eve, additional variants have been detected in mutants recently derived from the Eve strain; yet others have appeared in F_1 progeny of an Adm $\times$ Eve cross, apparently arising during meiosis. Interesting as the rapid evolution of the GT/CA family may be for other reasons, it makes the family unsuitable for use in establishment of a RFLP-based genetic map.

The family of sequences detected with GG/CC, by contrast, appears to have about the right level of variability to be useful as a genetic marker (as do the other families mentioned above). Although it is less unstable than the family just discussed, it is highly polymorphic. Because sequence data indicate that the G-rich and C-rich regions detected with this probe are segregated on opposite strands of the genomic DNA, we refer to members of this family as "C-rich sequences," for simplicity. C-rich sequences are as abundant as those detected with GT/CA, and they are highly dispersed in the genome, occurring about once every 100 kb, on the average. An estimate of the degree of homology between the probe and the C-rich sequences of the *Volvox* genome was obtained by comparison with a C-rich genomic sequence of known composition: the long tandem repeats of C_4A_2 present in *Tetrahymena* telomeres. In 0.3 $\times$ SSC at 65°C, no hybridization of the latter to GG/CC could be detected. In contrast, hybridization to the *Volvox* sequences was retained in 0.3 $\times$ SSC at 80°C. Thus, *Volvox* C-rich sequences appear to be extensive stretches of DNA with less than 33% mismatch with GG/CC.

Parallel studies reveal a similar abundance of C-rich sequences in *Chlamydomonas, Gonium,* and *Eudorina.* However, in contrast to the clear ladder of discrete bands seen with all these Volvocalean genera, DNA from all other organisms tested (various bacteria, *Tetrahymena, Drosophila,* corn, carrot, tobacco, chimpanzee, and others) exhibited a smear of weakly homologous material or none at all. Thus, C-rich sequences appear to be a unifying feature of the Volvocales and may ultimately prove of considerable value for phylogenetic studies. [It should be noted, however, that Tautz and Renz (1984) have detected relatively strong homology to this probe on Southern blots of sea urchin DNA.]

One C-rich genomic clone with strong homology to the synthetic probe has been isolated, mapped, subcloned, and partially sequenced. It has been found to have two long internally repetitive C-rich regions separated by about 2 kb of "normal" DNA. C-rich region 1 contains the following simple sequence repeats in a string: $(C_3T)_6C_6(C_3T)_6$; six repeats (with a one-base substitution in one repeat) of the "18-mer" $C_5TC_3TCTCT_2C_2T$, followed by numerous modified repeats; six perfect repeats of the "10-mer" GC_3TC_4T, flanked on either side by numerous modified repeats of the general formula $(C_{1-9}T)_n$; then follows a 76% C, nonrepetitive stretch. C-rich region 2 contains the longest simple sequence repeat of all: more than 30 repeats of the "13-mer" $C_4TCTC_2TC_2T$, containing only one variation (substitution of one C by a G in one repeat); other C-rich sequences are also present in this region. Note that with minor exceptions these are basically homopyrimidine sequences. The sequence has been confirmed by sequencing the complementary G-rich homopurine strand.

Regions 1 and 2 do not cross-hybridize; hence, probes derived by subcloning them (with 50–100 bases of flanking sequence) detect different subsets of C-rich sequences on genomic Southern blots. Between them they detect a family of genomic sequences and RFLPs that largely overlaps, but is not identical with the family detected using the synthetic GG/CC probe. These cloned fragments turn out to be more useful than the synthetic probe for detecting and analyzing polymorphisms. The nature of some of the other C-rich sequences is being examined.

At least some of the C-rich sequences are transcribed. GG/CC and the region 1 and 2 probes all detect at least four discrete bands and a general smear of hybridizing material on Northern blots of total cellular RNA. The nature of the translational products of these transcribed sequences, if any, is unknown. But the sequences predict that if translated, regions 1 and 2 together would code for peptides containing more than 50% proline, plus considerable amounts of serine and lesser amounts of leucine. This is a composition similar to that given by Wenzl *et al.* (1984) for SSG 185 and also is reminiscent of cell wall glycoproteins of *Chlamydomonas* and higher plants (Lamport, 1974).

There is also a striking similarity of the *Volvox* C-rich repeats to the repetitive sequences found in the telomeres of lower eukaryotes, all of which can be summarized by the formula (C_{1-8}(T or A)$_{1-4}$) (Blackburn, 1984). However, most of the *Volvox* C-rich repeats are located at internal chromosomal locations, and no data are yet available on their possible homology to *Volvox* telomeres. But the homopyrimidine–homopurine organization exhibited by some of them offers the potential for forming unusual DNA structures wherever they may be located in the chromosome. Cantor and Efstratiadis (1984) have shown that in supercoiled plasmids this class of DNA sequences forms unusual structures that are consistent with left-handed Z-DNA, and Lee *et al.* (1984) have evidence for the potential of such sequences to form triple helices (which they believe may be involved in transcriptional initiation) under *in vivo* conditions. Experimental evidence for the suggestion that some *Volvox* C-rich sequences may generate such unusual DNA structures is provided by the fact that all four C-rich regions we have tested are closely associated with S_1 nuclease hypersensitive sites when harbored in supercoiled plasmids.

There is, therefore, no shortage of potential roles for these C-rich sequences. It would not be surprising if different members of the C-rich family served many different functions, but for the moment at least, one of the major sources of interest the family holds for us is its potential—because of its associated polymorphisms—for aiding further analysis of the genome.

VI. Speculations on the Genetic Origins of Multicellularity

In a stimulating and penetrating recent study, Bell (1985) has argued from ecological considerations and detailed allometric analysis that colony formation, division of labor between somatic and reproductive cells, increasing emphasis on sexual versus asexual reproduction, and oogamous differentiation of gametes—all of which are seen in the order Volvocales—constitute a progression of natural sequelae to a basic selective advantage associated with increased size in environments populated with filter feeders. Others have made qualitatively similar suggestions before, but Bell has reinforced such ideas with a persuasive set of quantitative relationships derived from a large body of life-history data. Beyond this, he has made several detailed, testable predictions about pathways of nutrient assimilation and trophic relationships likely to be found in specific members of the order, and in the genus *Volvox* in particular. Efforts to test some of those predictions are now underway in his laboratory (V. Koufopanou, personal communication).

But if Bell's (or any similar) model turns out to account for the ultimate causes of all those features of Volvocalean life style that currently fascinate developmental biologists, it will not speak to the question of proximate causes—the genetic bases—of such features. It is to the latter type of question, the possible genetic origins of certain key features of the higher Volvocales, that we would now like to turn briefly. The two features to be considered are the evolution of a colonial or multicellular body and the division of labor between somatic and reproductive cells.

A. Origins of the Ties That Bind the Spheroid

The evolution of the higher Volvocales, containing thousands of cells per individual, from a unicellular *Chlamydomonas*-like ancestor has involved elaboration of the simple multilayered "cell wall" of the unicells into a complex extracellular matrix in which novel fibrous elements form cellular compartments that both separate the cells and bind them together and in which the spaces are inflated with mucilaginous materials. Structural considerations suggest that most of this elaboration has involved modifications of what in *Chlamydomonas* is a single, rather undistinguished looking layer of the wall, but the range of variations on the basic plan, even within the genus *Volvox,* are extraordinary (Kirk *et al.,* 1985). Chemical considerations (Lamport, 1974) suggest that many of these elaborations are based on a family of hydroxyproline-rich glycoproteins (HPRGs) already abundant in the unicells; but again the range of modifications extant is apparently great (Mitchell, 1980; U. W. Goodenough and J. Heuser, unpublished).

The possible genetic basis for this extensive evolutionary diversification was recently discussed by Goodenough (1985). With her colleagues she established that the sexual agglutinins of *Chlamydomonas* not only have extended rodlike structures like certain HPRGs of the wall, they have a similar hydroxyproline-rich amino acid composition and are almost certainly members of the same family, related by descent (Cooper *et al.,* 1983). This led her into a thought-provoking speculation about the features of fibrous glycoproteins that might facilitate rapid diversification and provide the basis for species-specific specializations that both prevent cell fusion most of the time and promote it in the case of compatible gametes. She observed that "A common feature of fibrous glycoproteins . . . is that they are internally repetitive . . . . Repetitive DNA sequences are prone to undergo addition and deletion events . . . . Therefore . . . fibrous glycoproteins might have a built-in propensity to change at a faster rate than the average gene . . ." (Goodenough, 1985).

Our studies, reviewed in the preceding section, show that the Volvocales have a large number of genes with the potential to code for proline-rich (hence hydroxyproline-rich) proteins and that at least some of these genes are indeed capable of rapid change. Although it is highly unlikely that all of these genes code for ECM and agglutinin molecules, it seems equally unlikely that none of them do. Direct test of Goodenough's suggestion should soon be possible.

We would like to go one step further, however, and suggest a possible origin for this family of internally repetitive, potentially ECM-coding genes. The sequences we have obtained to date also bear striking resemblances to known telomeric sequences which have been analyzed to date, all of which have numerous repeats of C-rich simple sequences (Blackburn, 1984). We suggest that recombinational fusion between telomeres followed by chromosome breakage and terminal addition of new telomeric sequences [which may be added in a nontemplated manner (Blackburn, 1984)] could have resulted in internalization of telomeric sequences. Once internalized, telomeric sequences would have been free to evolve, recombine, and generate the "evolutionary substrate" for elaboration of the HPRG genes that would eventually make possible construction of the multicellular spheroids of the higher Volvocales. In addition, as Goodenough suggested, once such an internally repetitive gene family was established, it would have a propensity for rapid evolution via recombination. Presumably, more extensive sequence data on actual HPRG-coding sequences, telomeres, and all the other C-rich genes of the Volvocales may permit future evaluation of the possible validity of this proposal.

B. Three Genes Required for Evolving Division of Labor

Our speculations into the genetic origins of division of labor in the Volvocales are derived from consideration of three types of *V. carteri* mutants with modified patterns of cellular differentiation. The first type are the well-known somatic regenerators, with lesions at the *regA* locus; these have been discussed at length in Section II,B,1,c. From consideration of the unrestricted reproductive potentials of somatic cells bearing a lesion at this locus, we have inferred that the function of the wild-type allele, *regA*$^+$, is to suppress all forms of reproductive development in somatic cells. The other two mutant categories have been much less studied. Huskey and Griffin (1979) described a mutant recovered from a *regA*$^-$ strain that "does not have gonidia but reproduces solely by somatic cell regeneration." We have recently isolated such a strain also. Although genetic analysis of such mutants is obviouly warranted, for the purposes of discussion we assume these gonidialess strains to have a lesion at a locus we will call *gls,* the wild-type allele of which would appear to control the asymmetric cleavage division that normally sets apart cells to be gonidia. (Obviously such a mutation would be lethal on a wild-type background, but is viable in combination with *regA*$^-$.) The third category was mentioned in Section II,B,1,c, but has not previously been described in print. These are the late gonidial, or *lag,* mutants; they have no obvious gonidial cells at the end of embryogenesis, but as development proceeds, it becomes obvious that gonidial specification has occurred because a predictably spaced set of cells enlarge, differentiate into gonidia, and cleave to yield a new generation that similarly lacks obvious gonidia at first (R. J. Huskey, personal communication; D. L. Kirk, unpublished). These late gonidia (like *reg* somatic cells) appear to differentiate first as somatic cells and then to redifferentiate as gonidia. The important difference from *reg* mutants is that it is only a defined subset of cells, in the characteristic locations of gonidia, that exhibit this behavior. We infer that the function of the *lag*$^+$ allele (that is lost in the mutants) is to shunt presumptive gonidia directly into the reproductive pathway, bypassing the vegetative pathway.

The significance of these three genes is best visualized by first considering the asexual life cycle of the lower Volvocales, both unicellular and colonial, which lack division of labor (Fig. 4). *Chlamydomonas* cells do not reproduce asexually by simple binary fission, contrary to statements in many textbooks. Rather, vegetative cells more than double in size, then redifferentiate into reproductive cells—nonmotile, flagellaless cells, with modified surface coats that cause them to stick to glass (Straley and Bruce, 1979), and reorganized organelles. Then these reproductive cells enter what is, by definition, cleavage: a series of rapid cell divisions in the absence of growth. In *Chlamydomonas,* the number of cleavage divisions can vary between two and four, to generate four, eight, or sixteen daughter cells that then differentiate as vegetative cells (Craigie and Cavalier-Smith, 1982). In colonial Volvocalean genera (including *Gonium, Pandorina,* and *Eudorina*), the number of cleavage divisions may be increased to as high as seven, and the cells remain together during their vegetative differentiation. But the basic cycle is the same as in the unicells: Each cell is first vegetative, then reproductive (Kochert, 1973).

In wild-type *Volvox* the two developmental pathways that are pursued sequentially in the lower forms are pursued separately by the two cell types (Fig. 4). But deletion of only two genetic functions reduces *Volvox* from a multicellular organism to colonial status: a *regA⁻/gls⁻* strain (described above) is essentially an overgrown *Eudorina* in which all cells first fulfill vegetative and then reproductive functions! Thus, these two genetic functions (*regA⁺* and *gls⁺*) appear to be the two most important to have been added to the ancient Volvocalean repertoire to achieve division of labor.

However, it would appear that if one were able to transform functional *regA⁺* and *gls⁺* genes into a colonial Volvocalean, the recipient would not thereby come to resemble *V. carteri,* for although it might have a subset of cells specified for reproduction and a subset prohibited from reproducing, the former would be expected to differentiate first as vegetative and then redifferentiate as reproductive cells. This type of incomplete division of labor is what is seen, of course, in the contemporary genus *Pleodorina* (Kochert, 1973). To achieve the complete division of labor seen in *V. carteri,* a third genetic function, the *lag⁺* function, would be required to shunt the presumptive gonidia directly into the reproductive pathway. The selective advantage of such an addition appears obvious: Forms with the *lag⁺* function can shorten their asexual life cycle greatly by bypassing the vegetative half of the ancient pathway.

Thus, these three functions appear by present analysis to hold as much potential interest from a phylogenetic as from an ontogenetic point of view. If and when the day comes that the DNA sequences encoding these three functions are identified, it should be possible to test by molecular methods our prediction that sequences comparable to *regA* and *gls* will be found in *Pleodorina,* but that the *lag* function will be possessed only by forms like *V. carteri,* in which the division of labor is complete.

VII. Coda

Since Haeckel's time *Volvox* has held particular fascination for evolutionary biologists. Since Weismann's time it has held equal fascination for many developmental biologists. The doorway to detailed laboratory exploration was opened by Darden and Kochert in the 1960s with the demonstration that reproduction and development of members of the genus could be studied under controlled conditions. The demonstration by Starr, later expanded by Huskey and collaborators, that mutants with instructive developmental variations could be isolated and analyzed was a giant step forward. Now that this attractive creature has begun to demonstrate its willingness to converse with us on a molecular level, it appears that the time will soon come (if it has not already) when, in the words of Kochert (1981), "*Volvox* will . . . finally achieve a measure of that potential to further aid our understanding of mechanisms of cellular differentiation which it so clearly proclaims to its every observer."

ACKNOWLEDGMENTS

We wish to thank all of our colleagues who have shared manuscripts and summaries of unpublished work with us for inclusion in this review and who have read this manuscript and made constructive suggestions for improvement of it. Our thanks also go to Jacqueline Hoffman for preparation of Figs. 2–4. Unpublished work from this laboratory was supported by grant GM 27215 from the National Institutes of Health. Predoctoral studies of J. F. H. were also supported by fellowships from The Division of Biology and Biomedical Sciences of Washington University (2 years) and the Samuel Roberts Noble Foundation (2 years).

REFERENCES

Baran, G. J. (1984). Ph.D. thesis, University of Virginia, Charlottesville.
Baran, G. J., Harper, J. F., Huskey, R. J., and Kirk, D. L. (1985). In preparation.
Bause, E., and Jaenicke, L. (1979). *FEBS Lett.* **106,** 321–324.
Bause, E., Muller, T., and Jaenicke, L. (1983). *Arch. Biochem. Biophys.* **220,** 200–207.
Bell, G. (1985). *In* "Origin and Evolution of Sex" (H. O. Halvorson and A. Monroy, eds.). Liss, New York, pp. 221–256.
Bendich, A., Anderson, R. S., and Ward, B. L. (1979). *In* "Genome Organization and Expression in Plants" (C. J. Leaver, ed.), pp. 31–32. Plenum, New York.
Birchem, R. (1977a). Ph.D. thesis, University of Georgia, Athens.
Birchem, R. (1977b). *Annu. EMSA Meet. Abstr., 35th* p. 346.
Birchem, R. (1978). *J. Phycol.* **14,** 30.
Birchem, R., and Kochert, G. (1979a). *Phycologia* **18,** 409–419.
Birchem, R., and Kochert, G. (1979b). *Protoplasma* **100,** 1–12.
Bisalputra, T., and Stein, J. R. (1966). *Can. J. Bot.* **44,** 1697–1710.
Blackburn, E. H. (1984). *Cell* **37,** 7–8.

Brunke, K. J., Young, E. E., Buchbinder, B. U., and Weeks, D. P. (1982). *Nucleic Acids Res.* **10,** 1295–1310.

Brunke, K. J., Anthony, J., Sternberg, E., and Weeks, D. P. (1984). *Mol. Cell. Biol.* **4,** 1118–1124.

Bryant, J. L., Green, K. J., and Kirk, D. L. (1980). *J. Cell Biol.* **87,** 133a.

Burr, F. A., and McCracken, M. D. (1973). *J. Phycol.* **9,** 345–346.

Callahan, A. M., and Huskey, R. J. (1980). *Dev. Biol.* **80,** 419–435.

Cantor, C. R., and Efstratiadis, A. (1984). *Nucleic Acids Res.* **12,** 8059–8072.

Caplan, H. S., and Blamire, J. (1980). *Cytobios* **29,** 115–128.

Catt, J. W., Hills, G. J., and Roberts, K. (1978). *Planta* **138,** 91–98.

Coggin, S., and Kochert, G. (1980). *J. Cell Biol.* **80,** 39a.

Coggin, S. J., Hutt, W., and Kochert, G. (1979). *J. Phycol.* **15,** 247–251.

Coleman, A. W., and Maguire, M. J. (1982). *Dev. Biol.* **94,** 441–450.

Collins, A., and Johnson, R. (1979). *Nucleic Acids Res.* **7,** 1311–1320.

Cooper, J. B., Adair, W. S., Mecham, R. P., Heuser, J. E., and Goodenough, U. W. (1983). *Proc. Natl. Acad. Sci. U.S.A.* **80,** 5898–5901.

Coss, R. A. (1974). *J. Cell Biol.* **63,** 325–329.

Cowan, N. J., and Dudley, L. (1983). *Int. Rev. Cytol.* **85,** 147–173.

Craigie, R. A., and Cavalier-Smith, T. (1982). *J. Cell Sci.* **54,** 173–191.

Darden, W. H., Jr. (1966). *J. Protozool.* **13,** 239–255.

Dauwalder, M., Whaley, W. G., and Starr, R. C. (1980). *J. Ultrastruct. Res.* **70,** 318–335.

Deason, T. R., Darden, W. H., and Ely, S. (1969). *J. Ultrastruct. Res.* **26,** 85–94.

Delsman, H. C. (1919). *K. Akad. Wet. Gew. Vergad. Wissen Natuurk. Afd.* **27,** 137.

Dobell, C. (1932). "Antony van Leeuwenhoek and His Little Animals." Staples Press, London.

Foster, K. W., Saranak, J., Patel, N., Zarilli, G., Okabe, M., Kline, T., and Nakanishi, K. (1984). *Nature (London)* **311,** 756–759.

Fulton, A. B. (1978). *Dev. Biol.* **64,** 236–251.

Fulton, C., and Simpson, P. A. (1976). *Cell Motil. Book A–C* 1976. pp. 987–1005.

Gilles, R., and Jaenicke, L. (1982). *Z. Naturforsch.* **37c,** 1023–1030.

Gilles, R., Bittner, C., Cramer, M., Mierau, R., and Jaenicke, L. (1980). *FEBS Lett.* **116,** 102–106.

Gilles, R., Bittner, C., and Jaenicke, L. (1981). *FEBS Lett.* **124,** 57–61.

Gilles, R., Gilles, C., and Jaenicke, L. (1983). *Naturwissenschaaften* **70,** 571–572.

Gilles, R., Gilles, C., and Jaenicke, L. (1984). *Z. Naturforsch.* **39c,** 584–592.

Gilles, R., Moka, R., Gilles, C., and Jaenicke, L. (1985). Submitted.

Goodenough, U. W. (1985). *In* "Origin and Evolution of Sex" (H. O. Halvorson and A. Monroy, eds.), pp. 123–140. Liss, New York.

Goodenough, U. W., and Heuser, J. (1985). *J. Cell Biol.* **101,** 1550–1568.

Green, K. J. (1982). Ph.D. thesis, Washington University, St. Louis, Missouri.

Green, K. J., and Kirk, D. L. (1981). *J. Cell Biol.* **91,** 743–755.

Green, K. J., and Kirk, D. L. (1982). *J. Cell Biol.* **94,** 741–742.

Green, K. J., Viamontes, G. I., and Kirk, D. L. (1981). *J. Cell Biol.* **91,** 756–769.

Hagen, G., and Kochert, G. (1980). *Exp. Cell Res.* **127,** 451–457.

Hamada, H., Petrino, M. G., and Kakunaga, T. (1982). *Proc. Natl. Acad. Sci. U.S.A.* **79,** 6465–6469.

Harper, J. F. (1982). *J. Cell Biol.* **95,** 353a.

Harper, J. F. (1984). *J. Cell Biol.* **99,** 327a.

Harper, J. F. (1986). In preparation.

Haseltine, W. A. (1983). *Cell* **33,** 13–17.

Hills, G. J., Phillips, J. M., Gay, M. R., and Roberts, K. (1975). *J. Mol. Biol.* **96,** 431–441.

Hoffman, J. L. (1983). *J. Cell Biol.* **97**, 63a.

Hoffman, J. L. (1984). *J. Cell Biol.* **99**, 243a.

Hoffman, J. L., and Kirk, D. L. (1986). In preparation.

Holtfreter, J. (1943). *J. Exp. Zool.* **94**, 261–318.

Huskey, R. J. (1979). *Dev. Biol.* **72**, 236–243.

Huskey, R. J., and Griffin, B. E. (1979). *Dev. Biol.* **72**, 226–235.

Huskey, R. J., Griffin, B. E., Cecil, P. O., and Callahan, A. M. (1979a). *Genetics* **91**, 229–244.

Huskey, R. J., Semenkovich, C. F., Griffin, B. E., Cecil, P. O., Callahan, A. M., Chace, K. V., and Kirk, D. L. (1979b). *Mol. Gen. Genet.* **169**, 157–161.

Hutt, W., and Kochert, G. (1971). *J. Phycol.* **7**, 316–320.

Ireland, G. W., and Hawkins, S. E. (1980). *Microbios* **28**, 185–201.

Ireland, G. W., and Hawkins, S. E. (1981). *J. Cell Sci.* **48**, 355–366.

Jaenicke, L. (1982). *Trends Biochem. Sci.* **7**, 61–64.

Jaenicke, L., and Gilles, R. (1982). *In* "Biochemistry of Differentiation and Morphogenesis" (L. Jaenicke, ed.), pp. 288–294. Springer-Verlag, Berlin and New York.

Jaenicke, L., and Waffenschmidt, S. (1979). *FEBS Lett.* **107**, 250–253.

Jaenicke, L., and Waffenschmidt, S. (1981). *Ber. Dtsch. Ges.* **94**, 375–386.

Janet, C. (1923). "Le *Volvox*. Troisieme Memoire." Protat Freres, Paris.

Johnson, U. G., and Porter, K. R. (1968). *J. Cell Biol.* **38**, 403–425.

Karn, R. C., Starr, R. C., and Hudock, G. A. (1974). *Arch. Protistenkd.* **116**, 142–148.

Kelland, J. L. (1964). Ph.D. Thesis, Princeton University, Princeton, New Jersey.

Kelland, J. L. (1977). *J. Phycol.* **13**, 373–378.

Kirk, D. L., and Kirk, M. M. (1976). *Dev. Biol.* **50**, 413–437.

Kirk, D. L., and Kirk, M. M. (1978a). *Plant Physiol.* **61**, 556–560.

Kirk, D. L., and Kirk, M. M. (1978b). *J. Phycol.* **14**, 198–203.

Kirk, M. M., and Kirk, D. L. (1978c). *Plant Physiol.* **61**, 549–555.

Kirk, D. L., and Kirk, M. M. (1983). *Dev. Biol.* **96**, 493–506.

Kirk, D. L., and Kirk, M. M. (1985a). *Science* **230**, in press.

Kirk, M. M., and Kirk, D. L. (1985b). *Cell* **41**, 419–428.

Kirk, D. L., Viamontes, G. I., Green, K. J., and Bryant, J. L. (1982). *In* "Developmental Order: Its Origin and Regulation" (S. Subtelny, ed.), pp. 247–274. Liss, New York.

Kirk, D. L., Birchem, R., and King, N. (1985). *J. Cell Sci.*, in press.

Kirk, D. L., Huson, K. S., and Zagris, N. (1986). In preparation.

Kochert, G. (1968). *J. Protozool.* **15**, 438–452.

Kochert, G. (1973). *In* "Developmental Regulation" (S. J. Coward, ed.), pp. 155–167. Academic Press, New York.

Kochert, G. (1975). *Dev. Biol. Suppl.* **8**, 55–90.

Kochert, G. (1981). *In* "Sexual Interactions in Eucaryotic Microbes" (D. H. O'Day and P. A. Horgen, eds.), pp. 73–93. Academic Press, New York.

Kochert, G., and Crump, W. J., Jr. (1979). *Gamete Res.* **2**, 259–264.

Kochert, G., and Olsen, L. W. (1970a). *Arch. Mikrobiol.* **74**, 19–30.

Kochert, G., and Olsen, L. W. (1970b). *Trans. Am. Micros. Soc.* **89**, 475–478.

Kochert, G., and Yates, I. (1970). *Dev. Biol.* **23**, 128–135.

Kochert, G., and Yates, I. (1974). *Proc. Natl. Acad. Sci. U.S.A.* **71**, 1211–1214.

Kurn, N. (1981). *Cell Biol. Int. Rep.* **5**, 867–875.

Kurn, N. (1982). *FEBS Lett.* **144**, 68–72.

Kurn, N., and Duskin, D. (1982). *Wilhelm Roux's Arch.* **191**, 169–175.

Kurn, N., and Sela, B.-A. (1979). *FEBS Lett.* **104**, 249–252.

Kurn, N., and Sela, B.-A. (1981). *Eur. J. Biochem.* **121**, 53–57.

Kurn, N., Colb, M., and Shapiro, L. (1978). *Dev. Biol.* **66**, 266–269.

Lamport, D. T. A. (1974). In "Macromolecules Regulating Growth and Development" (E. D. Hay, J. J. King, and J. Papaconstantinou, eds.), pp. 113–130. Academic Press, New York.

Lee, J. S., Woodsworth, M. L:, Latimer, L. J. P., and Morgam, A. R. (1984). Nucleic Acids Res. 12, 6603–6614.

L'Hernault, S. W., and Rosenbaum, J. L. (1983). J. Cell Biol. 97, 258–263.

L'Hernault, S. W., and Rosenbaum, J. L. (1985a). Biochemistry 24, 473–478.

L'Hernault, S. W., and Rosenbaum, J. L. (1985b). J. Cell Biol. 100, 457–462.

McCracken, M. D., and Barcellona, W. J. (1976). J. Histochem. Cytochem. 24, 668–673.

McCracken, M. D., and Barcellona, W. J. (1981). Cytobios 32, 179–187.

McCracken, M. D., and Starr, R. C. (1970). Arch. Protistenkd. 112, 262–282.

Marchant, H. J. (1976). In "Intercellular Communication in Plants: Studies on Plasmodesmata" (B. E. S. Gunning and A. W. Robards, eds.), pp. 59–80. Springer-Verlag, Berlin and New York.

Marchant, H. J. (1977). Protoplasma 93, 325–339.

Margolis-Kazan, H., and Blamire, J. (1976). Cytobios 15, 201–216.

Margolis-Kazan, H., and Blamire, J. (1977). Biochem. Biophys. Res. Commun. 76, 674–681.

Margolis-Kazan, H., and Blamire, J. (1979). Microbios Lett. 11, 7–13.

Matagne, R. F. (1978). Mol. Gen. Genet. 160, 95–99.

Mattson, D. (1985). M. S. thesis, Washington University, St. Louis, Missouri.

Mattson, D., Green, K. J., and Kirk, D. L. (1986). In preparation.

Miller, C. E., and Starr, R. C. (1981). Ber. Dtsch. Ges. 94, 357–372.

Mitchell, B. A. (1980). M. S. thesis, Michigan State University, East Lansing.

Moseley, K. R., and Thompson, G., Jr. (1980). Plant Physiol. 65, 260–265.

Muller, T., Bause, E., and Jaenicke, L. (1981). FEBS Lett. 128, 208–212.

Nikaido, T., Nakai, S., and Honjo, T. (1981). Nature (London) 292, 845–848.

Pall, M. L. (1973). J. Cell Biol. 59, 238–241.

Pall, M. L. (1975). In "Developmental Biology: Pattern Formation, Genetic Regulation" (D. McMahon and C. F. Fox, eds.), Vol. 2, pp. 148–156. Benjamin, Menlo Park, California.

Pickett-Heaps, J. D. (1970). Planta 90, 174–190.

Pickett-Heaps, J. D. (1975). "Green Algae: Structure, Reproduction and Evolution in Selected Genera." Sinauer, Sunderland, Massachusetts.

Pommerville, J. C., and Kochert, G. D. (1981). Eur. J. Cell Biol. 24, 236–243.

Pommerville, J. C., and Kochert. G. D. (1982). Exp. Cell Res. 140, 39–45.

Powers, J. H. (1908). Trans. Am. Microsc. Soc. 28, 141–175.

Provasoli, L., and Pintner, I. J. (1959). In "The Ecology of Algae" (C. A. Tryon and R. T. Hartman, eds.), pp. 84–96. Spec. Publ. 2, Pymatung Lab. of Field Biol., Univ. of Pittsburgh, Pittsburgh, Pennsylvania.

Roberts, K. (1974). Philos. Trans. R. Soc. London Ser. B 268, 129–146.

Schlesinger, M. J., Aliperti, G., and Kelley, P. M. (1982). Trends Biochem. Sci. 7, 222–225.

Schwartz, D. C., and Cantor, C. R. (1984). Cell 37, 67–75.

Sessoms, A. M. (1974). Ph.D. thesis, University of Virginia, Charlottesville.

Sessoms, A. M., and Huskey, R. J. (1973). Proc. Natl. Acad. Sci. U.S.A. 70, 1335–1338.

Smith, G. M. (1944). Trans. Am. Microsc. Soc. 63, 265–310.

Smith, L. D., and Williams, M. A. (1975). Dev. Biol. Suppl. 8, 3–24.

Starr, R. C. (1969). Arch. Protistenkd. 111, 204–222.

Starr, R. C. (1970). Dev. Biol. Suppl. 4, 59–100.

Starr, R. C. (1972a). Br. Phycol. J. 7, 284.

Starr, R. C. (1972b). Soc. Bot. Fr. Mem. 175–182.

Starr, R. C. (1975). Arch. Protistenkd. 117, 187–191.

Starr, R. C., and Jaenicke, L. (1974). *Proc. Natl. Acad. Sci. U.S.A.* **71**, 1050–1054.
Starr, R. C., O'Neill, R. M., and Miller, C. E., III. (1980). *Proc. Natl. Acad. Sci. U.S.A.* **77**, 1025–1028.
Straley, S. C., and Bruce, V. G. (1979). *Plant Physiol.* **63**, 1175–1181.
Stringer, J. R. (1982). *Nature (London)* **296**, 363–366.
Sumper, M. (1979). *FEBS Lett.* **107**, 241–246.
Sumper, M. (1984). *In* "Pattern Formation: A Primer in Developmental Biology" (G. M. Malacinski and S. V. Bryant, eds.), pp. 197–212. Macmillan, New York.
Sumper, M., and Wenzl, S. (1980). *FEBS Lett.* **114**, 307–312.
Tautz, D., and Renz, M. (1984). *Nucleic Acids Res.* **12**, 4127–4138.
Tucker, R. G., and Darden, W. H. (1972). *Arch. Mikrobiol.* **84**, 87–94.
Vande Berg, W. J., and Starr, R. C. (1971). *Arch. Protistenkd.* **113**, 195–219.
Viamontes, G. I., and Kirk, D. L. (1977). *J. Cell Biol.* **75**, 719–730.
Viamontes, G. I., Fochtmann, L. J., and Kirk, D. L. (1979). *Cell* **17**, 537–550.
Vig., B. K., and Lewis, R. (1978). *Mutat. Res.* **55**, 121–145.
Weinheimer, E. W. (1973). Ph.D. thesis, University of Georgia, Athens.
Weinheimer, T. (1983). *Cytobios* **36**, 161–173.
Weisshaar, B., Gilles, R., Moka, R., and Jaenicke, L. (1984). *Z. Naturforsch.* **39c**, 1159–1162.
Weismann, A. (1889). *In* "Essays upon Heredity" (E. B. Poulton, S. Schonland, and A. E. Shipley, eds.). Oxford Univ. Press (Clarendon), London and New York.
Wenzl, S., and Sumper, M. (1979). *FEBS Lett.* **107**, 247–249.
Wenzl, S., and Sumper, M. (1981). *Proc. Natl. Acad. Sci. U.S.A.* **78**, 3716–3720.
Wenzl, S., and Sumper, M. (1982). *FEBS Lett.* **143**, 311–315.
Wenzl, S., Thym, D., and Sumper, M. (1984). *EMBO J.* **3**, 739–744.
Willadsen, P., and Sumper, M. (1982). *FEBS Lett.* **139**, 113–116.
Wright, R., and Jarvik, J. (1985). *J. Cell Biol.*, in press.
Yates, I., and Kochert, G. (1976). *Cytobios* **15**, 7–21.
Yates, I., Darley, M., and Kochert, G. (1975). *Cytobios* **12**, 211–213.
Youngbloom, J., Schloss, J. A., and Silflow, C. D. (1984). *Mol. Cell. Biol.* **4**, 2686–2696.
Zagris, N., and Kirk, D. L. (1980). *J. Cell Biol.* **87**, 15a.
Zeikus, J. A., and Starr, R. C. (1980). *Arch. Protistenkd.* **123**, 127–161.

INTERNATIONAL REVIEW OF CYTOLOGY, VOL. 99

The Ribosomal Genes of *Plasmodium*

Thomas F. McCutchan

Laboratory of Parasitic Diseases, National Institute of Allergy and Infectious Diseases, National Institutes of Health, Bethesda, Maryland

I. Introduction

The ribosomal gene unit (rDNA unit) has been extensively studied and characterized in a wide variety of organisms. Although there are differences in the organization and composition of these genes from organism to organism, the most striking feature of eukaryotic rDNA units on an evolutionary scale is their similarity. In contrast, perhaps the most interesting features of the rDNA units of the *Plasmodium* species, and the subject of this review, are the sharp digressions made from what was thought to be standard features of the eukaryotic ribosomal gene. This review will focus on the *Plasmoidium* rDNA unit by first discussing briefly a feature of the rDNA units of other eukaryotic organisms and then comparing and contrasting this with what has been found about the *Plasmodium* rDNA unit. There are three main areas of interest: (1) the number of ribosomal genes in a haploid genome, (2) the structural variations among individual rDNA units, and (3) the arrangement of these genes with regard to each other.

II. Life Cycle of *Plasmodium berghei*

Since the biology of an organism affects the composition and organization of its genome, it is worthwhile discussing the life cycle of *Plasmodium* first. Figures 1 and 2 show a generalized life cycle of mammalian

295

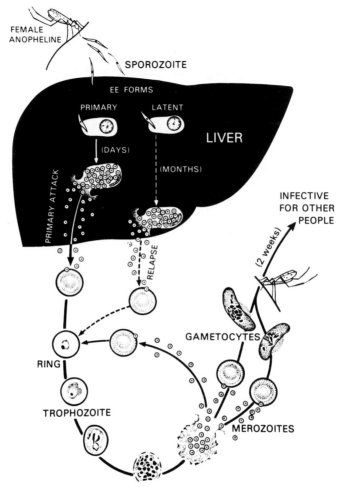

FIG. 1.   A generalized diagram of the malaria life cycle in the mammalian host (L. H. Miller, unpublished). The latent exoerythrocytic forms (EE forms) do not occur in many of the malarias. The time frame of the cycle varies considerably among the different malaria parasites.

malarias. Only the *Plasmodium berghei* life cycle will be discussed. This introduction is intended only to give the reader a sketch of the life cycle of one of the *Plasmodium* species, the time frame in which it occurs, and some idea of the stages where relatively rapid production of ribosomes appears to be occurring. The biology and biochemistry of the *Plasmodium* species have been extensively reviewed elsewhere (Garnam, 1966; Coatney 1971; Carter and Diggs, 1977; Sherman, 1979). It should also be noted

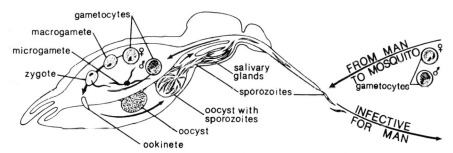

FIG. 2. The mosquito cycle and development of the malaria parasite in an *Anopheles* mosquito (Miller, 1974).

that the genus *Plasmodium* encompasses a large number of species of parasitic protozoa that vary widely with regard to host range, morphology, biological characteristics (Garnam, 1966; Coatney, 1971), and even DNA base composition (McCutchan *et al.*, 1984). Although many of the striking characteristics of the rDNA unit among these organisms are similar, interspecies differences may also exist.

The exocrythrocytic forms (EE forms in Fig. 1) of the parasite compose the first developmental stage in the mammalian host. This stage initiates with inoculation of sporozoites into the bloodstream and invasion of the liver. The parasites replicate asexually in the liver and are released as exoerythrocytic merozoites into the bloodstream (Fig. 1). In rodent malarias, this stage lasts 2–3 days and results in replication of the parasite up to 20,000-fold. The rodent malarias leave no latent forms in the liver, like those shown in Fig. 1 for some primate malarias, and hence are not relapsing. This stage is very active in cell replication and presumably in protein synthesis; many ribosomes appear in the cytoplasm. The entry of the exoerythrocytic merozoite into the bloodstream and infection of the red blood cell mark the start of the blood stages. Both asexual and sexual development occur at this point. In rodent malarias, a cycle of asexual development lasts 18–24 hours. Inside the merozoite-infected red blood cell the parasite develops through stages that are defined cytologically and referred to as rings, trophozoites, and schizonts. Approximately 12 new merozoites develop in one infected cell per cycle. These merozoites are released by lysis of the erythrocyte and reinvasion of new erythrocytes occurs. At this stage electron microscopy shows only thinly dispersed ribosomes. No defined nucleolus organizer, the site of ribosome synthesis, has been described in *P. berghei,* but a compact nucleolus has been described in the avian malaria *Plasmodium lophurae* (Aikawa *et al.*, 1969).

Sexual development of *Plasmodium* also occurs in the bloodstream. It is not clearly understood what even triggers sexual development. However, some infected cells develop into macrogametocytes and others into microgametocytes. The cytoplasm of the macrogametocyte is filled with developed endoplasmic reticulum and ribosomes while that of the microgametocyte is not (Fig. 3) (Sinden *et al.*, 1976). It has been suggested that differences in ribosome population and development of endoplasmic reticulum, where much protein synthesis occurs, relate to the subsequent fate of two different gametocytes (Sinden *et al.*, 1976). The microgametocyte, like the sperm cell, is a terminal cell which produces microgametes while the macrogametocyte, like the egg, continues to develop after fertilization. The time frame of gametocyte development is difficult to establish because no clear point of initiation has been established.

Gamete formation and fertilization initiates on ingestion of gametocytes in the blood meal of the mosquito (Fig. 2). In the gut of the mosquito the gametocytes emerge and become extracellular. The macrogametocyte becomes a macrogamete and the microgametocyte, after the process of exflagellation, yields microgametes. Approximately eight microgametes are produced per microgametocyte. Fertilization occurs within a few minutes, resulting in a diploid organism. (The point at which the organism again becomes haploid is not clear, but it does occur before sporozoites are formed.) In 12–18 hours the zygote develops into a motile cell called the ookinete. The ookinete then invades the wall of the mosquito midgut and continues to develop. The ookinete develops into an oocyst while in the wall of the midgut. The mature oocyst contains several thousand sporozoites. When the mature oocyst bursts, the sporozoites make their way to the salivary gland. Under optimal conditions a single zygote can result in several thousand sporozoites which start to reach the salivary gland of the mosquito 13–14 days after the ingestion of the blood meal.

In summary, the organism goes through at least 11 distinct stages in its life cycle, both haploid and diploid. Protein synthesis varies in rate and occurs at different temperatures throughout the cycle. Evidence obtained by microscopy indicates that certain stages are marked by developed endoplasmic reticulum and a high concentration of ribosomes, while others are not. It is clear that other eukaryotic organisms make adjustments in the rate of transcription of ribosomal RNA (McKnight and Miller, 1976; Foe, 1977) or actual changes in the number of genes involved in transcription (Engberg and Pearlman, 1972; Timmis and Ingle, 1973) to compensate for changes in their growth rate and need for protein synthesis. *Plasmodium* species may make similar adjustments, and hence the above information is important in the interpretation of results on the molecular biology of their rDNA units.

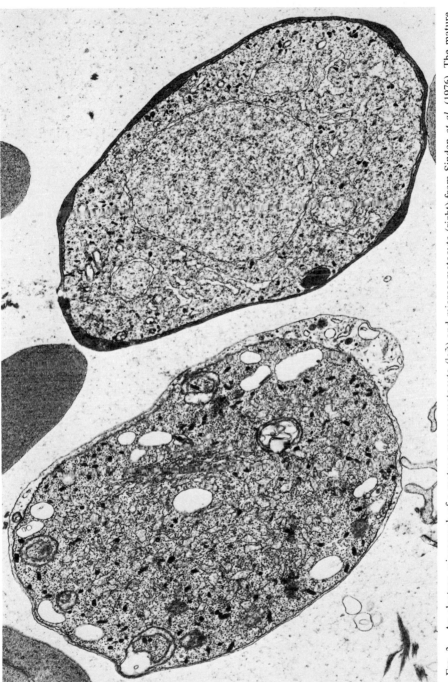

FIG. 3. An electron micrograph of a mature macrogametocyte (left) and microgametocyte (right) from Sinden *et al.* (1976). The mature macrogametocyte contains mitochondria and extensive endoplasmic reticulum while the microgametocyte does not.

III. Ribosomal Gene Copy Number

The number of copies of ribosomal genes per haploid genome in the
Plasmodium species is exceptionally low. The ribosomal genes of eukaryotic organisms are routinely characterized as exhibiting "dosage repeti-
tion" (Long and Dawid, 1980). This means that they occur in multiple
copies in the genome because the cell requires more ribosomal RNA than
a single gene could supply. The number of rDNA gene copies generally
ranges from about 50 to 10,000. The gene number of an organism is
usually fixed, but extrachromosomal amplification of the number of
rDNA units does occur in some organisms during stages of their life cycle
when a great amount of protein synthesis is occurring. For example, gene
amplification occurs during oogenesis in amphibians and some insects.
Amplification of rDNA also occurs in some protozoa such as *Tetrahymena* whose extrachromosomal gene number varies according to the
growth phase (Engbert and Pearlman, 1972). Thus, the ribosomal gene
number in most eukaryotic organisms is large, and there are occasionally
mechanisms in place to compensate for a temporary need for extra ribosomes by an increase in gene number.

The organism seems to need this large number of genes. What happens
phenotypically when some of the multiple copies of the rDNA unit are
deleted? This question has been studied using the bobbed mutants of
Drosophila (Tartof, 1975; Ritossa, 1976; Shermoen and Kiefer, 1975).
These mutants represent ribosomal gene deletions. Mutants that have
only 50% of their normal complement of ribosomal genes are phenotypically normal and, in fact, still produce the same amount of rRNA as the
wild-type strain. Mutants containing less than 50% of their normal complement of rDNA genes are phenotypically affected and are characterized
by slower development. Reduction of the gene level below 15% is lethal.
This work suggests that, although *Drosophila* may have some extra copies of the gene, an organism needs its multiple copies of the ribosomal
gene to develop properly and to compete in an evolutionary sense.

The rDNA copy number has been determined for a rodent malaria [*P.
berghei* (Dame and McCutchan, 1983a)], an avian malaria [*P. lophurae*
(Unnasch and Wirth, 1983a)], and a human malaria [*Plasmodium falciparum* (Langsley et al., 1983)]. The rodent malaria contains only four
ribosomal genes. This indicates that, of the organisms which do not amplify their DNA, *P. berghei* has fewer copies of these classic dosage
repeat genes than any other eukaryote (Dame and McCutchan, 1983a).
Further, it appears that two of the four *P. berghei* genes are not transcribed in the asexual form of the parasite. One of the four genes has been
shown by S_1 analysis to be transcribed, while another one is probably not

(Dame *et al.*, 1984). Further restriction map analysis has allowed us to separate the four genes into two structurally defined pairs, and it is therefore suggested that only two genes are actively transcribing RNA (Dame and McCutchan, 1983a). The other malaria species have between four and ten rDNA units (Unnasch and Wirth, 1983a; Langsley *et al.*, 1983; T. F. McCutchan and J. B. Dame, unpublished data).

We have attempted to calculate the maximal rate of rRNA synthesis in *P. berghei*. This has been approached by estimating the amount of ribosomes that could be synthesized by the *P. berghei* genes in a single asexual blood cycle. It was calculated that about 2.3×10^5 ribosomes could be synthesized by four rDNA units in a 24-hour cycle and that there are approximately 0.65×10^5 ribosomes in a mature *P. berghei* schizont (Dame and McCutchan, 1983a). In making this calculation we assumed that the RNA polymerase I of *Plasmodium* functions approximately as other eukaryotic polymerases (Coupar *et al.*, 1979; Cox, 1976; Greenberg and Penman, 1966; Darnell *et al.*, 1967; Miller and Bakken, 1972) with regard to the rate of synthesis and the density to which enzyme molecules can pack on their template. We also assumed that newly replicated DNA functions for transcription almost immediately after formation. During a 24-hour growth cycle, a single *P. berghei* merozoite containing one genome equivalent goes to a fully developed schizont containing an average of eight genome equivalents. We came to the conclusion that two active genes could supply the protein-synthesizing machinery of *P. berghei* asexual blood stages if the equivalent of four genomes were working over the entire cycle. The need of the metabolically more active stages of the parasite's life cycle to supply its protein-synthesizing machinery seems less likely to be satisfied.

Another way to consider the gene number in *Plasmodium* species is to relate the ribosomal gene number to total genome size. Unnasch and Wirth point out that, although the absolute number of ribosomal genes is quite low, this number, taken in context of the size of the *Plasmodium* genome, is not unique (Unnasch and Wirth, 1983a). *Plasmodium* species have about one rDNA unit for every 8.3×10^6 bp of genome (Table I). While this number is quite low for other lower eukaryotic organisms (e.g., *Dictyostelium discoideum* has one rDNA unit for 7.0×10^5 bp; *Leishmania donovani* has one rDNA unit per 3.3×10^5 bp), it is quite similar to some higher eukaryotic organisms, such as mice and humans, which have approximately one rDNA unit per 1.6×10^7 bp of genome.

The unusually low number of rDNA units in *Plasmodium* species raises the question whether there is as yet undetected amplification at points during development of the parasite. The asexual blood stages of the rodent malaria, *P. berghei,* show no amplification of gene number (Dame

TABLE I

SIZE OF GENOME AND ARRANGEMENT OF rDNA UNITS OF *Plasmodium* SPECIES COMPARED WITH SEVERAL OTHER ORGANISMS

Organism[a]	Haploid genome size (bp)	Number of rDNA units	Presence of intron		Arrangement of rDNA units
			Large rRNA	Small rRNA	
Escherichia coli	4.2×10^6	8	−	−	Interspersed in chromosome
Saccharomyces cerevisiae	1×10^7	140	−	−	Tandem repeats in chromosome
Dictyostelium discoideum	5×10^7	200	−	−	Amplified extra-chromosomally
Plasmodium berghei	$2–5 \times 10^{7\,b}$	4^c			Interspersed in the genome
Plasmodium lophurae		8^d	$+^e$	$+^e$	Interspersed in the genome
Plasmodium falciparum		8^f	$+^f$		Interspersed in the genome
Leishmania donovani	6.5×10^7	170	+	−	Tandem repeats in the genome
Tetrahymena	1×10^8	200	+	−	Amplified extra-chromosomally
Drosophila melanogaster	1.8×10^8	200	+	−	Tandem repeats in the genome
Xenopus laevis	2.2×10^{10}		−	−	Tandem repeats both in the genome and in extra-chromosomal elements
Chicken	8×10^8	200	−	−	Tandem repeats in the genome
Man	2.8×10^9	160	−	−	Tandem repeats in the genome

[a] Information about the eukaryotic organisms other than for those specified is from Lewin (1980) and Long and Dawid (1980).

[b] Dore *et al.* (1980).

[c] Dame and McCutchan (1983a).

[d] Unnasch and Wirth (1983a).

[e] Unnasch and Wirth (1983b).

[f] Langsley *et al.* (1983).

and McCutchan, 1983a). However, these may not be the most metabolically active stages of the parasite's life cycle. We have also studied DNA
from the male and female gameteocytes, zygotes, and ookinetes of the
avian malaria, *Plasmodium gallinaceum*. In these studies both chromosomal and extrachromosomal amplifications would have been detected,
but not was found (T. F. McCutchan and J. B. Dame, unpublished data).
This leaves sporozoite development and liver exoerythrocytic stages
which have not been studied. These are the two points in the parasite's
life cycle that are the most active in replication as well as the most difficult
materials to obtain for study.

IV. Arrangement of rDNA Units

The ribosomal genes in *Plasmodium* are not tandemly arranged, but are
interspersed among other genes, like the ribosomal genes of *Escherichia
coli*. Further, the various genes do not have identical primary sequences.
Organisms that have their complement of ribosomal genes integrated into
a chromosome have these units clustered in one or a few regions of DNA.
Units are connected in a head to tail fashion in what is called a tandem
array (Fig. 4). An array can be recognized as the dark-staining region of
the nucleus called the nucleolus. The individual units on ribosomal gene
tandem arrays and even other types of tandem arrays have the characteristic of maintaining the same sequence as their siblings. This may be a
genetic characteristic of the array, and the mechanisms involved have
been discussed (Lewin, 1980). The process of maintaining sequence
homogeneity in rRNA transcripts is important because variability in ribosome types does not appear to be biologically tolerated. Other organisms
have many fewer rDNA units integrated into their genome, but expand
their number by amplification of extrachromosomal elements containing

rDNA unit

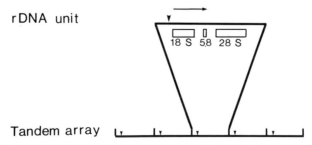

Tandem array

FIG. 4.   A diagram of the structure of a eukaryotic rDNA unit (above) and of several units
in a tandem arrangement (below).

rDNA units. Sequence identity in transcribed rRNA may be in part maintained in these organisms because organisms only amplify one of their members.

Restriction mapping data on the genes of *P. berghei* (Dame and McCutchan, 1983a), *P. lophurae* (Unnasch and Wirth, 1983a), and *P. falciparum* (Langsley *et al.*, 1983) indicate that each has variability in the sequence of their ribosomal genes. There are clearly restriction site differences internal to mature coding regions, indicating sequence variation. *P. berghei* contains four genes which can be divided into two pairs by restriction mapping. *P. lophurae* DNA contains from six to eight different genes which can be divided into four different groups by restriction analysis. *P. falciparum* has eight different genes divided into at least two classes. These data would argue that sequence variation of mature rRNAs occurs in *Plasmodium* if several genes are being transcribed at any one time.

The ribosomal gene of the three species of *Plasmodium* that have been tested appear not to be arranged in a tandem fashion. The exact boundaries of the rDNA units beyond the rRNA-coding regions is difficult to define, and so the exact size of the unit remains a mystery. The locale of different rDNA units has been studied both directly and indirectly. Two different units from the *P. berghei* genome have been studied by DNA heteroduplex analysis and show homology only in mature RNA-coding regions (Dame *et al.*, 1984). The regions surrounding each gene and those in the spacer regions between small and large rRNA-coding regions are completely different. Further, if there were a distinct unit size and the units followed in a tandem arrangement, this should be detectable by finding restriction enzymes which cut one time per unit and yield single size restriction fragments of the internal rDNA units (Fig. 4). This is not the case for *Plasmodium* DNA. Further, Dame and McCutchan (1983a) have done restriction mapping of a large area surrounding each of the genes of *P. berghei* and find no linkage where ~150 kb of the genome are accounted for. We have further data suggesting that the rDNA units of *Plasmodium cynomolgi* are on at least three different chromosomes (T. F. McCutchan, unpublished).

The 5 S RNA is also a ribosomal RNA. The 5 S genes of most eukaryotic organisms are not linked to the other ribosomal RNA genes even though all the different ribosomal RNAs are produced in approximately equimolar ratios. The 5 S gene is transcribed by RNA polymerase III, while the rDNA unit is transcribed by RNA polymerase I. Yeast and *Dictyostelium* 5 S genes are exceptional in that they are linked to the other rRNA genes, but as in other eukaryotic organisms, they are under separate control and are transcribed by RNA polymerase III. The 5 S genes of bacteria are different in that they not only are linked to the rDNA unit, but are transcribed as a part of the rDNA unit.

The position of this gene in *P. berghei* is interesting because, although organization of its rDNA unit is similar to that found in *E. coli* (e.g., low copy number and nontandem arrangement of genes), the 5 S gene is unlinked to the main rDNA unit as in other eukaryotic organisms. The gene number is limited to only a few copies and is not within 5 kb of any rDNA unit (Dame and McCutchan, 1984).

V. The Structure of the rDNA Unit

The structure of the individual units of *Plasmodium* rDNA is not unusual except for the presence of a novel type of intron in some units. In most eukaryotic organisms the rDNA unit is like the one diagrammed in Fig. 4 (Long and Dawid, 1980). Each unit contains a pre-rRNA gene and a spacer. Mature 5.8, 18, and 28 S RNA is made by processing the pre-RNA molecule. The polarity of transcription is from the small, 18 S, rRNA gene through the 5.8 S RNA gene and internal spacer region and to the large, 28 S, rRNA gene. The regions that code for mature RNAs are remarkably conserved. No variations in the sequence of transcribed RNA are seen in an organism, even though its transcripts are being produced from many different rDNA units. Most metazoan rDNA units do not have introns. Dipterian insects are an exception, but the intron-containing units are not transcribed to any extent. Some unicellular eukaryotic organisms such as *Tetrahymena* also have introns in the large rRNA gene, and here they are transcribed and later removed by splicing.

The general organization of an rRNA transcription unit of *Plasmodium* species is like the rDNA of other organisms (Fig. 4) with the small rRNA-coding region located at the 5' end and the large rRNA-coding region at the 3' end. This has been shown in both *P. berghei* (Dame and McCutchan, 1983a) and *P. lophurae* (Unnasch *et al.*, 1985). A putative 5.8 S gene has also been detected between the coding region of the small rRNA and the large rRNA (Langsley *et al.*, 1983; Dame *et al.*, 1984; Dame and McCutchan, 1984). The presence of large, apparently unprocessed rRNA in *Plasmodium knowlesi* (Trigg *et al.*, 1975) suggests that the gene is transcribed into a pre-rRNA and then processed in a manner seen in other eukaryotic organisms. The sizes of the mature RNAs have been determined and, when compared with the sizes of mature rRNAs from other organisms, show nothing unusual (Table II). The large rRNA of the rodent malaria is cleaved or nicked *in vivo* while large rRNAs of the other *Plasmodium* species are not.

Perhaps the most interesting finding with regard to gene structure is described by Unnasch and Wirth in *P. lophurae* (Unnasch and Wirth,

THOMAS F. McCUTCHAN

TABLE II
Ribosomal RNAs of the *Plasmodium* Species Compared with
Several Other Organisms[a]

Organism[a]	Precursor RNA S value	Length of[b]		Natural nick[c]
		Large rRNA	Small rRNA	
Escherichia coli		3000	1541	−
Saccharomyces cerevisiae	37	3750	2000	−
Dictyostelium discoideum		4075	1800	−
Plasmodium berghei		4300[d]	2500[d]	+[d]
		4000[e]	2200[e]	
Plasmodium knowlesi	48, 38, 32[f]			−[g]
Plasmodium lophurae				−[h]
Plasmodium falciparum		3720[i]	2100[i]	−[i]
Leishmania donovani		3750	2100	−
Tetrahymena	35, 26	3720	1970	+
Xenopus laevis	40	4475	1925	+
Drosophila melanogaster	34	4100	2000	+
Chicken	45	4626	1800	−
Man	45	5000	1950	−

[a] Information about the organisms other than for those specified is from Lewin (1980).

[b] The lengths of the RNAs are expressed in the number of nucleotides.

[c] A natural nick refers to the specific cleavage of the large rRNA *in vivo*.

[d] Miller and Ilan (1978).

[e] Dame and McCutchan (1983b).

[f] Trigg *et al.* (1975).

[g] J. B. Dame and T. F. McCutchan, unpublished.

[h] Unnasch and Wirth (1983a).

[i] Vezza and Trager (1981).

1983b). There, both the small and large rRNA genes have introns (Fig. 5). The small rRNA gene contains an intron of 230 bp, the large rRNA gene contains two introns of 240 and 110 bp. This is confirmed by sequence analysis and is the first observation of an intron in the small rRNA (Unnasch and Wirth, 1983b). The question remains as to whether these intron-containing genes are transcribed. A single rDNA unit has been found in *P. berghei* that contains an uninterrupted coding sequence by S_1 analysis (Dame *et al.*, 1984). This unit is apparently responsible for some, if not all, of the active transcription; thus, transcripts of some genes are not spliced. Yet, a different rDNA unit was also isolated from the *P. berghei* genome. It is apparently interrupted in both the small and large rRNA genes as determined by S_1 nuclease analysis (Dame *et al.*, 1984). When the two *P. berghei* rDNAs were compared by heteroduplex analysis, it appeared that they varied in the large rRNA gene because of sequence differences in particular areas rather than by the addition of introns in one

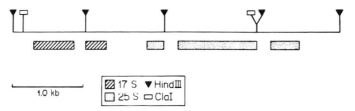

FIG. 5. A diagram of the structure of the *P. lophurae* rDNA unit. Coding regions for the small rRNA, 27 S, and large rRNA, 25 S, are as indicated. From Unnasch and Wirth (1983b).

particular unit. Differences between the two small rRNA genes were too small to see by electron microscopic techniques. Although the internal structure of *P. lophurae* and *P. berghei* rDNA units so far analyzed is different, it must also be remembered that only a subset of the genes in each organism has been analyzed so far. As more genes from each system are analyzed some clear pattern may emerge.

Finally, the structure of the intron in the large rRNA gene of *P. lophurae* deserves mention because it has features that are characteristic of a transposable element (Unnasch *et al.*, 1985). The intron contains several regions of dyad symmetry, including a 12-bp terminal repeat characteristic of transposable elements. The intron also has very significant sequence homology (47%), with the terminal repeat of the *Drosophila* transposable element copia.

The question remains as to whether these interrupted rDNA units are transcribed. Two interesting studies suggest that gene splicing does occur in *Plasmodium*. Ravetch *et al.* (1984) have shown that the gene for the histidine-rich protein of *P. lophurae* has a small intron which is processed out in the mature message. Further, Francoeur *et al.* (1985) have found small ribonucleoprotein particles in *P. falciparum*. They appear to be similar to and possibly performing the same function as the small ribonucleoprotein particles of higher eukaryotic cells which are involved in the processing and transport of mRNA. However, DNA sequence data that include the intron and its boundaries do not detect the usual splice sites found in lower eukaryotic organisms (Unnasch *et al.*, 1985). Therefore, if these genes are transcribed and the rRNA processed, the mechanism must be different from that shared by organisms, such as *Tetrahymena* and *Physarum polysephalum* (Waring *et al.*, 1983; Cech, 1983).

VI. Conclusion

The rDNA units of the *Plasmodium* species are clearly unusual. There are a very few copies of the gene in the genome and none of these appears to be amplified in number during growth of the organism. The copies are not arranged tandemly, but are interspersed in the genome. The individual units do not have identical DNA sequence, and some units have a novel type of intron in the small rRNA gene. The reason for breaking the established rules of rDNA structure may be related to the unusual biological pressures that face this parasitic protozoa. The results described above raise several questions: Is there a reason to maintain different rDNA units whose sequences are evolving independently? Could the different rDNA units be producing different ribosomes which function at different times during the life cycle or translate different subsets of messenger RNAs? Are the mechanisms involved in maintaining sequence homology between genes not working in the *Plasmodium* species or simply not working to maintain homology between rDNA units? Answering questions like these may tell us as much about the *Plasmodium* genome as a whole as it does about ribosome genes in particular. It is also possible that studying these genes that are apparent exceptions to the rules of standard genomic arrangement may help us focus on the relationship between gene organization and the biology of an organism.

ACKNOWLEDGMENTS

I would like to thank F. C. Eden, L. A. McNicol, J. Hansen, and J. A. Welsh for their help with this manuscript. I would also like to thank Tom Unnasch, Dyann Wirth, Robert Sinden, Christine Gritzmacher, Michelle Francoeur, and Louis Miller for making their unpublished data available to me.

REFERENCES

Aikawa, M., Huff, C. G., and Sprinz, H. (1969). *J. Ultrastruct. Res.* **26,** 316–331.
Carter, R., and Diggs, C. L. (1977). *In* "Parasitic Protozoa" (J. Kreier, ed.), Vol. III, pp. 359–465. Academic Press, New York.
Cech, T. R. (1983). *Cell* **34,** 713–716.
Coatney, G. R. (1971). "The Primate Malarias." National Institutes of Health, Bethesda, Maryland.
Coupar, B. E. H., Davies, J. A., and Chesterton, C. J. (1979). *Eur. J. Biochem.* **84,** 611–623.
Cox, R. F. (1976). *Cell* **7,** 455–465.
Dame, J. B., and McCutchan, T. F. (1983a). *J. Biol. Chem.* **258,** 6984–6990.
Dame, J. B., and McCutchan, T. F. (1983b). *Mol. Biochem. Parasitol.* **8,** 263–279.
Dame, J. B., and McCutchan, T. F. (1984). *Mol. Biochem. Parasitol.* **11,** 301–307.
Dame, J. B., Sullivan, M., and McCutchan, T. F. (1984). *Nucleic Acids Res.* **12,** 5943–5952.
Darnell, J. E., Girard, M., Baltimore, D., Sumners, D. F., and Maizel, J. V. (1967). *In* "The Molecular Biology of Viruses" (J. Colter, ed.), pp. 375–401. Academic Press, New York.

Dore, E., Birago, C., Frontali, C., and Battaglia, P. A. (1980). *Mol. Biochem. Parasitol.* **1,** 199–208.

Engberg, J., and Pearlman, R. E. (1972). *Eur. J. Biochem.* **26,** 393–400.

Foe, V. E. (1977). *Cold Spring Harbor Symp. Quant. Biol.* **42,** 723–739.

Francoeur, A. M., Pebbles, C. L., Gritzmacher, C. A., Reese, R. T., and Tan, E. M. (1985). *Proc. Natl. Acad. Sci. U.S.A.,* in press.

Garnam, P. C. C. (1966). "Malaria Parasites and Other Haemosporidia" pp. 60–84. Blackwell, Oxford.

Greenberg, H., and Penman, S. (1966). *J. Mol. Biol.* **21,** 527–535.

Hyde, J. E., Zolg, J. W., and Scaife, J. G. (1981). *Mol. Biochem. Parasitol.* **4,** 283–290.

Langsley, G., Hyde, J. E., Goman, M., and Scaife, J. G. (1983). *Nucleic Acids Res.* **11,** 8703–8717.

Lewin, B., ed. (1980). "Gene Expression-2," pp. 563–564. Wiley, New York.

Long, E. O., and Dawid, I. B. (1980). *Annu. Rev. Biochem.* **49,** 727–764.

McCutchan, T. F., Dame, J. B., Miller, L. H., and Barnwell, J. (1984). *Science* **225,** 808–811.

McKnight, S. L., and Miller, O. L., Jr. (1976). *Cell* **8,** 305–319.

Miller, F. W., and Ilan, J. (1978). *Parasitology* **77,** 345–365.

Miller, L. H. (1974). *In* "Tranmissible Disease and Blood Transfusion" (T. J. Greenwalt, M. D. Jamieson, and G. A. Jamieson, eds.), pp. 244. Grune & Stratton, New York.

Miller, O. L., and Bakken, A. H. (1972). *In* "Gene Transcription in Reproductive Tissue" (E. Discfalusy, ed.), pp. 155–177. Karolinska Institute, Stockholm.

Ravetch, J. V., Feder, R., Pavlovec, A., and Blobel, G. (1984). *Nature (London)* **312,** 616–620.

Ritossa, F. (1976). *In* "The Genetics and Biology of Drosophila" (M. Ashberner and L. Novitski, eds.), Vol. 1A, pp. 801–846. Academic Press, New York.

Sherman, I. W. (1979). *Microbiol. Rev.* **43,** 453–495.

Sherman, I. W., Cox, R. A., Higginson, B., McLaren, D. J., and Williamson, J. (1975). *J. Protozool.* **22,** 568–572.

Shermoen, A. W., and Kiefer, B. I. (1975). *Cell* **4,** 275–280.

Sinden, R. E., Canning, E. U., and Spain, B. (1976). *Proc. R. Soc. London Ser. B* **193,** 55–76.

Tartof, K. D. (1975). *Annu. Rev. Genet.* **9,** 355–385.

Timmis, J. N., and Ingle, J. (1973). *Nature (London)* **244,** 235–236.

Trigg, P. I., Shakespeare, P. G., Burt, S. J., and Kyd, S. I. (1975). *Parasitology* **71,** 199–209.

Unnasch, T. R., and Wirth, D. F. (1983a). *Nucleic Acids Res.* **11,** 8443–8459.

Unnasch, T. R., and Wirth, D. F. (1983b). *Nucleic Acids Res.* **11,** 8460–8472.

Unnasch, T. R., McLafferty, M., and Wirth, D. F. (1985). Submitted.

Vezza, A. C., and Trager, W. (1981). *Mol. Biochem. Parasitol.* **4,** 149–162.

Waring, R. B., Scazzocchio, C., Brown, T. A., and Davies, R. W. (1983). *J. Mol. Biol.* **167,** 595–605.

INTERNATIONAL REVIEW OF CYTOLOGY, VOL. 99

Molecular Biology of DNA in *Acanthamoeba*, *Amoeba*, *Entamoeba*, and *Naegleria*

THOMAS J. BYERS

Department of Microbiology, The Ohio State University, Columbus, Ohio

I. Introduction

Amoebae have been favorite models for the study of problems in cell biology for years. There have been comparatively few studies, however, on the molecular biology of these organisms. One reason for this lack of interest is that the larger free-living amoebae, which are particularly useful for nuclear transfers and other kinds of micromanipulation, have never been cultured axenically. Consequently, studies of molecular biology in

311

these organisms potentially are complicated by the presence of food organisms. Many of the smaller amoebae can be cultured axenically, but the absence of evidence for sexual reproduction and, therefore, classical genetic analysis, has dampened the enthusiasm of molecular biologists for studies on these organisms. With the advent of recombinant DNA technology, however, the genomes of asexual organisms have become much more accessible to study. Several laboratories already have begun investigating the structure and function of amoeba genes and others have indicated an interest in starting. There are a number of interesting problems that might be examined in the amoebae. I have tried to summarize here what is known about amoeba DNA in order to encourage others to explore the molecular biology of these organisms.

The four genera of amoebae that have been studied most extensively will be discussed. I will only consider amoebae that remain as single cells, but four distinctly different life cycles are found in these groups. Interested molecular biologists should be aware that there are many additional genera of amoebae that are essentially untouched with respect to any significant biochemistry or molecular biology (Singh, 1981; Levine *et al.,* 1980). Some of these may prove to be very interesting subjects for future studies.

The genus *Amoeba* includes large free-living organisms. Work on the biology of these amoebae was last reviewed by Jeon (1973). The cell cycle is simple; replication is by binary fission and differentiation is unknown. These amoebae are large enough to manipulate by various surgical procedures, and this attribute has been exploited extensively (Jeon, 1973). Cultures typically are grown using the ciliated protozoan *Tetrahymena* as the major food organism.

The genus *Acanthamoeba* includes small amoebae that are ordinarily free-living, but can be opportunistic pathogens of animals and humans (Martinez, 1980, 1983; Kadlec, 1978; Visvesvara, 1980). Such organisms are considered amphizoic (Page, 1974). In a few cases, infection has been lethal, but these organisms are ubiquitous in nature and most humans appear to have good resistance to infections. Unfortunately, the taxonomy of *Acanthamoeba* is confused, and it is unclear whether pathogenicity is restricted to certain species or whether all strains are capable of it. Acanthamoebae replicate by binary fission. In addition, they differentiate into dormant cysts (i.e., they encyst) during adverse environmental conditions. A number of reviews of this process are available (Byers, 1979; Weisman, 1976; Griffiths, 1970; Krishna Murti, 1975). Cultures are readily grown axenically in complex or chemically defined media.

The genus *Naegleria* includes small amoebae that can differentiate either into a cyst or a flagellated stage (Fulton, 1970, 1977). In *Naegleria,* it is clear that certain species are pathogenic and that other species are not (John, 1982; Schuster, 1979). The pathogenic *Naegleria fowleri* is highly virulent and fatal to humans. Human infections are rare, but still, there has been little interest in the molecular biology of this species. The non-pathogenic species *Naegleria gruberi* has been the main organism studied. As in the case of *Acanthamoeba*, cultures of *Naegleria* can be grown axenically in complex or chemically defined media (Fulton *et al.*, 1984).

The genus *Entamoeba* is distinct from the rest because it includes anaerobes and parasites. The life cycle includes a phase of replication by binary fission and encystment, which occurs in response to environmental factors (Martinez-Palomo, 1982; Albach and Booden, 1978). Nuclear division can continue in the cyst in the absence of cytoplasmic division. Species are typically identified according to their hosts. *Entamoeba histolytica*, the human parasite, and *Entamoeba invadens*, a parasite of reptiles, are the two most commonly studied species. Both can be grown axenically under anaerobic conditions. *Entamoeba moshkovskii* is interesting because it appears to be free-living.

My objective is to discuss what is known about amoeba DNA during the vegetative replication cycle and during differentiation. I have attempted to present a reasonably current review of what is known about gene structure and the overall organization of the DNA and its metabolism. In the past, such a review probably would have had a reasonable lifetime. Now, however, an increasing number of good molecular biologists are becoming interested in these organisms, and it can be expected that knowledge about amoeba genomes and the molecular biology of the cell cycle and cell differentiation will increase rapidly in the near future.

II. Characteristics of Nuclear DNA and Gene Structure

A. Nuclear DNA Content and Chromosome Numbers

The nuclear DNA content of growing amoebae varies widely among species for which data are available (Table I). The lowest contents are found in the *E. histolytica*-like Laredo and Huff strains, which have 0.08–0.09 pg/amoeba, and in *E. moshkovskii*, which has 0.06 pg/amoeba. Reported values for *E. invadens* and *E. histolytica* are higher, ranging up to about 2 pg/amoeba. It is unknown whether the large differences in DNA content found in different laboratories for the HK 9 strains of *E. histolytica* are real or due to differences in methodology. The differences would suggest changes in ploidy levels or in gene amplification if substantiated. Two species of *Naegleria* have intermediate values around 0.2–0.3

TABLE I

CHARACTERISTICS OF AMOEBA NUCLEAR DNA

Species/strain	DNA/amoeba (pg)[a]	GC (%) nuclear main-band DNA[b]	GC (%) nuclear satellite DNA[b]	References
Acanthamoeba				
A. castellanii/Neff	1.28	57–61		Adam et al. (1969); Bohnert and Herrmann (1974); Hettiarachchy and Jones (1974); L. E. King and T. J. Byers, unpublished; Jantzen (1973)
A. astronyxis		50		Band and Mohrlok (1973)
A. palestinensis		62		Adam et al. (1969)
A. polyphaga		57		Adam et al. (1969)
Amoeba				
A. indica		59.8	28.4	Fritz (1982)
A. proteus/F		57.4	23.5	Fritz (1982)
A. proteus/A/D/$_T$D/$_x$D/T$_1$		23.4–29.5	44.7–61.8	Fritz (1982)
A. proteus/Bk	42.8	34.0	55	Spear and Prescott (1980)
A. proteus	34			Tautvydas (1971)
Amoeba hybrids				
A. proteus (nucleus) + A. indica (cytoplasm)		63.5	28.9	Fritz (1982)

A. indica (nucleus) + A. proteus (cytoplasm)		25.3	48.6	Fritz (1982)
Entamoeba				
E. histolytica/200:NIH/F-22	0.5, 0.5	29.2, 30.5		Gelderman et al. (1971b); Reeves et al. (1971)
E. histolytica/HK9-1/HK9-2	0.45–2.21	27.6		Gelderman et al. (1971a,b); Reeves et al. (1971); Lopez-Revilla and Gomez (1978)
E. histolytica-like/Laredo	0.09	30.8		Gelderman et al. (1971b); Reeves et al. (1971)
E. histolytica-like/Huff	0.08	30.7		Gelderman et al. (1971b); Reeves et al. (1971)
E. invadens/PZ/IP/165	0.32–0.82	26.1–33.49 (T_m)		Gelderman et al. (1971b); Lopez-Revilla and Gomez (1978)
E. invadens/IP-1	0.15			Sirijinktakarn and Bailey (1980)
E. moshkovskii/FIC	0.06	31.7 (T_m)		Gelderman et al. (1971b); Lopez-Revilla and Gomez (1978)
Naegleria				
N. fowleri/Lee	0.2			Weik and John (1978)
N. gruberi/NEG	0.17	34		Fulton (1977)
N. gruberi/NB-1	0.34	43		Fulton (1977)

[a] Includes cytoplasmic DNA in Acanthamoeba and Naegleria. All values are for mononucleated populations except for Entamoeba, where the averages ranged from 1 to 1.2 nuclei per amoeba in different strains (Lopez-Revilla and Gomez, 1978).

[b] Percentage of GC is based on buoyant density except for E. invadens and E. moshkovskii, where it is based on melting temperature (T_m).

pg/amoeba, whereas *Acanthamoeba castellanii* Neff most commonly has 1–2 pg/amoeba. Substantially higher values have been reported for this strain (Marzzoco and Colli, 1974; Coulson and Tyndall, 1978), but these values seem inconsistent with measurements from other laboratories (Byers, 1979). The largest amoeba DNA values are for *Amoeba proteus*, at about 34–43 pg/nucleus.

Entamoebae are anaerobic and lack mitochondria; therefore, the nuclear DNA and whole cell DNA contents are the same. Very little is known about mitochondrial DNA contents of the aerobic amoebae. Our measurements indicate that mitochondrial DNA can be as high as 20% of total cell DNA in early log-phase *A. castellanii* (King and Byers, 1985). Earlier work by Adam *et al.* (1969) indicated that the mitochondrial DNA could be as high as 30% of the total, but that it was variable. In *N. gruberi*, the mitochondrial DNA was 14% of the cell total (Fulton, 1977). No figures are available for *Amoeba*, but in general, amoebae appear to have relatively high proportions of mitochondrial DNA.

Both nuclear and mitochondrial DNA contents vary during culture growth. The nuclear DNA content of unagitated cultures of *A. castellanii* decreases during the log to post-log-phase transition (Byers *et al.*, 1969; King and Byers, 1985). This decrease is associated with preparations for encystment (see Section V,A), but probably is not essential for differentiation, since suspension cultures induced to encyst by $MgCl_2$ double their DNA content during encystment (Chagla and Griffiths, 1974). Since these authors have demonstrated elsewhere that G_1 is the predominant cell-cycle phase under their culture conditions (Chagla and Griffiths, 1978), the increase may simply be due to arrest of amoebae in G_2 phase after the DNA has been replicated. Variations in DNA content also were noted in *N. gruberi* where the DNA/nucleus also increased post-log phase (Weik and John, 1977). In contrast, the DNA/nucleus was relatively constant during the log- and post-log-phase transition in *N. fowleri* (Weik and John, 1978). No explanation is currently available for this difference.

Our experience with *A. castellanii* indicates that mitochondrial DNA is extensively degraded during late log- and post-log-growth phases (see Section V,A). No evidence about this matter is available for the other amoeba genera.

Amoeba chromosomes are small, and estimates of chromosome numbers are difficult to obtain. As many as 500–1000 small chromosomes have been observed in *A. proteus* (Andresen, 1973; Ord, 1973). Early authors reported counts of 3–16 chromosomes for *Naegleria*, but Fulton (1977) concluded that the chromosomes of *N. gruberi* were too small and too tightly packed to count. In *Acanthamoeba* also, chromosomes were small and too tightly clustered to permit accurate counts (Pussard, 1964,

1972). The picture is even more confused in *Entamoeba* where chromosomes have yet to be identified (Martinez-Palomo, 1982).

Until recently, the prospect of obtaining information about numbers, sizes, and linkage groups of amoeba chromosomes was bleak. With the development of pulse-gradient electrophoresis, however, the prospect has brightened considerably (Schwartz and Cantor, 1984; Carle and Olsen, 1984). With this technique, which has been used successfully to separate small chromosomes from trypanosomes and yeast, it should be possible to study the chromosomes of many lower eukaryotes.

B. Unique and Repetitious Sequences

The sequence organization of amoeba nuclear DNA is interesting because of the primitive nature of these organisms. Very limited data are available on the proportions of unique and repetitious nucleotide sequences in amoeba DNA. Estimates of the unique fraction range from 0.74 to 1.0 for *A. castellanii* Neff, but the lower value seems more likely because it has been obtained in more than one laboratory (references in Byers, 1979). The values for three strains of *Entamoeba*, estimated from a published graph (Gelderman *et al.*, 1971a), range from 0.4 to 1.0. Since these data are all from the same laboratory, the variability may be real. The *E. histolytica*-like Laredo strain, which appears to have the least amount of repetitious sequences, has an estimated genome size of 0.01 pg. A variety of fungi, which are the main group of eukaryotic organisms having very little repetitious DNA, also have genomes of less than 0.05 pg (references in Pellegrini *et al.*, 1981). *E. moshkovskii*, which seems to have the largest repetitious fraction, has an estimated genome size of 0.1 pg. Further studies of *Entamoeba* genomes would be particularly significant because they currently are evolving in the absence of any possible interaction with organelle DNA. It would be interesting to determine whether they carry any organelle-like sequences indicative of interactions with symbionts during their previous evolutionary history.

C. Genome Size and Ploidy Levels

There is no definitive evidence for sexual reproduction in any of the species examined here, and ploidy levels are unknown. Fulton (1970) suggested that haploid (NEG) and diploid (NB-1) strains of *N. gruberi* occur in the laboratory. The strongest evidence favoring this view is the observation that strains usually contain DNA in relative amounts of $1\times$ or $2\times$ (Table I) and that a $1\times$ strain (NEG) is readily mutagenized, whereas a $2\times$ strain (NB-1) is not. Knowledge about ploidy levels and possible sexual processes may be available from analyses of interstrain variations in the electrophoretic mobility of selected enzymes. This approach has been applied with some success in studies of the same problems in the kinetoplastida (Tait, 1983). Ward (1985) has briefly reported on variations in the mobility of lactic dehydrogenase with respect to its genetic coding in *Acanthamoeba,* but the data raise more questions than they answer. Nevertheless, a thorough genetic analysis of isozyme variants in each of the four amoeba genera certainly would be worthwhile.

The likelihood of a polyploid state in *A. proteus* is suggested by the very large amounts of nuclear DNA (40–50 times the amount in a human cell) and by evidence that some strains may have 500–1000 chromosomes (Andresen, 1973; Ord, 1973). More direct evidence for polyploidy is available for *Acanthamoeba* and *Entamoeba* from comparisons of total nuclear DNA and estimates of genome size based on measurements of kinetic complexity. The haploid genome of *A. castellanii* Neff is 0.045 pg based on the best estimates of the kinetic complexity of the unique fraction (2×10^{10} D), the proportion of the total nuclear DNA that is unique (74%), and the proportion of the cell DNA that is nuclear (85%) (Bohnert and Herrman, 1974; Jantzen, 1973; King and Byers, 1985). The average amoeba in a log-phase population has 1.28 pg DNA (King and Byers, 1985) and, therefore, would be around $25n$. The Laredo strain of *Entamoeba,* with an estimated haploid genome size of 0.01 pg (10^7 nucleotide pairs) and a nuclear content of 0.09 pg (Gelderman *et al.,* 1971a,b), would be about $9n$. *E. moshkovskii,* with an estimated genome size of 0.1 pg (10^8 nucleotide pairs) and an estimated nuclear content of 0.06 pg, would be about $1n$. *E. histolytica* HK 9, with a genome size of 0.033 pg ($10^{7.5}$ nucleotide pairs) and a nuclear content of at least 0.45 pg (Table I), would be $14n$ or greater. (See also Section V,B for *E. invadens.*) Of course, there is a potentially large error in these calculations. Nevertheless, they provide the best estimates of ploidy levels that we currently have, and they suggest that several strains are likely to be polyploid.

D. NUCLEAR SATELLITES

The nuclear DNAs of a number of *Amoeba* strains and at least two strains of *Naegleria* (NEG and NB-1) separate into two bands on CsCl gradients (Fritz, 1981, 1982; C. Fulton, 1977, personal communication), whereas only one band has been reported for *Entamoeba* and *Acanthamoeba* (Gelderman *et al.,* 1971a; Adam *et al.,* 1969). The main band of *A. proteus* typically was 75% or more of the total nuclear DNA and had an average GC content of about 30% in 10 strains (Fritz, 1981, 1982). The heavier satellite band included the balance of the nuclear DNA and had an average GC content of 54% in six strains. The heavier band, however, was not always the quantitatively minor band. In the closely related *Amoeba indica* and in *A. proteus* F, the heavy band included the majority of the DNA and, therefore, was referred to as the main band (Table I). Experiments with hybrid amoebae, which contained a nucleus from one strain and cytoplasm from another, suggested that the main and satellite bands could change in relative proportions under the influence of cytoplasmic conditions (Fritz, 1982). For example, in a hybrid including a nucleus from *A. proteus* and cytoplasm from *A. indica,* the GC contents of the main and satellite bands were 63.5% and 28.9%, respectively (Table I). In the reciprocal hybrid, the main band was 25.3% GC and the satellite was 48.6%. In each case, the distribution of DNA between the two bands tended to be characteristic of the strain that was the source of cytoplasm. Clearly, a study of the regulation of the relative proportions of DNA in the main and satellite bands might contribute to our understanding of roles that cytoplasmic factors can play in differential replication of nuclear DNA. Unfortunately, the present lack of axenic culture conditions weighs against the use of *Amoeba* for these studies, but it should at least be possible to characterize differences between donor strains and reciprocal hybrids in main band and satellite DNA. In addition, the possiblity that variations in proportions of main band and satellite DNA might occur under different conditions should be explored with *Naegleria,* which can be grown axenically.

It is unknown whether the nuclear satellites contain unique or repetitious sequences or both. The possibility has been considered that the nuclear satellite in *A. proteus* includes the genes coding for ribosomal RNA (Spear and Prescott, 1980). This proposal is appealing because the amoeba nucleus has an unusually large number of nucleoli (Andresen, 1973) and because DNA associated with the nucleoli replicates late in the cell cycle and, therefore, somewhat independently of the main body of DNA (Minassian and Bell, 1976). Furthermore, starved amoebae in G_2 reinitiate DNA synthesis in the satellite upon refeeding (Spear and Prescott, 1980). Since rDNA replication could be induced by refeeding of starved *Tetrahymena* (Engberg *et al.,* 1974), it was possible that the same might be true for *A. proteus.* However, satellite DNA failed to hybridize with rRNA from *Tetrahymena,* seemingly ruling out the possibility that satellites are amplified rDNA (Spear and Prescott, 1980).

Microspectrophotometric analysis of fluorescent-stained nuclei indicated that nuclear DNA in *A. proteus* more than doubles during the cell growth-duplication cycle (Makhlin *et al.,* 1979). It would be interesting to know whether the excess DNA is associated preferentially with either of the nuclear DNA bands. It has been suggested that the excess might be rDNA (Makhlin *et al.,* 1979), but there is no evidence for this.

E. Structure of Actin, Myosin, and Ribosomal
RNA Genes in *Acanthamoeba*

At present, data on nucleotide sequences are available for one gene, for portions of two others, and for two RNA products in *A. castellanii* Neff. Actin is the most abundant polypeptide in *Acanthamoeba*. Two-dimensional electrophoresis revealed at least three different actin molecules (Jantzen, 1981). A gene for one of these molecules, actin I, has been cloned and totally sequenced (Nellen and Gallwitz, 1982). Actin I is a protein of 374 amino acids and, therefore, requires a coding sequence of 1122 bp (base pairs). In addition, the gene has a single intron of 129 bp. Common eukaryotic signal sequences are found in the promoter region. A modified TATA box 5'-CATAAAA-3' begins 30–31 nucleotide pairs upstream from the mRNA start site. The sequence 5'-AACCAAGCG-3', which is similar to the eukaryotic consensus sequence 5'-GGC/TCAATCT-3', begins 70–71 nucleotide pairs upstream from the start site. The exact 3' end of the mRNA has not been identified, but the sequence 5'-AATACA-3' located 44 nucleotide pairs downstream from the translation termination codon may be a modification of the more common polyadenylation signal 5'-AATAAA-3'. The overall amino acid sequence of actin I is closely related to the predominant actins from *Physarum* and *Dictyostelium*, but is very different from yeast actin. The amino acid sequence is unusual in having an N-terminal glycine. The codon usage is very biased; there is a strong preference for C and an avoidance of A in the third position. Nellen and Gallwitz (1982) found evidence for at least three different actin genes, an observation consistent with previous electrophoretic data (Jantzen, 1981).

At the time of this writing, the sequencing of an actin gene of *N. gruberi* is nearly complete (C. Fulton, personal communication). It will be interesting to compare the two amoeba genes with respect to coding biases, signal sequences, and other organizational features.

Three different myosin molecules, myosin IA, IB, and II, have been found in *A. castellanii* Neff (Gadasi *et al.,* 1979). The gene for the heavy chain of myosin II has now been cloned and partially sequenced (Hammer *et al.,* 1984; J. A. Hammer, personal communication). The putative gene was isolated from a genomic library of *Acanthamoeba* DNA in phage λ using a portion of the myosin heavy-chain gene from the nematode *Caenorhabditis elegans* as a probe. Most or all of the gene is contained in a 6.5-kb restriction fragment. The fragment is complementary to a 5300-bp mRNA which codes *in vitro* for a protein of 185 kDa. The protein comigrated in SDS–polyacrylamide gel electrophoresis with purified *Acanthamoeba* myosin II heavy chain and is quantitatively and selectively precipitated by antiserum to the heavy chain. The identity of the gene has been confirmed by antiserum to the heavy chain. The identity of the gene has been confirmed by sequencing of the 3' end. Preliminary results suggest that introns are absent or very short and that there may be more than one myosin II heavy-chain gene in *Acanthamoeba*.

The genes for ribosomal RNA in *Acanthamoeba* include the typical 18–5.8–26 S repeat unit plus 5 S genes which are unlinked to the repeat unit. The organization of the 5 S genes is unknown. The repeat unit has been cloned and partially sequenced (D'Allessio *et al.*, 1981; M. R. Paule, personal communication). The number of repeat units is unknown, but all seem to be the same. There is no evidence for introns, but an unusual transcribed spacer occurs within the 26 S gene. Stevens and Pachler (1972) discovered that the 26 S rRNA in the large ribosomal subunit actually consisted of two large fragments held together by noncovalent bonds. Paule and his associates have shown that the coding region for these fragments includes two sequences, one about 2 kb and the other about 2.4 kb, separated by a 200-bp transcribed spacer. An *in vitro* transcription system for the ribosomal repeat unit in which transcription consistently begins at the authentic initiation site has been developed (Paule *et al.*, 1984a). The site is located within a 74-bp sequence and is recognized by purified RNA polymerase I plus a single species of initiation factor, designated TIF-1. Correct initiation is species specific for both the polymerase and the initiation factor (Grummt *et al.*, 1982; M. R. Paule, personal communication).

The promoter used for transcription of the ribosomal genes can be subdivided into three regions, referred to as motifs (Kownin *et al.*, 1985; Iida *et al.*, 1985). The *start motif* overlaps the initiation start site, defined as nucleotide pair 0; it extends from upstream position -26 to downstream position $+8$. This region is necessary for transcription of the 18 S rRNA. It is very AT rich, which might facilitate strand separation for initiation, but this probably is not essential, since start motifs in other organisms are not especially AT rich. The *Acanthamoeba* start motif does not bind TIF-1. The second region, *motif A*, extends further upstream from approximately -20 to -31; the boundaries are somewhat uncertain. This motif does bind TIF-1. Together, motif A and the start motif constitute the *core promoter*, extending from -31 to $+8$, which is essential for faithful *in vitro* translation. The third region of the promoter, *motif B*, extends from -32 to -47. It is not essential for transcription, but the efficiency of the process is greater when motif B is left intact. The region is necessary for the formation of a stable preinitiation complex that includes one or more initiation factors plus part of the core promoter sequence. The complex does not include RNA polymerase I. Although transcription of rRNA is species-specific, motif B contains a sequence ACTTT, extending from −40 to −45, that closely resembles the ATCTTT sequence found in mouse, rat, and human rRNA promoters. Paule's group suggests that the sequence may have a general function in promoting transcriptional efficiency (Kownin *et al.*, 1985).

Ribosomal RNA synthesis is shut down during the encystment of *Acanthamoeba* (Stevens and Pachler, 1973). Paule and his colleagues have determined that the down-regulation of the rRNA genes is due to a modification of RNA polymerase I (Paule *et al.*, 1984b). This is the first example of eukaryotic gene regulation by polymerase modification. The nature of the modification has not been determined, but a phosphorylation, adenylation, or similar modification is suspected.

Indirect information about organization of the 5 and 5.8 S rRNA genes has been obtained from sequencing of the two RNAs (MacKay and Doolittle, 1981). The 5 S RNA consists of 119 nucleotides, and the base sequence fits a secondary structure model that is consistent with other eukaryotic 5 S rRNAs. The nucleotide sequence was compared to 5 S rRNA sequences for two other protozoans, *Tetrahymena* and *Crithidia*, and a variety of other organisms. The protozoan:protozoan comparisons gave sequence homologies averaging 64%. In contrast, 5 S rRNA from the insect *Drosophila melanogaster* is 76% homologous with human 5 S rRNA and 85% homologous with 5 S rRNA from the echinoderm *Lytechinus variegatus*. Furthermore, the *Acanthamoeba* sequences were as divergent from those of other protozoans as from organisms in other kingdoms (i.e., wheat, human, and yeast 5 S rRNAs are all 65% homologous with the amoeba RNA).

The 5.8 S rRNA of *Acanthamoeba* consists of 162 nucleotides, including four pseudouridine and four methylated bases (Mackay and Doolittle, 1981). The sequence only partially fits either of two secondary structure models previously proposed, even though the models were both consistent with 5.8 S rRNA from organisms as diverse as yeast and vertebrates. The indications of extensive molecular diversity in the amoeba 5 and 5.8 S rRNAs are fully consistent with evidence from other sources which indicate an unusually high level of molecular diversity in the protozoans, probably explained by a long evolutionary history (Nanney, 1984; Williams, 1984).

F. RIBOSOMAL RNA GENES IN *Amoeba*

No information is available for the nucleotide sequences of any genes in the genus *Amoeba*. However, Murti and Prescott (1978) probably have observed transcriptionally active rRNA genes in *A. proteus* by using electron microscopy and the Miller technique to obtain spreads of nucleolar material. Of the putative rRNA genes observed, 90% occurred as single units (matrix units) averaging 6.07 μm in length and decorated with 100–200 lateral fibers that were 20 nm in diameter and ranged up to an average of 0.85 μm in length. The central axis, presumably DNA plus associated protein, was 4 nm in diameter. The lateral fibers, presumably ribonucleoproteins, increased in length from one end of the unit to the other, but the length gradient was not nearly as great as is typical for transcriptionally active rRNA genes isolated from other organisms. There was no evidence for the typical tandem repeating arrangement of the basic units. Of the units, 10% were seen in aggregates of two or three, but it was impossible to determine whether the central axes of the aggregated units were continuous with each other.

The 25 and 19 S rRNA molecules of *A. proteus* have molecular weights of 1.5 and 0.84 × 10^6, respectively (L. Goldstein, quoted in Murti and Prescott, 1978). Similar molecular-weight values have been reported for *A. discoides* (1.55 and 0.8 × 10^6; Hawkins and Hughes, 1973) and *A. castellanii* (1.53 and 0.89 × 10^6; Stevens and Pachler, 1972). Approximately 2.2–2.3 μm of DNA is required to code for the two rRNA molecules of *A. proteus*. If the basic ribosomal gene repeat unit in *A. proteus* is typical and includes coding sequences for 29, 19, and 5.8 S rRNA plus transcribed spacers, then a single matrix unit would be large enough to contain one or two repeat units. Information about the size of rRNA precursors might help distinguish between these two possibilities, but so far no precursor has been identified for *A. proteus*. However, the single gradient of lateral fiber lengths suggests that the matrix unit is transcribed as a single unit rather than as two units.

III. Mitochondrial DNA Characteristics

Acanthamoeba, Amoeba, and *Naegleria* are genera of obligate aerobes and, consequently, all possess mitochondria. Entamoebae are anaerobes and do not have mitochondria or cytochromes. With the exception of *Acanthamoeba,* very little is known about amoeba mitochondrial DNA. Several laboratories are currently investigating the mitochondrial DNA of *Naegleria,* but the studies are only in preliminary stages. It is only known that the mitochondrial DNA of *N. gruberi* NEG has a buoyant density of 1.683 (23% GC), compared to the main-band nuclear DNA density of 1.693 (34% GC), and is 14% of the total DNA (Fulton, 1970, 1977).

Mitochondrial DNA has been isolated from *Acanthamoeba* in a number of laboratories (reviewed in Byers, 1979). In our own laboratory, DNA from more than 20 strains has been examined (Bogler *et al.*, 1983; T. J. Byers, S. A. Bogler, V. Stewart, and E. Hugo, unpublished). The mitochondrial genome occurs as a circular molecule (Bohnert, 1973; Bogler *et al.*, 1983). The average size that we have measured for 19 strains is 41.5 ± 1.5 (SD) kb. Sizes ranged from 38.7 to 44.0 kb. A twentieth strain, HOV.6, has a significantly larger genome of 49 kb. The G + C compositions have been determined from CsCl buoyant density measurements by Adam and Blewett (1974) for *A. castellanii* Neff (34%), *A. castellanii* Reich (*A. palestinensis*) (37%), *A. polyphaga* (43%), and *A. astronyxis* Ray (30%). Clearly, as in *Naegleria*, all mtDNAs are rich in A + T. A more extensive effort to study sequence heterogeneity in *A. castellanii* Neff examined the following: the buoyant density, the differential melting profile, the elution of mtDNA from hydroxyapatite by a gradient of sodium phosphate, and the percentage GC of fragments retarded on Sephadex G-25 in samples progressively degraded by micrococcal nuclease (Mery-Drugeon *et al.*, 1981). The overall percentage GC was 32.9, essentially the same as for the fungus *Ustilago cyanodontis*, which was examined in the same study. Although both organisms were equally rich in A + T, the fungal DNA appeared to have long nonalternating A–T regions, such as previously discovered in *Saccharomyces*, whereas the amoeba DNA did not.

We have digested mtDNA obtained from 20 strains of *Acanthamoeba* with 5–17 different restriction endonucleases (see Section VI,B) and are currently constructing physical (restriction site) maps from these data. In addition, we have begun to identify putative gene locations by use of heterologous gene probes from *Neurospora* (R. A. Akins, S. A. Bogler, E. Hugo, V. Stewart, and T. J. Byers, unpublished). The first physical map of the mitochondrial genome was constructed for *A. castellanii* Neff by H. J. Bohnert and A. von Gabain (Fig. 1, unpublished results). The techniques used to order the DNA fragments included double digestions using combinations of enzymes (restriction fragment sizes have been published by Mery-Drugeon *et al.*, 1981, and Bogler *et al.*, 1983), partial digestions, redigestion of isolated DNA fragments with other restriction endonucleases, and DNA–DNA hybridization. Bohnert and von Gabain also localized the rRNA genes. The large ribosomal RNAs are very similar in size to the 16 and 23 S rRNAs of *Escherichia coli*. One set of rRNA genes was found in *Acanthamoeba*, and the two genes were separated by about 3.5 kb.

Log-phase amoebae of the Neff strain contained an average of about 0.15 pg of mitochondrial DNA per amoeba (King and Byers, 1985). With a genome size of ~2.7 × 10^7 Da, there should be ~3300 mitochondrial DNA molecules per cell. During encystment, King and Byers found mitochon-

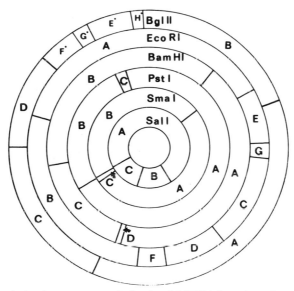

FIG. 1. Restriction fragment map of mitochondrial DNA from *Acanthamoeba castellanii* Neff. The map is from unpublished results of H. J. Bohnert and A. von Gabain (used with permission). These authors located the 16 S rRNA gene on *Sal*I fragment C at the end adjacent to fragment A. The 23 S rRNA gene was located on *Bam*HI fragment B overlapping the ends of *Eco*RI fragments A and B, but mostly B. The order of *Bgl*II fragments E, F, G, and H is uncertain. Fragment sizes for *Eco*RI, *Bam*HI, *Pst*I, *Sal*I, and their double digests have been published (Mery-Drugeon *et al.*, 1981). Unpublished fragment sizes for *Sma*I and *Bgl*II are 13.9, 13.1, and 0.25 mDa and 9.40, 5.7, 4.9, 3.0, 2.4, 1.0, 0.52, and 0.48 mDa, respectively.

drial DNA levels dropping to between 0.01 and 0.02 pg per amoeba; this would be enough for 223–446 molecules per cyst. No rigorous estimates of mitochondrial numbers are available, but rough estimates can be made from data available for suspension cultures of the Neff strain (Bowers and Korn, 1969). According to these workers, cyst volumes averaged about 2200 μm^3, and the mitochondrial volume was about 6.8% of the total cell volume. The cyst mitochondria in their pictures appeared to be roughly spherical with an average diameter of about 0.5 μm; thus, the average volume for a mitochondrion was about 0.065 μm^3. Cysts would have about 2300 mitochondria of this size. This number is much larger than the estimated number of mitochondria based on our data. However, for amoebae grown in monolayer cultures in our laboratory, the average volume of trophozoites was only about 29% of the trophozoite volume in the suspension cultures of Bowers and Korn (Byers *et al.*, 1969). Assuming a constant mitochondrial size and fraction of total cell volume, the smaller cell might have about 670 mitochondria per cyst (0.29 × 2300). Because of the uncertainties in this estimate, this number probably is not significantly different from the number of molecules of DNA estimated from King's data. Unfortunately, we do not have any reasonable estimate of the numbers of mitochondria in trophozoites. However, in the work of Bowers and Korn, the total mitochondrial volume of log-phase cells is 5.3 times that of cysts. If this factor is used for the monolayer cultures and mitochondrial sizes remain unchanged, the trophozoites should have about 3600 mitochondria (5.3 × 670); this agrees well with the estimate of 3300 mitochondrial DNA molecules. Thus, it seems likely that the number of mtDNA molecules per mitochondrion is relatively small and may even be one.

IV. DNA Replication during the Growth-Duplication Cycle

A. DNA POLYMERASES OF *Amoeba*

Very little is known about the enzymatic machinery of DNA replication in amoebae. To date, only the DNA polymerases of *Amoeba discoides* have been examined (Abbott and Hawkins, 1980). These studies were stimulated by the observation that certain phenotypic traits, e.g., strepto-mycin resistance, can be heritably transmitted by microinjection of frac-tions containing small-molecular-weight RNA (Hawkins, 1973; Abbott and Hawkins, 1980). The authors wondered whether this phenomenon might be explained by the presence of an RNA-dependent DNA poly-merase capable of synthesizing DNA from the injected RNA. Conse-quently, template preferences for thymidine triphosphate ([³H]TTP) incorporation into acid-insoluble material were examined using a variety of deoxyribonucleotide and ribonucleotide templates. The DNA poly-merase activities from whole cell homogenates were pooled after passage through Sephadex G-200 and had maximum activity when either heat-denatured calf thymus DNA or synthetic poly[d(A-T)] were used as tem-plates. They also had respectable activity with *E. coli* rRNA and *A. discoides* microsomal RNA as templates. The template activity of amoeba RNA was reduced 81% by treatment with RNase. Three peaks of nuclear polymerase activity were identified using nicked double-stranded calf thy-mus DNA as a template. Two of these activities also were able to utilize heat-denatured calf thymus DNA as a template. When short columns were used for separations, the void volume activity was able to efficiently utilize RNA templates and poly(A)·d(pT)10, a synthetic riboadenylic polymer base paired to decadeoxythymidylate. The evidence is consistent with the possibility that an RNA-dependent DNA polymerase is present. When longer columns were used, the ability to utilize the synthetic polymer disappeared; possibly a necessary cofactor was removed under these conditions. DNA polymerase activity able to utilize nicked double-stranded calf thymus DNA was found in the cytosol, but was not characterized further. The synthetic polymer poly(A·d(pT)10 could not be used by enzymes present in the cytosol. Relationships between the DNA polymerases of *A. discoides* and the higher eukaryotes are unclear. However, the low-molecular-weight polymerase β seems to be missing, as in a number of other lower eukaryotes (Abbott and Hawkins, 1978).

B. The Cell Cycle in *Acanthamoeba*

The DNA replication cycle of *Acanthamoeba* has been reviewed previously (Byers, 1979). The review should be consulted for additional information. Based on [³H]thymidine pulse labeling, Neff (1971) concluded that the cell cycle of *A. castellanii* Neff was 2% M, 10% G_1, 3% S, and 85% G_2. We reexamined this problem in the Neff strain, combining [³H]thymidine pulse labeling with microspectrophotometric measurements of nuclear DNA content (King and Byers, 1985). The DNA measurements revealed a unimodal distribution characteristic of a population primarily in one phase of the cell cycle. Autoradiographs of the same cells indicated that the amoebae with the lowest DNA contents were in S phase. Thus, we agree that G_2 is the predominant phase, but find no evidence for G_1. We conclude that the cycle under our conditions is 2% M, 9% S, and 89% G_2. Evidence for the absence of G_1 also was obtained for *A. castellanii* HR (*A. rhysodes;* Band and Mohrlok, 1973). Colchicine blocked the cell cycle and the nuclear, but not cytoplasmic, incorporation of [³H]thymidine. Incorporation resumed following cell division after removal of the drug, but before complete reformation of nuclei in the daughter cells. Thus, G_1 would have been very brief if it existed at all. A third approach to analyzing the cycle utilized synchronous cultures of the Neff strain (Chagla and Griffiths, 1978). In this study, [³H]thymidine incorporation occurred just before cell division, suggesting that G_1 was both present and the predominant phase. It seems likely that the difference was due to differences in culture methods. It appears that G_1 can be present in the Neff strain, but it may not be essential.

C. The Cell Cycle in *Amoeba*

The DNA replication cycle of *Amoeba* was most recently reviewed by Prescott (1973). The large size of *A. proteus, A. discoides,* and *Amoeba indica* has made it possible to obtain small synchronous populations by using micropipettes to select mitotic cells (Prescott and Goldstein, 1967; Ord, 1968; Rao and Chatterjee, 1974). Studies of [^{3}H]thymidine incorporation into nuclear DNA in these cultures generally have demonstrated an S phase beginning immediately after cell division, as in *Acanthamoeba*. The G_1 phase is either very brief or absent. One report of evidence for a G_1 phase could not be substantiated by Prescott and Lauth (quoted in Prescott, 1973). Microspectrophotometric measurements revealed a unimodal distribution of nuclear DNA contents (Makhlin *et al.*, 1979), suggesting that amoebae were mostly in a single phase, presumably G_2. In *A. indica,* S phase lasted 3 hours out of a total generation time of 24 hours (Rao and Chatterjee, 1974). In *A. proteus* the S phase typically was 5–6 hours out of a cycle time of 36–48 hours (Prescott and Goldstein, 1967; Ord, 1968; Rao and Chatterjee, 1974). Ord demonstrated that the S period in *A. proteus* is biphasic, and Minassian and Bell (1976) demonstrated that the second burst of synthesis was associated with nucleoli and presumably included replication of DNA coding for ribosomal RNA. It would be interesting to know when the nuclear satellite DNA (see Section II,D) replicates, especially since one suggestion is that it might include rDNA (Spear and Prescott, 1980), but no evidence seems to be available on this point. Mitochondrial DNA continues to replicate throughout G_2 in *A. proteus,* but Rao and Chatterjee (1974) found very little [^{3}H]thymidine incorporation in the G_2 phase of *A. indica*.

D. The Cell Cycle in *Naegleria*

It has been difficult to determine the durations of cell-cycle phases in *Naegleria;* it has not been possible to obtain synchronous populations by selection of mitotic cells, the uptake rate of [^{3}H]thymidine is low and internal pools appear to saturate slowly, the stages of mitosis are difficult to discern, and effective inhibitors of mitosis have not been identified (Fulton, 1977). In spite of these difficulties, estimates of cell-cycle phase durations for axenic cultures have been obtained (Fulton, 1977). The duration of mitosis was determined using estimates of the mitotic index. The duration of S phase was determined from the accumulation of labeled nuclei and from the percentage of labeled mitoses during continuous labeling of asynchronous cultures with [^{3}H]thymidine. The latter method also was used to determine the duration of G_2. The best estimates of the cell-cycle parameters were as follows: M, 28 minutes (6%); G_1, 180 minutes (38%); S, 180 minutes (38%); and G_2, 90 minutes (19%), for a total cell-cycle time of 8 hours. Somewhat shorter estimates of S were obtained for monoxenic cultures. No published distributions of nuclear DNA content are available to confirm the conclusions based on labeling studies, but these should be easy to obtain using fluorescence-activated cell analysis (Coulson and Tyndall, 1978). There also is no evidence on the timing of replication of the nuclear satellite relative to the main band.

E. Nucleocytoplasmic Interactions and the Regulation of DNA Synthesis

Regulation of DNA synthesis has only been studied in *Amoeba* where the large size of these organisms has facilitated nuclear transplantation operations. Much of this work has been reviewed by Prescott (1973). The most obvious question to ask was what would happen to nuclear activities if nuclei from one phase of the cell cycle were transferred to cytoplasm at another phase. Consequently, S phase nuclei were transferred into G_2 phase cytoplasm and G_2 nuclei were placed into S cytoplasm. The results from short-term experiments are conflicting. In *A. proteus,* Goldstein and Prescott (1967) (Prescott, 1973) found that the [^{3}H]thymidine incorporation decreased in S nuclei transplanted into G_2 cytoplasm and increased in G_2 nuclei transplanted into S cytoplasm. Thus, they proposed that the cytoplasm had immediate control of nuclear DNA synthesis. In contrast, following an extensive series of careful experiments, Ord (1969) concluded that the nuclear synthetic activity was unchanged in both types of transfers and, therefore, that DNA synthesis was not under the immediate control of cytoplasmic factors. This conflict remains unresolved. However, in a related experiment, Rao and Chatterjee (1974) found that S phase nuclei from *A. proteus* continued to incorporate labeled thymidine at an undiminished rate when transferred to G_2 cytoplasm of *A. indica*. These results may be consistent with those of Ord, but, of course, they also could indicate that cytoplasmic regulatory factors are species-specific. Rao and Chatterjee (1974) and Chatterjee and Rao (1974) clearly demonstrated, however, that the cytoplasm controlled nuclear diameter and the duration of S in the progeny of cells that survived nuclear implants and produced viable clones. When the smaller nuclei of *A. indica* were placed in *A. proteus* cytoplasm, they increased in size prior to the first cell division. However, when the larger nuclei of *A. proteus* were placed in cytoplasm of *A. indica,* the size adjustment did not occur until after the first division. It was technically impossible to follow changes in S phase in the same way; therefore, it is unknown when changes might have occurred. It is even possible that the S period did not change. This could be true if survival of hybrid cells only occurred when nuclei with S phase lengths at the appropriate extreme of the normal range, i.e., nuclei of *A. proteus* with the shortest S phases and nuclei of *A. indica* with the longest

S phases, were transferred into foreign cytoplasm. This would be consistent with the observation that only 22% of the hybrids with an *A. proteus* nucleus and *A. indica* cytoplasm and 3% of the reciprocal hybrids produced clones. Even if this explanation is correct, however, it is clear that the cytoplasm controls the duration of S phase in the long term. Goldstein and Ron (1969) suggested that nuclear proteins that are released to the cytoplasm at mitosis might include the factors controlling the duration of S. Their conclusions were based on the observation that S phase was prolonged in the daughters of cells from which 30–50% of the cytoplasm had been amputated during mitosis. It was presumed that the amputations resulted in daughter nuclei that were deficient in these particular proteins. As these authors recognized, however, other cytoplasmic factors that might be important also would be removed by the amputations.

V. DNA Metabolism during Differentiation

A. ENCYSTMENT IN *Acanthamoeba castellanii*

Several inhibitors of nucleic acid synthesis induce cyst formation in *Acanthamoeba* (reviewed in Byers, 1979). Since this fact was first discovered by Neff and Neff (1969, 1972), there has been considerable effort to understand why this phenomenon occurs. One approach has been to examine changes in DNA content and metabolism. The DNA content of log-phase amoebae is quite variable. The reason is unknown, but might include the possibility of a polyploid state (see Section II,C). Whole cell DNA content decreases during post-log phase or during starvation-induced encystment (Neff and Neff, 1969; Byers *et al.,* 1969; King and Byers, 1985). On the average, cysts, which have single nuclei, have about 50% as much nuclear DNA as early log-phase trophozoites, which also are mononucleated. The decrease would be explained if encystment medium stimulated a final cell division and daughter cells arrrested without entering S (Rudick, 1971). Two observations make this unlikely. First, all available data indicate that cell multiplication ceases as soon as amoebae are transferred to encystment medium (King and Byers, 1985). In contrast, a final division does seem to occur when *Entamoeba* is incubated in encystment medium. Second, the decrease in DNA content during post-log phase probably is associated with preparations for encystment (Byers *et al.,* 1969) and, therefore, may occur for the same reasons that a decrease occurs in encystment medium. If the post-log amoebae arrest prior to S phase after dividing, then cultures diluted into fresh medium should exhibit a lag of nearly 8 hours before resuming multiplication. This does not occur. Rather, King and Byers observed a small initial burst of division that was more consistent with a cell-cycle arrest near the end of the cycle. This result would be consistent with a reduction division in which amoebae divided and then continued to a late stage in the cycle without replicating DNA. The likelihood that acanthamoebae are polyploid is consistent with this possibility. Alternatively, amoebae might divide and complete S phase, but simultaneously or subsequently degrade approximately half of the total DNA; again, this would be more likely if the cells were polyploid. Either of these models would produce nuclei with about half the log-phase amount of DNA. Measurements of nuclear cross-sectional area indicate that the cyst nuclear volume averages 43% of the log-phase trophozoite nuclear volume (King and Byers, 1985). This decrease is consistent with the decreased DNA content, but does not help choose between the two models.

King and Byers have obtained evidence for nuclear and mitochondrial DNA degradation during post-log phase and starvation-induced encystment. Log-phase cultures were labeled with [³H]thymidine, and then the loss of label was observed by examining the DNA on CsCl gradients during encystment. Label was lost from both nuclear and mitochondrial DNA. When unlabeled amoebae were incubated in encystment medium plus labeled thymidine, only the mitochondrial DNA became labeled. The simplest explanation for all the data is that net losses of both nuclear and mitochondrial DNA occur during encystment, nuclear loss occurring by degradation in the absence of synthesis and mitochondrial loss occurring during turnover. The nuclear DNA loss must be around 50%, but the mitochondrial loss can be 90% or greater (King and Byers, 1985).

B. ENCYSTMENT IN *Entamoeba invadens*

Little work has been done on the molecular biology of encystment in *Entamoeba*. The earliest work on [³H]thymidine incorporation was complicated due to the use of monoxenic cultures (Albach *et al.*, 1966). The only work with axenic cultures that has been reported explored the relationship between DNA synthesis, nuclear division, and cell division in *E. invadens* (Sirijintakarn and Bailey, 1980). Cultures growing in a nutrient-rich medium were induced to encyst in a hypoosmotic encystment medium supplemented with labeled thymidine or were prelabeled with thymidine and then induced to encyst in unlabeled medium. Data on changes in DNA content per amoeba and per nucleus as well as observed decreases in levels of incorporated thymidine suggested the following. Uninucleate amoebae with 0.151-pg DNA divided once upon transfer to encystment medium, presumably reducing the nuclear DNA content in half. DNA synthesis and cyst wall formation were initiated and duplication of the DNA most likely occurred. One or two nuclear divisions followed, resulting in cyst populations with approximately one-half binucleates and one-half tetranucleates. The average DNA content was then 0.155 pg per amoeba. If the binucleates and the tetranucleates had the same whole cell DNA content, as seems likely, then the tetranucleate nucleus contained 0.039 pg. This nucleus, which was visibly smaller than the binucleate or mononucleate nuclei, resulted from three nuclear divisions, the first associated with cell division followed by DNA duplication and the latter two without cell division or DNA synthesis. If this interpretation of the results is correct, then the amoebae must have had nuclei that were at least $4n$ at the time of transfer to encystment medium. However, a few octanucleate cysts were observed, and amoebae became octanucleate when first induced to excyst. If the octanucleates were formed without DNA synthesis, then the trophozoites initially induced to encyst could have been $8n$. This value agrees well with estimates of ploidy levels for other species of *Entamoeba* (see Section II,C). Sirijintakarn and Bailey observed that the total [^{3}H]thymidine content of prelabeled cells decreased more than 50% during the course of differentiation from trophozoite to mature cyst. A 50% decrease would be expected if there was only one cell division following transfer of trophozoites to encystment medium. The authors attributed the additional loss to continued cell multiplication by some cells. However, considering the observations for *Acanthamoeba* (see Section IV,A), nuclear DNA turnover should be considered a good alternative possibility.

C. FLAGELLATION IN *Naegleria gruberi*

Essentially nothing has been reported on DNA metabolism during encystment in *Naegleria*. Interest has focused on the amoeba–flagellate transformation that is induced by replacing growth medium with a nutrient-free transformation buffer (Fulton, 1970, 1977). Flagellation is accompanied by a decrease in DNA synthesis, and the factors that regulate this decrease have been the subjects of several studies. Under conditions used by Yuyama and Corff (1978), flagellum formation begins about 60 minutes after incubation in transformation buffer. These authors utilized CsCl gradients to examine the accompanying changes in [^{3}H]thymidine incorporation and observed that there was a marked decrease in the incorporation into nuclear DNA. Mitochondrial DNA synthesis continued and became increasingly dominant as differentiation progressed. Some label continued to be incorporated into nuclear DNA, but this might be explained by the fact that a few amoebae failed to become flagellated.

The decrease in thymidine incorporation was accompanied by a drop in thymidine uptake and in the activity of thymidine kinase (Bols *et al.*, 1977). The extent of the decrease in enzyme activity was strongly correlated with the proportion of the culture forming flagella. In contrast to thymidine kinase, the activity of nucleotide phosphotransferase, which phosphorylates thymidine to dTMP, remained essentially unchanged throughout differentiation. The thymidine kinase activity in equal mixtures of extracts from log phase and differentiating amoebae was the sum of the two activities mixed; thus, the differences are not due to activators or inhibitors. The decrease in activity during differentiation was only slightly slowed by concentrations of cycloheximide and actinomycin D that block protein and RNA synthesis. Therefore, the lost activity does not result from the concomitant synthesis of a degradative enzyme.

Most cell types arrest in G_1 or sometimes G_2 phase when differentiation is initiated. It was concluded, however, that *N. gruberi* arrests in mid-S phase during flagellum formation (Corff and Yuyama, 1976; Yuyama and Corff, 1978). Since the amoebae are in an asynchronously multiplying culture when first induced to differentiate, it would take about 3 hours for the amoebae in the earliest stages of S phase to complete DNA synthesis (S = 3 hours). Since nearly 100% differentiation occurs in 2 hours, some amoebae must initiate differentiation while in mid-S phase. However, that proportion of the population that was in the first hour of G_1 (G_1 − 3 hours) would not reach S phase within the 2-hour period of differentiation and, therefore, must initiate flagellation from G_1. In the expected age-frequency distribution of a growing asynchronous population, the number of cells having any particular age decreases continuously from the youngest to the oldest groups (James and Cook, 1958; Steel, 1973). In *N. gruberi*, where the duration of G_1 equals the duration of S, substantially more cells should be in G_1 than in S. Thus, a relatively large proportion of the population could be in G_1 when they differentiate. Additional amoebae must differentiate from G_2 phase (G_2 = 1.5 hours) unless nuclear division occurs and they enter G_1 during the induction period. Thus, the rapid nature of the flagellation process makes it unlikely that differentiating cells could be blocked at a common point with respect to DNA replication. It would be interesting to determine whether the same conclusion could be drawn for encystment.

VI. DNA and Phylogeny

A. Interstrain Variation in Nuclear DNA Base Composition

Variations in base composition have been used to examine interstrain relationships in *Amoeba, Acanthamoeba, and Entamoeba*. In all cases, the compositions have been determined indirectly from buoyant densities or melting temperatures. Reeves *et al.* (1971) attempted to distinguish strains of *E. histolytica* that were tolerant of high temperatures from *E. histolytica*-like strains that were intolerant of these temperatures. Although some data indicate that these strains differ very significantly in DNA content and in the presence or absence of repetitious DNA sequences (Gelderman *et al.*, 1971a,b), the estimated percentage of G + C was very similar for both types: 28.79 ± 0.98% for 10 high-temperature strains and 29.86 ± 0.97% for 5 low-temperature strains. There are some statistically signficant pairwise differences among strains, but the overall intraspecific diversity is relatively low. For example, the variations among strains are less than the variations of repetitive measurements on DNA from a single strain of *Amoeba* (Fritz, 1981). In two other species, *E. moshkovskii* and *E. invadens,* the G + C values were 31.7 and 31.0%, respectively. Thus, diversity is relatively low even at the interspecific level.

Intraspecific diversity in overall base composition also is low in *Acanthamoeba*. Seven strains of *A. castellanii* all were 61% G + C (Adam and Blewett, 1974). *Acanthamoeba palestinensis,* which some classify as a strain of *A. castellanii,* was 62% G + C. The values for these species were clearly distinct, however, from *Acanthamoeba polyphaga, Acanthamoeba culbertsoni,* and *Acanthamoeba astronyxis,* which were 57, 56, and 50% G + C, respectively. The *Acanthamoeba* relationships were tested further by DNA–DNA hybridization with the same conclusion that intraspecific diversity was substantially less than interspecific diversity (Adams and Blewett, 1974). At the extreme, *A. astronyxis* Ray showed no hybridization with *A. castellanii* Neff. The significance of the hybridization studies is lessened, however, by the fact that the comparisons were not restricted to unique sequences.

Intraspecific diversity was higher in *Amoeba* where 10 strains of *A. proteus* averaged 34.4 ± 5.6% G + C (Fritz, 1981). However, these values are for the main-band DNA and do not take into account the large nuclear satellite. The evidence suggesting that the main band and the satellite might vary in relative proportions (see Section II,D) creates a problem. This is most apparent in a comparison of *A. proteus* with *A. indica* (Table I) where it is unclear whether one should compare main bands, bands of similar density, or total nuclear DNA. If the variation in relative amounts means that differential replication of selected subfractions of nuclear DNA occurs, then the higher diversity indicated by the calculated main-band base compositions may only reflect physiological differences. As Fritz (1981) has recognized, comparisons of base composition have only limited value in phylogenetic studies. Properly executed DNA–DNA hybridization studies (Ahlquist and Sibley, 1983) or studies of ribosomal RNA sequences (Pace *et al.*, 1985) have more promise for the elucidation of phylogenetic relationships.

B. Interstrain Variation in Mitochondrial DNA

There have been several reports on the possible use of variations in mitochondrial DNA to study the taxonomy of *Acanthamoeba* (Byers *et al.*, 1983; Bogler *et al.*, 1983; Costas *et al.*, 1983). The initial approach has been to compare electrophoretic patterns for mitochondrial DNA fragments obtained by digestion with restriction endonucleases. In our laboratory, 21 strains representing several different species have been examined with at least five different restriction enzymes each. Sixteen different phenotypes have been identified by electrophoresis. The most striking observation is that intraspecific and interspecific differences both are relatively large. The largest difference is between *A. astronyxis* and the other strains. This is consistent with differences in nuclear DNA base composition and DNA–DNA hybridization results (see Section VI,A). Evidence also was obtained for one cluster of strains in which no polymorphisms were detected with up to 19 different restriction enzymes. These strains may form clusters of closely related organisms, but it has been difficult to rule out the possibility that the similarity may have arisen historically through problems with cross contamination.

Although we have attempted to quantitate interstrain relationships based on the differences in electrophoretic patterns (Bogler *et al.*, 1983), it is not known whether the basic assumptions underlying the model used for the calculations are valid for *Acanthamoeba*. Thus, we have begun to examine sequence differences more directly by comparing restriction site and gene distribution maps. The latter are being located using heterologous probes from *Neurospora* (R. A. Akins and T. J. Byers, unpublished results).

VII. Bacterial Endosymbionts and Viruses

Both bacterium-like endosymbionts and viruses are found in various strains of amoebae. These are important because of influences that their presence may have on studies of nuclear and organelle DNA function. In addition, the existence of viruses offers the possibility of developing useful vectors for recombinant DNA approaches to the study of gene structure and function. Unfortunately, relatively little is known about the nucleic acids of these entities.

Evidence for rhabdovirus-like RNA viruses (Bird and McCaul, 1976) and filamentous and polyhedral DNA viruses (Mattern *et al.*, 1972; Hruska *et al.*, 1973) has been obtained for *Entamoeba*. Particles resembling rhabdoviruses were observed in several strains of *E. histolytica,* in a Laredo strain, and in a strain of *E. invadens* (Bird and McCaul, 1976). Filamentous and polyhedral types of DNA viruses were produced when uninfected cultures of *E. histolytica* were treated with media from infected cultures, but these viruses were not ordinarily seen in healthy axenic cultures. Therefore, it was suggested that the viral DNA might be integrated into the amoeba genome (Diamond and Mattern, 1976). Virus-like particles also have been described for *N. gruberi* EGs (Schuster and Dunnebacke, 1971, 1976). These elements developed in the nucleus, passed into the cytoplasm, and, eventually, into the medium. Particles were not ordinarily seen in axenic cultures, but could be induced by bacteria in the medium or by BUdR (Schuster and Clemente, 1977). Thus, these particles also may have been present in a latent form. It is unknown, however, whether nucleic acids are associated with the particles. I am unaware of any evidence for virus particles in either *Amoeba* or *Acanthamoeba,* although there have been reports of tritiated thymidine incorporation into the cytoplasm of *A. castellanii* in regions other than over mitochondria (Ito *et al.,* 1969; McIntosh and Chang, 1971). Since no particles could be found associated with the incorporation, it was suggested that defective viruses might be present. I am inclined to believe that the incorporation was more likely not into DNA (Byers and King, 1985), but the author's conclusions cannot be ruled out.

Bacterium-like endosymbionts have been described for three of the amoeba genera. An intracellular diptheroid form was observed in *N. fowleri* CJ (Phillips, 1974). The organism could be eliminated by antibiotic treatment, but amoebae tended to grow better in its presence, thus suggesting a symbiotic relationship. Proca-Ciobanu *et al.* (1975) described a bacterium-like symbiont in axenic cultures of *A. castellanii* SN. This organism, which was present free in the cytoplasm, possibly was an obligate symbiont because it never was possible to culture bacteria from the medium during several years of axenic growth. A bacterial parasite which originally was isolated from the soil was able to infect *A. castellanii* and to multiply internally (Drozanski, 1956; Drozanski and Chmielewski, 1979). The amoebae eventually were destroyed. Although the bacterium was able to survive for several months outside of the amoebae, it could not replicate under these conditions. Another bacterium-like parasite was described in one strain of *A. astronyxis,* but details were not discussed (Page, 1967). There have been several reports of *Acanthamoeba* harboring mycobacteria (e.g., Jadin, 1976) and legionella (Holden *et al.,* 1984).

Although it is clear that a few strains of *Naegleria* and *Acanthamoeba* can harbor bacteria, most strains show no signs of any infectious agent. In contrast, numerous strains of *A. proteus* and *A. discoides* contain bacteria or self-replicating DNA-containing bodies (Andresen, 1973; Hawkins, 1973). In some cases, these particles number in the thousands. I am unaware, however, of any studies of the DNA in these symbionts, except for X bacteria described by Jeon (1983) and his colleagues. The X bacteria are obligate symbionts that initially arose from a spontaneous infection of *A. proteus* (Jeon and Lorch, 1967). In the course of time, the amoebae became dependent on the symbionts. It is now possible to deliberately infect cultures and the symbiotic dependency develops over about 200 cell generations. Currently, amoebae carry an average load of about 42,000 symbionts per cell (Jeon, 1983). The X bacteria have circular plasmids of 11 and 39 million Da. To date, nothing is known about the functions or organization of the plasmids or of relationships between bacterial and amoeba DNA.

VIII. Conclusions

We have only begun to explore the molecular biology of amoeba DNA, and a number of very interesting problems await the attention they deserve. Some of the more interesting questions are the following. Are the nuclear genes polyploid? If so, how is gene expression regulated? Are there any natural mechanisms for gene recombination? How does the overall organization of coding and noncoding DNA sequences in these primitive organisms compare with that of other organisms? Are there really some species of *Entamoeba* that have very little repetitious nuclear DNA and other species with very much? If so, what are the signficant differences? What are the nuclear satellites in *Amoeba* and *Naegleria?* How do they relate to the main-baind DNAs and how are the relative proportions of the main-band and satellite DNAs regulated? Do amoebae have transposable elements? Are there any viruses in *Amoeba* or *Acanthamoeba?* Do the nuclear genomes of *Entamoeba* species that have no mitochondria differ in any significant way from the genomes of the aerobic amoebae? If so, what is the evolutionary signficance of the differences?

In addition to the aforementioned problems in basic molecular biology, there are two practical problems that deserve more attention. Taxonomic studies, especially of *Acanthamoeba* and *Entamoeba,* could benefit from the use of additional molecular approaches to help elucidate the relationships between pathogenic and nonpathogenic or between parasitic and

nonparasitic strains. A beginning has been made with studies of mitochondrial DNA in *Acanthamoeba,* but interstrain variability is high, and
more conservative DNA sequences need to be examined. Also, this approach will not be useful for *Entamoeba* where mitochondria are absent.
Another practical problem is the development of a library of specific
probes that readily could be used to identify living amoebae isolated from
human infections or organisms in tissue sections.

A few years ago, a number of the problems cited would have been
beyond reach. Today, they are mostly well within reach and only need to
attract the attention and interest of creative molecular biologists. I hope
that this review will help to stimulate the interest that these organisms
deserve.

ACKNOWLEDGMENTS

The author thanks Drs. H. J. Bohnert, C. Fulton, J. A. Hammer III, L. E. King, and M.
R. Paule for permission to discuss their unpublished work. The research work of the author
and his associates that is described in this article was partially supported by grants from the
National Institutes of Health and the Ohio State University.

REFERENCES

Abbott, G. F., and Hawkins, S. E. (1978). *Experientia* **34,** 427.
Abbott, G. F., and Hawkins, S. E. (1980). *Cytobios* **27,** 135.
Adam, K. M. G., and Blewett, D. A. (1974). *Ann. Soc. Belge Med. Trop.* **54,** 163.
Adam, K. M. G., Blewett, D. A., and Flamm, W. G. (1969). *J. Protozool.* **16,** 6.
Ahlquist, J. E., and Sibley, C. G. (1983). *In* "Current Ornithology" (R. F. Johnson, ed.),
    Vol. 1, p. 245. Plenum, New York.
Albach, R. A., and Booden, T. (1978). *In* "Parasitic Protozoa" (J. P. Kreier, ed.), Vol 2, p.
    455. Academic Press, New York.
Albach, R. A., Shaffer, J. G., and Watson, R. H. (1966). *J. Protozool.* **13,** 349.
Andresen, N. (1973). *In* "The Biology of Amoeba" (K. Jeon, ed.), p. 99. Academic Press,
    New York.
Band, R. N., and Mohrlok, S. (1973). *Nature (London)* **227,** 379.
Bird, R. G., and McCaul, T. F. (1976). *Ann. Trop. Med. Parasitol.* **70,** 81.
Bogler, S. A., Zarley, C. D., Buriaek, L. L., Fuerst, P. A., and Byers, T. J. (1983). *Mol.
    Biochem. Parasitol.* **8,** 145.
Bohnert, H. J. (1973). *Biochim. Biophys. Acta* **324,** 199.
Bohnert, H. J., and Herrmann, R. G. (1974). *Eur. J. Biochem.* **50,** 83.
Bols, N. C., Corff, S., and Yuyama, S. (1977). *J. Cell. Physiol.* **90,** 271.
Bowers, B., and Korn, E. D. (1969). *J. Cell Biol.* **41,** 786.
Byers, T. J. (1979). *Int. Rev. Cytol.* **61,** 2834.
Byers, T. J., Rudick, V. L., and Rudick, M. J. (1969). *J. Protozool.* **16,** 693.
Byers, T. J., Bogler, S. A., and Burianek, L. L. (1983). *J. Protozool.* **30,** 198.

Carle, G. F., and Olson, M. V. (1984). *Nucleic Acids Res.* **12**, 5647.
Chagla, A. H., and Griffiths, A. J. (1974). *J. Gen. Microbiol.* **85**, 139.
Chagla, A. H., and Griffiths, A. J. (1978). *J. Gen. Microbiol.* **108**, 39.
Chatterjee, S., and Rao, M. V. N. (1974). *Exp. Cell Res.* **84**, 235.
Corff, S., and Yuyama, S. (1976). *J. Protozool.* **23**, 587.
Costas, M., Edwards, S. W., Lloyd, D., Griffiths, A. J., and Turner, G. (1983). *FEMS Microsc. Lett.* **17**, 231.
Coulson, P. B., and Tyndall, R. (1978). *J. Histochem. Cytochem.* **26**, 713.
D'Allessio, J. M., Harris, G. H., Perna, P. J., and Paule, M. R. (1981). *Biochemistry* **20**, 3822.
Diamond, L. S., and Mattern, C. F. T. (1976). *Adv. Virus Res.* **20**, 87.
Drozanski, W. (1956). *Acta Microbiol. Pol.* **5**, 315.
Drozanski, W., and Chmielewski, T. (1979). *Acta Microbiol. Pol.* **28**, 123.
Engberg, J., Nilsson, J. R., Pearlman, R., and Leick, V. (1974). *Proc. Natl. Acad. Sci. U.S.A.* **71**, 894.
Fritz, C. T. (1981). *Biochem. Syst. Ecol.* **9**, 207.
Fritz, C. T. (1982). *Comp. Biochem. Physiol.* **72B**, 641.
Fulton, C. (1970). *In* "Methods in Cell Physiology" (D. M. Prescott, ed.), p. 341. Academic Press, New York.
Fulton, C. (1977). *Annu. Rev. Microbiol.* **31**, 597.
Fulton, C., Webster, C., and Wu, J. S. (1984). *Proc. Natl. Acad. Sci. U.S.A.* **81**, 2406.
Gadasi, H., Maruta, H., Collins, J. H., and Korn, E. D. (1979). *J. Biol. Chem.* **254**, 3631.
Gelderman, A. H., Keister, D. B., Bartigis, I. L., and Diamond, L. S. (1971a). *J. Parasitol.* **57**, 906.
Gelderman, A. H., Bartigis, I. L., Keister, D. B., and Diamond, L. S. (1971b). *J. Parasitol.* **57**, 912.
Goldstein, L., and Prescott, D. M. (1967). *In* "Control of Nuclear Activity" (L. Goldstein, ed.), p. 3. Prentice-Hall, Englewood Cliffs, New Jersey.
Goldstein, L., and Ron, A. (1969). *Exp. Cell Res.* **55**, 144.
Griffiths, A. J. (1970). *Adv. Microb. Physiol.* **4**, 105.
Grummt, I., Roth, E., and Paule, M. R. (1982). *Nature (London)* **296**, 173.
Hammer, J. A., III, Korn, E. D., and Paterson, B. M. (1984). *J. Cell Biol.* **99**, 33a.
Hawkins, S. E. (1973). *In* "The Biology of Amoeba" (K. W. Jeon, ed.), p. 525. Academic Press, New York.
Hawkins, S. E., and Hughes, J. M. (1973). *Nature (London) New Biol.* **246**, 199.
Hettiarachchy, N. S., and Jones, I. G. (1974). *Biochem. J.* **141**, 159.
Holden, E. P., Winkler, H. H., Wood, D. D., and Leinbach, E. D. (1984). *Infect. Immun.* **45**, 18.
Hruska, J. F., Mattern, C. F. T., Diamond, L. S., and Keister, D. B. (1973). *J. Virol.* **11**, 129.
Iida, C. T., Kownin, P., and Paule, M. R. (1985). *Proc. Natl. Acad. Sci. U.S.A.* **82**, 1668.
Ito, S., Chang, R. S., and Pollard, T. D. (1969). *J. Protozool.* **16**, 638.
Jadin, J. B. (1976). *Pathol. Biol.* **24**, 171.
James, T. W., and Cook, J. R. (1958). *In* "Synchrony in Cell Division and Growth" (E. Zeuthen, ed.), p. 485. Wiley (Interscience), New York.
Jantzen, H. (1973). *Arch. Mikrobiol.* **91**, 163.
Jantzen, H. (1981). *Dev. Biol.* **82**, 113.
Jeon, K. W., ed. (1973). "The Biology of Amoeba." Academic Press, New York.
Jeon, K. W. (1983). *Int. Rev. Cytol. Suppl.* **14**, 29.
Jeon, K. W., and Lorch, J. (1967). *Exp. Cell Res.* **48**, 236.
John, D. T. (1982). *Annu. Rev. Microbiol.* **36**, 101.

Kadlec, V. (1978). *J. Protozool.* **25**, 235.

King, L. E., and Byers, T. J. (1985). Submitted.

Kownin, P., Iida, C. T., Brown-Shimer, S., and Paule, M. (1985). *Nucleic Acids Res.* **13**, 6237.

Krishna Murti, C. R. (1975). *Indian J. Med. Res.* **63**, 757.

Levine, N. D., Corliss, J. O., Cox, F. E. G., Deroux, G., Grain, J., Honigberg, B. M., Leesdale, G. F., Leoblick, A. R., Lom, J., Lynn, D., Merinfeld, E. G., Page, F. C., Poljansky, G., Sprague, V., Vavra, J., and Wallace, F. G. (1980). *J. Protozool.* **27**, 37.

Lopez-Revilla, R., and Gomez, R. (1978). *Exp. Parasitol.* **44**, 243.

McIntosh, A. H., and Chang, R. S. (1971). *J. Protozool.* **18**, 632.

Mackay, R. M., and Doolittle, W. F. (1981). *Nucleic Acids Res.* **9**, 3321.

Makhlin, E. E., Kudryavtseva, M. V., and Kudryavtseva, B. N. (1979). *Exp. Cell Res.* **118**, 143.

Martinez, A. J. (1980). *Neurology* **30**, 567.

Martinez, A. J. (1983). *Commonw. Inst. Parasitol. Protozool. Abstr.* **7**, 293.

Martinez-Palomo, A. (1982). "The Biology of *Entamoeba histolytica*," p. 161. Research Studies Press, New York.

Marzzoco, A., and Colli, W. (1974). *Biochim. Biophys. Acta* **374**, 292.

Mattern, C. F. T., Diamond, L. S., and Daniel, W. A. (1972). *J. Virol.* **9**, 342.

Mery-Drugeon, E., Crouse, E. J., Schmitt, J. M., Bohnert, H. J., and Bernardi, G. (1981). *Eur. J. Biochem.* **114**, 577.

Minassian, I., and Bell, L. G. E. (1976). *J. Cell Sci.* **22**, 521.

Murti, K. G., and Prescott, D. M. (1978). *Exp. Cell Res.* **112**, 233.

Nanney, D. L. (1984). *Proc. Int. Congr. Protozool. 6th,* p. 242.

Neff, R. J. (1971). *In Vitro* **6**, 300.

Neff, R. J., and Neff, R. H. (1969). *Symp. Soc. Exp. Biol.* **23**, 51.

Neff, R. J., and Neff, R. H. (1972). *C. R. Trav. Lab. Carlsberg* **39**, 11.

Nellen, W., and Gallwitz, D. (1982). *J. Mol. Biol.* **159**, 1.

Ord, M. J. (1968). *J. Cell Sci.* **3**, 483.

Ord, M. J. (1969). *Nature (London)* **221**, 964.

Ord, M. J. (1973). *In* "The Biology of Amoeba" (K. W. Jeon, ed.), p. 349. Academic Press, New York.

Pace, N. R., Stahl, D. A., Lane, D. J., and Olsen, G. J. (1985). *ASM News* **51**, 4.

Page, F. C. (1967). *J. Protozool.* **14**, 709.

Page, F. C. (1974). *Acta Protozool.* **13**, 143.

Paule, M. R., Iida, C. T., Perna, P. J., Harris, G. H., Brown-Shimer, S., and Kowin, P. (1984a). *Biochemistry* **23**, 4167.

Paule, M. R., Iida, C. T., Perna, P. J., Harris, G. H., Knoll, D. A., and D'Aleesio, J. M. (1984b). *Nucleic Acids Res.* **12**, 8161.

Pellegrini, M., Timerlake, W. E., and Goldberg, R. B. (1981). *Mol. Cell. Biol.* **1**, 136.

Phillips, B. P. (1974). *Am. J. Trop. Med. Hyg.* **23**, 850.

Prescott, D. M. (1973). *In* "The Biology of Amoeba" (K. W. Jeon, ed.), p. 467. Academic Press, New York.

Prescott, D. M., and Goldstein, L. (1967). *Science* **155**, 469.

Proc-Ciobanu, M., Lupascu, G., Petrovici, A., and Ionescu, M. D. (1975). *Int. J. Parasitol.* **5**, 49.

Pussard, M. (1964). *Rev. Ecol. Biol. Sol.* **4**, 587.

Pussard, M. (1972). *J. Protozool.* **19**, 557.

Rao, M. V. N., and Chatterjee, S. (1974). *Exp. Cell Res.* **88**, 371.

Reeves, R. E., Lushbaugh, T. S., and Montalve, F. E. (1971). *J. Parasitol.* **57**, 939.

Rudick, V. L. (1971). *J. Cell Biol.* **49,** 498.

Schuster, F. L. (1979). *In* "Biochemistry and Physiology of Protozoa" (M. Levandowsky and S. Hutner, eds.), 2nd Ed., Vol. 1, p. 215. Academic Press, New York.

Schuster, F. L., and Clemente, J. S. (1977). *J. Cell Sci.* **26,** 359.

Schuster, F. L., and Dunnebacke, T. II. (1971). *J. Ultrastruct. Res.* **36,** 659.

Schuster, F. L., and Dunnebacke, T. H. (1976). *Cytobiologie* **14,** 131.

Schwartz, D. C., and Cantor, C. R. (1984). *Cell* **37,** 67.

Singh, B. N. (1981). *Indian J. Parasitol.* **5,** 133.

Sirijintakarn, P., and Bailey, G. B. (1980). *Arch. Invest. Med.* **11** (Suppl. 1), 3.

Spear, B. B., and Prescott, D. M. (1980). *Exp. Cell Res.* **130,** 387.

Steel, G. G. (1973). *In* "The Cell Cycle in Development and Differentiation" (M. Balls and F. S. Billet, eds.), p. 13. Cambridge Univ. Press, London and New York.

Stevens, A. R., and Pachler, P. F. (1972). *J. Mol. Biol.* **66,** 225.

Stevens, A. R., and Pachler, P. F. (1973). *J. Cell Biol.* **57,** 525.

Tait, A. (1983). *Parasitology* **86,** 29.

Tautvydas, K. J. (1971). *Exp. Cell Res.* **68,** 299.

Visvesvara, G. S. (1980). *Int. Conf. Biol. Pathogen. Small Free-living Amoebae 2nd,* Centers for Disease Control, U.S. Public Health Service.

Ward, A. (1985). *J. Protozool.* **32,** 357.

Weik, R. R., and John, D. T. (1977). *J. Protozool.* **24,** 196.

Weik, R. R., and John, D. T. (1978). *J. Parasitol.* **64,** 746.

Weisman, R. A. (1976). *Annu. Rev. Microbiol.* **30,** 189.

Williams, N. E. (1984). *Evolution* **38,** 25.

Yuyama, S., and Corff, S. (1978). *J. Protozool.* **25,** 408.

Index